T4-AHD-357

ADVANCED SUMMER SCHOOL IN PHYSICS 2008

ADVANCED SUMMER SCHOOL IN PHYSICS 2008

Frontiers in Contemporary Physics
4th Edition

EAV08

Cinvestav, México City, México *7 - 11 July 2008*

EDITORS
Luis Manuel Montaño Zetina
Gabino Torres Vega
Miguel García Rocha
Luis F. Rojas Ochoa
Ricardo López Fernández
Cinvestav
México City, México

All papers have been peer reviewed

SPONSORING ORGANIZATIONS
Centro de Investigación y de Estudios Avanzados del IPN - (Cinvestav)
Consejo Nacional de Ciencia y Tecnología - (CONACyT)
Academia Mexicana de Ciencias - (AMC)
Fundación México-Estados Unidos para la Ciencia - (FUMEC)

Melville, New York, 2008
AIP CONFERENCE PROCEEDINGS ■ VOLUME 1077

Editors

Luis Manuel Montaño Zetina
Gabino Torres Vega
Miguel García Rocha
Luis F. Rojas Ochoa
Ricardo López Fernández

Departamento de Física
Cinvestav
Av. IPN 2508, Col. San Pedro Zacatenco
GAM 07360, México D.F.
México

E-mail: lmontano@fis.cinvestav.mx
gabino@fis.cinvestav.mx
mrocha@fis.cinvestav.mx
lrojas@fis.cinvestav.mx
lopezr@fis.cinvestav.mx

L.C. Catalog Card No. 2008939548
ISBN 978-0-7354-0608-7
ISSN 0094-243X

Printed in the United States of America

CONTENTS

V. SOLID STATE PHYSICS

FOREWORD

The *Advanced Summer School in Physics* of the *Center of Research and Advanced Studies* (Cinvestav by its acronym in Spanish) celebrates its 39th anniversary. This event became an important date in the annual schedule of the Physics international scientific community because of the cutting edge themes dealt and given by renowned scientific hosts, for example, Abdus Salam (Physics, 1979) and Pierre-Gilles de Gennes (Physics, 1991) both Nobel Prize recipients, in addition to other remarkable scientists who have attended this event over the years.

The 1969 school constitutes a landmark for the Mexican Physics scientists because it was one the first international meetings in the country in regards to this discipline. The aim of this event is to offer a forum for graduate students to discuss their own work with leading scientists from all over the world and, for the undergraduate students, to gain insight on the frontiers of contemporary physics. Both objectives have been fully fulfilled since the beginning.

The 2008 *Advanced Summer Schools in Physics* is the fourth of the present series, and is characterized by some relevant issues, for example, efforts has been devoted to celebrate the School every year and to include areas currently under development in the Cinvestav´s Physics Department, such is the case of mathematical physics, gravitation, high energy physics, statistical and medical physics, solid-state and condensed matter physics, among others.

The 2008 event hosted more than 100 participants from 20 different institutions and from five countries. They attended six courses and seven plenary talks which were given by international experts, including professors of our Physics Department. In addition, post graduated students from Cinvestav and other research centers delivered nine parallel talks. The conferences program was a tight scheduled developed within a week of intense activity.

The *Advanced Summer School of Physics* reflects the strong development of the physics community in Cinvestav. Nowadays the Physics Department is integrated by 50 professors and is one of the main contributors of new Physics scientists in Mexico: it has graduated more than 300 M. Sc. and 200 Ph.D. in physics. Also it is evident that Cinvestav has contributed successfully to the decentralization of Physics in the country and to the formation of successful research groups. In particular, the Physics Department has been the source of scientist for several groups in Mexico and abroad.

In summary, the 2008 *Advances Summer School od Physics* created an environment where close interaction between students and researchers was achieved. The enthusiasm of the Organizing Committee, Lecturers and Plenary Conferences Speakers during the event had an influence on students to acquire a scientific way of thinking by carrying out research. We are sure that the *Cinvestav´s Advanced Summer School in Physics* will preserve its position of privilege as a remarkable event that encourages students to follow a scientific career in Physics in Mexico and will continue to have an impact in both Mexican and foreign students.

Arnulfo Albores
Provost
Cinvestav
México City, 2008

PREFACE

The *Advanced Summer School* in Physics started in 1969 with the primary goal of offering academic high quality lectures. The one-month courses lectured by M.B. Beg, A. Sirlin, and G. Altarelli drew the attention of students to the frontiers of high energy physics and mathematical physics at those times. Over the years, a number of recognized experts in diverse disciplines of physics have lectured in the Summer School: D. Finley, M.A. del Olmo, R. Klein, D.A. McQuarrie, B. Mielnik, L. Woronowicz, and quite recently D.E. Aspnes, D.B. Cline, M.O. de la Cruz, J.F. Gracia-Bondia, J. Greene, F. Halzen, S. Parke, J. Stavans, and J.W.F. Valle, among others. The School has also been honored by hosting distinguished scientists like Abdus Salam (1979 Nobel Prize in Physics) and Pierre-Gilles de Gennes (1991 Nobel Prize in Physics).

The *Advanced Summer School 2008* (EAV08) was held at Cinvestav, México City, on July 7-11, 2008. This was the fourth in a new epoch of Schools. As in the three previous years, the aim of the School was to bring together graduate and advanced undergraduate students and researchers from the different areas developed at the Physics Department of Cinvestav. A general overview of the current state of the art was presented on topics related to mathematical physics and gravitation, high energy physics, statistical, medical and solid state physics. The School was attended by 20 researchers from Germany, Italy, Spain, Switzerland, USA, and Mexico, and by more than 100 students from twenty different institutions in Mexico. The courses in the School consisted of a starting introductory general session and continued with a series of lectures at increasing levels to reach applications on special topics.

During the School, six courses were presented. The first one was delivered by Georg Raffelt who lectured on cosmology and black matter and energy. An overview on the workings of the LHC detector was covered by Andreas Hoecker. Ralf Menk discussed the principles and applications of phase contrast radiology. In regard of statistical physics, Eduardo R. Hernández lectured about molecular dynamics simulations in computers and Maria C. Tamargo presented an interesting overview on some semiconductor devices. Luis Viña lectured on Bose-Einstein condensation in solids.

As a complement of the courses, seven invited talks were given by colleagues from Cinvestav. There were also nine short talks given mainly by students. We thank all of them for the professional attitude in preparing their speeches; these contribute to establish an intense and friendly environment of work during the School.

The proceedings are organized in five chapters according with our Physics Department's groups of research. In all cases, the chapter starts with the lecture notes of the courses while contributions are arranged to similarity in subject.

We are indebted to The *United States-Mexico Foundation for Science* and *Academia Mexicana de Ciencias*. The support of the *Mexican Council for Science and Technology* CONACyT is also acknowledged. The Chair of the Physics Department, Isaac Hernández gave us important help to solve many organizational and financial difficulties. Finally, we want to acknowledge all the support that our Institution, *Cinvestav*, gave to the organizing committee for accomplishing all the objectives which made our school a succesful event.

We also would like to express our gratitude to all the people who dedicated many hours of their time to help in the organization of the School, particularly to our support team: Luz del Carmen Cortés, Sara Cruz, Claudia Celia Díaz Huerta and Enrique Campos González. Daniel Pérez gave technical support to speakers, solving all the problems that usually appear in any presentation. We have to thank Miriam Lomelí, secretary of the School, for all the effort and patience during the organization of the event. Her experience in all the administrative duties was indispensable for the organization of the School since its earlier stages.

Luis Manuel Montaño Zetina
Gabino Torres Vega
Miguel García Rocha
Luis Fernando Rojas Ochoa
Ricardo López Fernández

Mexico City, October 2008

SPONSORS

Centro de Investigación y de Estudios Avanzados del IPN (Cinvestav)

Consejo Nacional de Ciencia y Tecnología (CONACyT)

Academia Mexicana de Ciencias (AMC)

Fundación México-Estados Unidos para la Ciencia (FUMEC)

ORGANIZING COMMITTEE

Luis Manuel Montaño Zetina (Chairman) lmontano@fis.cinvestav.mx
Physics Department, Cinvestav
México

Gabino Torres Vega gabino@fis.cinvestav.mx
Physics Department, Cinvestav
México

Miguel García Rocha mrocha@fis.cinvestav.mx
Physics Department, Cinvestav
México

Ricardo López Fernández lopezr@fis.cinvestav.mx
Physics Department, Cinvestav
México

Luis Fernando Rojas Ochoa lrojas@fis.cinvestav.mx
Physics Department, Cinvestav
México

SUPPORT TEAM

Sara Cruz y Cruz (UPIITA/Cinvestav)
Claudia Celia Díaz Huerta (Depto. de Física, Cinvestav)
Luz del Carmen Cortés Cuautli (ESIME)
Enrique Campos González (Depto. de Física, Cinvestav)
Milriam Lomelí, secretary (Depto. de Física, Cinvestav)

I. MATHEMATICAL PHYSICS AND GRAVITATION

Dark Side of the Universe

Georg G. Raffelt

Max-Planck-Institut für Physik (Werner-Heisenberg-Institut), Föhringer Ring 6, 80805 München, Germany

Abstract. The dynamics on scales from dwarf galaxies to the entire universe is dominated by gravitating mass and energy that can not be accounted for by ordinary matter. We review the astrophysical and cosmological evidence for dark matter and dark energy. While there is no obvious interpretation for dark energy, dark matter presumably consists of some new form of weakly interacting elementary particles. We discuss certain candidate particles and search strategies.

Keywords: Dark Matter, Dark Energy
PACS: 95.35.+d, 95.36.+x

INTRODUCTION

The question of what makes up the gravitating mass of the universe is as old as extragalactic astronomy which began with the insight that nebulae such as M31 in Andromeda are actually galaxies like our own. Sometimes they form gravitationally bound clusters. From Doppler shifts of the spectra of the galaxies in the Coma cluster, Zwicky derived in 1933 their velocity dispersion and could thus estimate the cluster mass with the help of the virial theorem, concluding that there was far more gravitating than luminous matter [1]. It took until the 1970s and 1980s to become broadly acknowledged that on scales from dwarf galaxies to the entire universe, luminous matter (stars, hydrogen clouds, x-ray gas in clusters, etc.) can not account for the observed dynamics.

Moreover, dark matter can not consist of ordinary ("baryonic") matter in some class of dark astrophysical objects such as compact stars. The amount of baryons is well measured to be about 4% of the cosmic critical density by the observed cosmic deuterium abundance in the context of big-bang nucleosynthesis and by the acoustic peaks in the angular power spectrum of the cosmic microwave background temperature fluctuations. So, most of the cosmic baryons are actually dark, yet most of the dark matter can not consist of baryons.

In 1973 Cowsik and McClelland proposed that some of the dark matter might consist of neutrinos, the only known examples of weakly interacting particles [2]. Today we know that neutrinos indeed have small masses and provide some dark matter (no less than 0.1% of the cosmic critical density and no more than about 1–2%), but the bulk of dark matter must be something else. As far as we understand today, cosmology is demanding new weakly interacting particles beyond the particle-physics standard model. The experimental search for the proposed candidates is arguably the most important task on the path to answering the question first raised by Zwicky 75 years ago.

Since 1998 the problem has become more mysterious still because yet another dark component became apparent in the global dynamics of the universe. The measured

CP1077, *Advanced Summer School in Physics 2008, Frontiers in Contemporary Physics—EAV08*
edited by L. M. Montaño Zetina, G. Torres Vega, M. García Rocha, L. F. Rojas Ochoa, and R. López Fernández

distance-redshift relation of thermonuclear supernovae (SNe Ia) as standard candles showed that the expansion of the universe is accelerating, in contrast to the deceleration expected if the dynamics is governed by matter or radiation, in whichever physical form. The new component ("dark energy") acts like an ideal fluid with negative pressure, a behavior expected for the vacuum energy predicted by quantum field theory. However, a convincing physical interpretation of this dominating component is missing.

The "cosmic pizza plot" shown in Fig. 1 illustrates that most of the cosmic dynamical mass and energy components are dark. They lack an established physical interpretation and can not be accommodated by standard-model microscopic physics. One may well worry that we are completely missing some crucial insight about how our universe works. However, here we take the point of view that the dark elements of the universe provide us with opportunities to make new discoveries, for example in the form of new particles, that may well show up in the next round of dark matter searches or at the Large Hadron Collider that is about to take up operation at CERN. The dark universe is providing us with windows of opportunity for new and exciting discoveries!

In these brief notes I will present only some of the key ideas about the dark universe. Excellent and detailed recent reviews are provided by Gondolo (2003) [3], by Trodden and Carroll (2004) [4], and by Bertone, Hooper and Silk (2004) [5]. For the original ideas about particle dark matter and experimental search strategies see Primack, Seckel and Sadoulet (1988) [6], for supersymmetric dark matter Jungman, Kamionkowski and Griest (1995) [7]. An early review on dark energy is by Peebles and Ratra (2002) [8], a more recent one by Copeland, Sami and Tsujikawa (2006) [9]. A recent summary with references and resources is given by Ratra and Vogeley in their "Resource letter on the beginning and evolution of the universe" (2007) [10]. A set of concise reviews on issues at the interface of particle physics with cosmology is found in the Review of Particle Physics [11]. And of course, there are numerous cosmology textbooks, each taking its own perspective on the dark universe [12, 13, 14, 15, 16, 17].

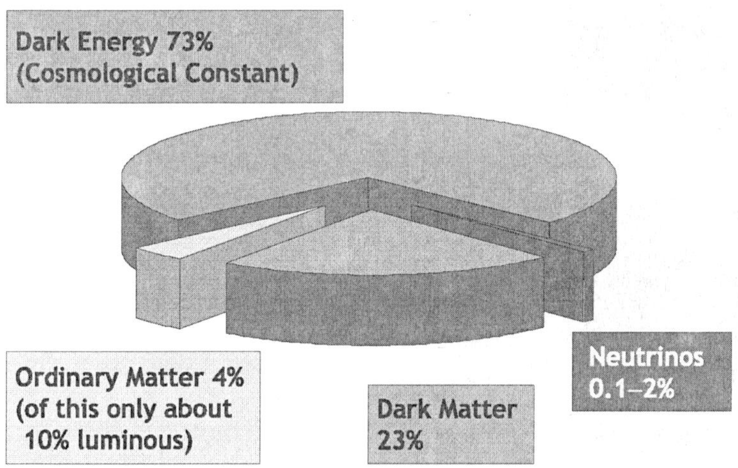

FIGURE 1. Cosmic pizza plot: Components of gravitating mass and energy.

EXPANDING UNIVERSE

Hubble Expansion and Robertson-Walker Metric

Two cosmological key observations were made in the early 1930s, the existence of large amounts of dark matter in galaxy clusters by Fritz Zwicky and the expansion of the universe by Edwin Hubble. Most of the luminous matter is concentrated in spiral galaxies such as the one shown in Fig. 2 that, as we will see, contain large amounts of dark matter. Moreover, the measured spectral Doppler shifts suggest that on average all other galaxies recede from us, with the more distant ones at larger velocities. Edwin Hubble formulated the relationship $v_{\rm recession} = H_0 \times \text{distance}$ named after him where the present-day Hubble parameter is usually written in the form

$$H_0 = h\,100~\text{km s}^{-1}~\text{Mpc}^{-1}\,. \tag{1}$$

Today we know that $h = 0.701 \pm 0.013$, but h used to be very uncertain so that it is often kept as an explicit "fudge factor" that appears in many cosmological quantities. (Astronomical distances are usually measured in parsec (pc) where 1 pc = 3.26 light-years = 3.08×10^{18} cm. Note that 1 pc is a typical distance between stars, 10 kpc is a typical scale for a galactic disk—the Sun is at 8.5 kpc from the center of the Milky Way—and the visible universe has a radius of about 3 Gpc.) If one extrapolates the Hubble expansion linearly into the past one finds that all galaxies must have begun their race about $H_0^{-1} = 14 \times 10^9$ years ago ("Hubble time").

Hubble's linear regression applies only for recession velocities safely below the speed of light. To interpret the cosmic expansion consistently we must transcend the Newtonian concept of space and time. If we assume homogeneity and isotropy on large scales,

FIGURE 2. *Left:* Spiral galaxy NGC 2997. *Right:* Edwin Hubble (1889–1953) who formulated the law of systematic galactic recession velocities.

the space-time geometry is fully described by the Robertson-Walker metric

$$ds^2 = dt^2 + a(t)^2 \left[\frac{dr^2}{1-kr^2} + r^2 \left(d\theta^2 + \sin^2 \theta d\varphi^2 \right) \right] , \tag{2}$$

where ds is the differential of the invariant distance, r, θ and φ are co-moving spherical coordinates with an arbitrarily chosen origin, t is the clock time of an observer that is co-moving with the cosmic expansion, and $a(t)$ is the time-dependent cosmic scale factor. The curvature parameter k takes on the value 0 if space is Euclidean (flat), $k = +1$ if space is positively curved, and $k = -1$ for negative curvature. The time-dependent Hubble expansion parameter is defined as

$$H = \frac{\dot{a}}{a} . \tag{3}$$

H_0 denotes the present-time value ("Hubble constant").

Within general relativity, the Hubble expansion is an expansion of the space between galaxies that remain locally at rest. For small distances we can interpret the same observations in terms of galaxies moving in a static space. Their spectral redshift is then interpreted as a Doppler effect (source and detector moving relative to each other). The general interpretation of the cosmic redshift is that the wavelength gets stretched along with the cosmic expansion. If the wavelength of a photon at emission is $\lambda_{\rm E}$ and at absorption $\lambda_{\rm A}$, then the corresponding ratios of cosmic scale factor is $a_{\rm E}/a_{\rm A} = \lambda_{\rm E}/\lambda_{\rm A}$. Every epoch of the universe can be characterized by its scale factor relative to the present epoch, or equivalently by the redshift $z + 1 = \lambda_0/\lambda$ of some radiation that was emitted at epoch z with wavelength λ and today has wavelength λ_0,

$$\frac{a}{a_0} = \frac{1}{1+z} \quad \text{or} \quad z = \frac{a_0}{a} - 1 . \tag{4}$$

The cosmic expansion affects the distance between galaxies, but not the size of galaxies themselves, or of the solar system, people, atoms and so forth. All of these objects are dominated by forces that are much stronger than cosmic effects. Likewise, the galaxies in clusters, which are gravitationally bound, are not subject to the cosmic expansion. All astrophysical objects tend to have "peculiar velocities" relative to the cosmic expansion. These peculiar motions are determined by local dynamics.

For positive curvature ($k = +1$) space is closed in analogy to a sphere. For flat geometry or negative curvature space is infinite. However, for nontrivial topologies it is also possible that these spaces are finite. In particular, flat geometry allows for simple periodic boundary conditions, corresponding to a torus-like topology. Periodic boundary conditions are used in numerical simulations of structure formation.

In the context of the inflationary picture of the early universe, space was "stretched" by an early phase of exponential expansion, making it seem almost perfectly flat today. Possible topological effects would matter only at distances very much larger than a Hubble volume (the region of space that we can observe). One can speculate that other parts of the universe, not causally connected to ours, could have inflated differently and could even exhibit different physical laws, different dimensions of space, different properties of elementary particles, and so forth [13].

Dynamics: Friedmann Equation

The maximally symmetric cosmological models considered here (homogeneous and isotropic) are fully described by their curvature parameter k, topology, and the function $a(t)$. It is determined by the effect of gravitating mass and energy. For our case the Einstein equations reduce to the Friedmann equation

$$H^2 = \left(\frac{\dot{a}}{a}\right)^2 = \frac{8\pi}{3} G_{\rm N} \rho - \frac{k}{a^2} + \frac{\Lambda}{3}, \tag{5}$$

where $G_{\rm N}$ is Newton's constant, ρ the gravitating density of mass and energy, and Λ is Einstein's cosmological constant.

If we assume flat geometry ($k = 0$) and ignore Λ, then the Friedmann equation establishes a unique relation between H and ρ. Evaluated at the present epoch, this defines the "critical density"

$$\rho_{\rm crit} = \frac{3H_0^2}{8\pi G_{\rm N}} = h^2\, 1.88 \times 10^{-29}\ {\rm g\ cm^{-3}} = h^2\, 10.5\ {\rm keV\ cm^{-3}}\,. \tag{6}$$

The contribution ρ_i of various mass or energy components is usually expressed by

$$\Omega_i = \frac{\rho_i}{\rho_{\rm crit}}\,. \tag{7}$$

Space is flat for $\Omega_{\rm tot} = 1$, the curvature is negative for $\Omega_{\rm tot} < 1$, and positive for $\Omega_{\rm tot} > 1$.

One contribution to ρ is matter, including ordinary and dark matter. One can view it as an ideal fluid with density ρ and vanishing pressure p or on the microscopic level as particles that are essentially at rest relative to the cosmic expansion. The term "particle" could encompass an entire galaxy, but also a single dark matter particle. As the universe expands, matter trivially dilutes as a^{-3}. In a matter-dominated universe, the scale factor evolves as $a(t) \propto t^{1/2}$ (Table 1).

Another contribution is radiation (relativistic particles) where $p = \rho/3$. Here, $\rho \propto a^{-4}$ where one factor a^{-3} comes from volume dilution, one factor a^{-1} from redshift by the expansion. All particles with mass eventually become nonrelativistic and then contribute to matter so that today the remaining radiation consists of photons, some gravitational waves, and perhaps the lightest neutrino. The relative importance of radiation decreases because of redshift so that radiation dominates at early times, matter at late times. Today $\Omega_{\rm M} \approx 0.27$ whereas the cosmic microwave background contributes about $\Omega_\gamma \approx 5 \times 10^{-5}$. Therefore, matter-radiation equality occurs at a redshift of about 3–5 $\times$ 10^3, depending in detail on the neutrino masses.

In the absence of Λ and for $k = +1$, the universe recollapses. For $k = -1$ the curvature term eventually dominates because it scales as a^{-2}. Therefore, at late times curvature dominates. In this "empty universe" the scale factor grows linearly with time and the Hubble parameter remains constant.

The cosmological constant is the most puzzling term. When Einstein derived the general theory of relativity, two parameters needed to be specified. One is Newton's

TABLE 1. Evolution of the cosmic scale factor for generic terms on the r.h.s. of the Friedmann equation.

Type	Equation of state	Behavior under cosmic expansion	Evolution of cosmic scale factor
Radiation	$p=\rho/3$	$\rho \propto a^{-4}$	$a(t) \propto t^{1/2}$
Matter	$p=0$	$\rho \propto a^{-3}$	$a(t) \propto t^{2/3}$
Negative Curvature	$p=0$	$\rho=0$	$a(t) \propto t$
Vacuum energy	$p=-\rho$	$\rho=\Lambda/8\pi G_\mathrm{N}=\mathrm{const.}$	$a(t) \propto \exp(\sqrt{\Lambda/3}\,t)$

constant that is determined by experiments. The other is Λ that has no impact on local physics. At that time the cosmic expansion was not yet observed and it seemed preposterous that the equations implied an expanding or contracting universe. A suitable Λ allows the r.h.s. of the Friedmann equation to be finely balanced to give zero. Later Einstein reportedly called the cosmological term his "biggest blunder."

On the contrary, Zeldovich realized that Λ seems unavoidable in the framework of quantum field theory. Quantum fields are viewed as collections of harmonic oscillators, one for each wave number, each of which must have a zero-point energy of $\frac{1}{2}\hbar\,\omega$. As there are infinitely many modes, the zero-point energy (vacuum energy) of quantum fields seems infinite, or at least very large if there is an ultraviolet cutoff. This vacuum energy, being a property of "empty space," must be invariant so that its energy-momentum tensor should be proportional to the metric, implying $p=-\rho$. As a consequence, ρ does not dilute under cosmic expansion as behooves "empty space" that should be invariant under cosmic expansion. Therefore, vacuum energy behaves exactly like Λ with $8\pi G_\mathrm{N}\rho_\mathrm{vac} \leftrightarrow \Lambda$. Since ρ_vac does not dilute with cosmic expansion, it always dominates at late times and then leads to exponential expansion. The early-universe inflationary phase is thought to be driven precisely by such a term, which however disappears later when the underlying quantum fields decay, thereby reheating the universe.

In summary, the evolution $a(t)$ after the radiation epoch is dominated by matter, curvature, and Λ. A given model has three parameters: The Hubble constant H_0, the fraction of matter today Ω_M, and the fraction of vacuum energy today Ω_Λ. The curvature follows from whether $\Omega_\mathrm{tot}=\Omega_\mathrm{M}+\Omega_\Lambda$ is larger, smaller, or equal to 0. The best-fit parameters from observations to be discussed later are shown in Table 2.

TABLE 2. Best-fit parameters for a cosmic ΛCDM model.

Parameter	Best-fit value
Expansion rate	$H_0=(70.1\pm1.3)$ km s^{-1} Mpc^{-1}
Spatial curvature	$\lvert R_\mathrm{curv}\rvert > 33$ Gpc
Age	$t_0=(13.73\pm0.12)\times10^9$ years
Vacuum energy	$\Omega_\Lambda=0.721\pm0.015$
Cold dark matter (CDM)	$\Omega_\mathrm{CDM}=0.233\pm0.013$
Baryons	$\Omega_\mathrm{B}=0.0462\pm0.0015$

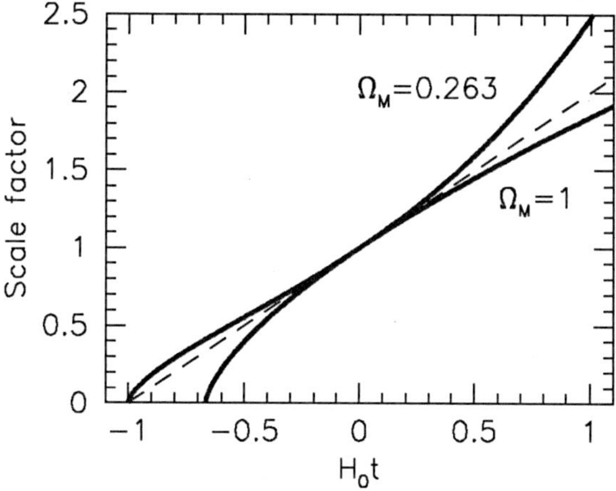

FIGURE 3. Cosmic scale factor as a function of time for geometrically flat models with matter only (lower curve) and one with $\Omega_M = 0.2629$ and $\Omega_\Lambda = 1 - \Omega_M$ (upper curve). The dashed line is a linear extrapolation of the present-day Hubble law, corresponding to an empty universe with negative curvature.

The evolution $a(t)$ for such a three-parameter model is given by the inverse of the "look-back time"

$$H_0 t = \int_1^a \frac{dx}{\sqrt{1 + (x^{-1} - 1)\,\Omega_M + (x^2 - 1)\,\Omega_\Lambda}}, \tag{8}$$

where the normalization $a_0 = 1$ was used. We take the present time to be $t_0 = 0$ and past times negative. The choice $\Omega_\Lambda = 0$ and $\Omega_M = 1$ corresponds to a flat, matter-dominated model where we show $a(t)$ as the lower solid curve in Fig. 3. We also show, as a dashed line, the linear extrapolation of the Hubble law, corresponding to an empty universe with $\Omega_M = \Omega_\Lambda = 0$. By definition the dashed line is the tangent to the solid line at the present epoch. The matter-dominated evolution is decelerating: the expansion rate slows down. The time to the big bang is here $\frac{2}{3} H_0^{-1}$. We show another flat example with $\Omega_M + \Omega_\Lambda = 1$ and $\Omega_M = 0.2629$, chosen such that the big bang occurs at exactly one Hubble time in the past. The universe first decelerates, then accelerates when the vacuum energy takes over. This model is very close to how our universe appears to be evolving.

Accelerated Expansion

In 1998 a revolution swept through cosmology in that accelerated expansion was discovered in the measured brightness-redshift relation of SNe Ia. These exploding stars can be used as cosmic standard candles and for a few weeks are so bright that they can be measured throughout the observable universe. Once automated searches had turned up enough of them, one could construct a Hubble diagram, displaying their redshift vs. their apparent brightness that translates into their distance. The latest "union compilation" of

307 SNe reaches to redshifts up to about 1.6 and thus traces a significant portion of the cosmic evolution [18]. Fitting the observations to a Robertson-Walker-Friedmann-Lemaître model of the universe, confidence contours in the parameter plane of Ω_M and Ω_Λ are shown in the left panel of Fig. 4. Together with other observations to be discussed later, one is restricted to a very small region consistent with a flat universe and the parameters shown in Table 2. However, the crucial point is that even the first useful SN data showed the need for a Λ term because an expansion like the upper curve in Fig. 3 provided a much better fit than one corresponding to the lower curve. In other words, the SN data pointed to an accelerating rather than decelerating expansion.

The quick acceptance of this result was not only due to the SN data alone, but the new "concordance model" also relaxed a number of other tensions that had plagued the previous situation. For example, the expansion age could now easily accommodate the globular cluster ages that had always pointed to a relatively old universe. Moreover, a flat geometry had seemed natural for a long time, yet the dark matter inventory of the universe had always seemed too small to allow for $\Omega_{tot} = 1$.

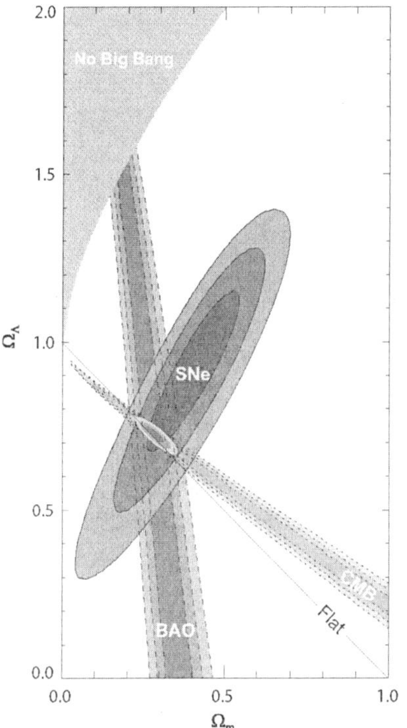

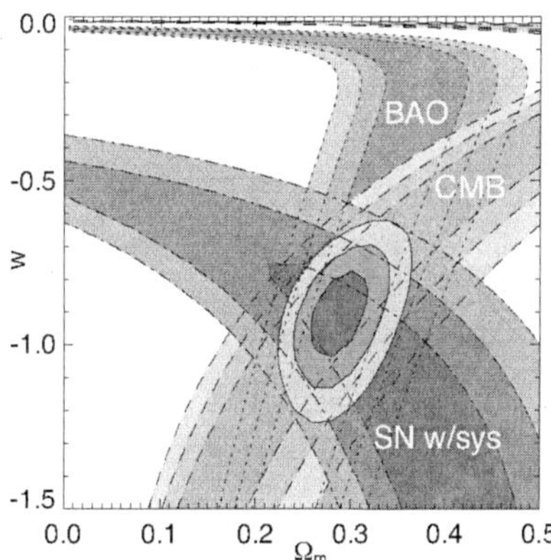

FIGURE 4. *Left:* Allowed regions of Ω_M and Ω_Λ at 68.3, 95.4 and 99.7% confidence level obtained from the cosmic microwave background (CMB) temperature fluctuations, baryon acoustic oscillations (BAO) in large-scale structure data, and from the supernova (SN) brightness-distance relation. Also shown is the combined allowed region. (Figure from Ref. [18].) *Right:* Same for Ω_M and the equation-of-state parameter w for dark energy, assuming a flat geometry. (Figure from Ref. [19].)

Accelerated expansion alone does not prove that it is caused by a Λ term. To consider hypothetical other equations of state (EoS), one may assume that the r.h.s. of the Friedmann equation is governed by an ideal fluid with pressure

$$p = w\rho\,. \tag{9}$$

All examples of Table 1 are of this form. If the EoS parameter obeys $w < -\frac{1}{3}$, the expansion is accelerated, otherwise decelerated. If $w < -1$ ("phantom energy"), the scale factor becomes infinite in a finite time ("big rip").

However, analyzing the data with w a free parameter shows that it is quite close to the "boring" vacuum-energy value $w = -1$ (right panel of Fig. 4). The primary goal of future SN surveys with far better data is to find a deviation from $w = -1$ that would show a deviation from "simple" vacuum energy. For example, a slowly evolving scalar field permeating the universe ("quintessence") would typically produce different w values. Moreover, the simple EoS considered here is but one parametrization of what could be a far more complicated function.

In summary, the SN Ia Hubble diagram and other data, when interpreted in the framework of Friedmann-Lemaître-Robertson-Walker cosmology, requires a "dark energy" component that, on present information, acts exactly like a cosmological constant. While it dominates the mass-energy inventory of the universe, it is very small by particle-physics standards where quantum field theory would naively point to a much larger value. For the time being, the observations are best fit with a simple Λ as originally formulated by Einstein. If there is any positive Λ it eventually dominates the expansion, but why is this happening "now," i.e., why do we live in an epoch when $\Omega_{\rm M}$ and Ω_Λ are comparable? The apparent fine-tuning is even more striking if we note that the age of the best-fit universe is almost exactly H_0^{-1}, i.e., the upper solid curve in Fig. 3 begins almost exactly where the dashed curve begins. Be that as it may, a theoretical understanding of "dark energy" is largely lacking. For further reading see Peebles and Ratra [8] and Copeland, Sami and Tsujikawa [9].

STRUCTURE FORMATION AND PRECISION COSMOLOGY

Matter distribution

While the SN Ia Hubble diagram provides the most direct evidence for accelerated expansion and thus for something like dark energy, far more precise information on the overall model and its detailed parameters derive from other data (Fig. 4). The Hubble diagram uses information from the overall expansion, whereas a huge amount of additional information comes from deviations from perfect homogeneity. While the Friedmann models assume homogeneity on average (in practice on scales exceeding about 100 Mpc), on smaller scales the matter distribution is structured. The formation, evolution and observation of these structures leads to what is called "precision cosmology."

Even a sky map of the observed galaxies shows that they are not uniformly distributed, but a true revolution began in the mid 1980s when the first systematic redshift surveys were performed. Here one measures the redshifts of galaxies in a slice on the sky and

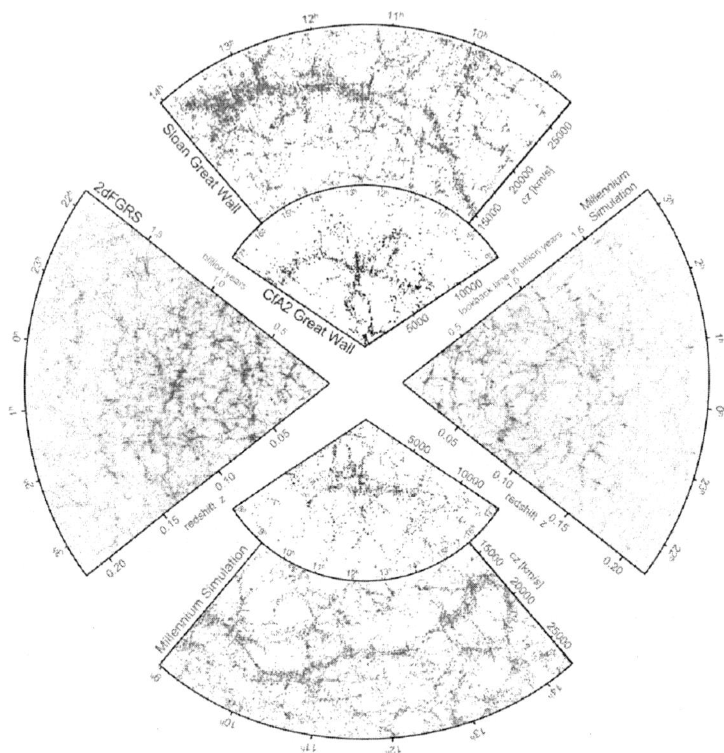

FIGURE 5. Galaxy redshift surveys juxtaposed with numerical simulations. (Figure from Ref. [20].)

shows their redshift as a function of angular location. The redshift serves as a measure of distance through the overall Hubble flow. So one constructs a 3-dimensional map of the universe, the first one being the CfA slice shown in Fig. 5. The other blue images are from more recent surveys that reach to far larger redshifts and encompass far more galaxies. The distribution is indeed uniform on large scales, so the picture of overall homogeneity and isotropy appears justified. However, on small scales the galaxy distribution forms sheets and filaments as well as nodes and clusters. There are voids and large coherent structures called "walls."

How did these structures form? The standard cosmological picture holds that the universe was almost perfectly homogeneous after the early phase of inflation, with tiny density fluctuations imprinted on it. What remains is the action of gravity. Regions of slight overdensity attract more matter, those of underdensity lose, so density contrasts enhance themselves by gravitational instability. Moreover, if the matter distribution is dominated by collisionless dark matter, the action of gravity is all that shapes the structures once an initial fluctuation spectrum has been specified. The red slices in Fig. 5 result from the largest-scale numerical simulations, showing a striking similarity between observed and simulated examples.

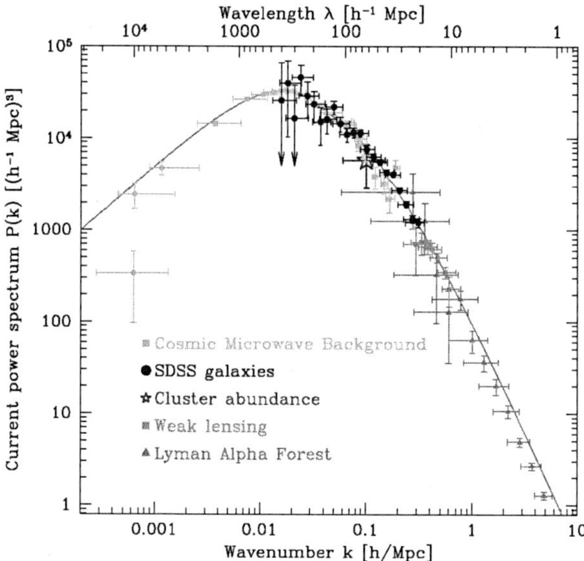

FIGURE 6. Power spectrum of cosmic density fluctuations measured by different probes on different scales. The solid line represents the best-fit cold dark matter (CDM) scenario including a cosmological constant Λ. (Figure from Ref. [21].)

Going beyond a visual comparison requires quantitative measures. The simplest is the power spectrum of density fluctuations. Assume a matter density $\rho(\mathbf{r})$ with an average $\bar{\rho}$. The field of density fluctuations is $\delta(\mathbf{r}) = \rho(\mathbf{r})/\bar{\rho} - 1$, its spatial Fourier transform is $\delta(\mathbf{k})$ with $\mathbf{k}$ the wave vector of a given mode. One expects the fluctuations to be a Gaussian random field, meaning that the $\delta(\mathbf{k})$ have no phase correlation. The random field is then fully characterized by $|\delta(\mathbf{k})|^2$, where no phase information is preserved, a quantity that is the power spectrum $P(\mathbf{k})$. With the additional assumption of isotropy, it only depends on $k = |\mathbf{k}|$. $P(k)$ is identical with the two-point correlation function that is relatively straightforward to extract from redshift surveys.

In Fig. 6 we show observations on different scales overlaid with a best-fit theoretical curve. It assumes an initial power spectrum of the form

$$P(k) \propto k^n \tag{10}$$

where $n = 1$ is the classic Harrison-Zeldovich spectrum, but in general n is a cosmic fit parameter, today found to be $n = 0.960 \pm 0.014$. In addition one allows for a Λ term in the Friedmann equation and assumes that matter is mostly collisionless "cold dark matter" (CDM). These are particles so massive that they become nonrelativistic ("cold") early enough to avoid washing out the initial fluctuations.

While the assumed initial spectrum is a power law and thus featureless, the observed spectrum has a distinct kink at intermediate scales. It can be understood from the way inhomogeneities grow. If space were not expanding, gravitational instabilities would grow exponentially. The cosmic expansion reduces the growth rate and one finds for a matter-dominated situation, where the scale factor grows as $a(t) \propto t^{2/3}$, that density contrasts grow linearly with scale factor: $|\delta(k)| \propto a$. On the other hand, when radiation dominates, the situation is more complicated. Small-scale modes do not grow at all or at most logarithmically, whereas large-scale modes grow as $|\delta(k)| \propto a^2 \propto t$. The distinction between "small" and "large" wavelengths is given by the horizon scale H^{-1} which limits scales that are in causal contact with themselves in view of the limited distance traveled by any signal since the big bang. The universe began radiation dominated and later became matter dominated. In the earlier phase, the growth of small scale (large wave number) modes was suppressed. After matter domination, all modes grow together. Therefore, the kink in the present-day power spectrum is an imprint of the transition between radiation and matter domination in the early universe.

In a phase of exponential expansion caused by a Λ term, structures do not grow at all. Therefore Λ must be small enough to allow for the formation of galaxies, stars, and so forth. The observed Λ is "typical" in the sense that a much larger value would have prevented structures from forming and observers from existing. Such anthropic arguments make only sense if we picture Λ and/or other properties of our universe as numbers that could have been different and are not fixed by the laws of nature. We noted earlier that one way of looking at our universe is that it is but one realization of many possibilities that may actually exist in causally disconnected regions.

The agreement between this simple theory of structure formation and observations over a wide range of scales is stunning and one of the main pillars of support for the standard ΛCDM model. In the inflationary phase of the early universe, the metric was smoothed on large scales, providing for an almost flat geometry on our Hubble scale and far beyond. Another consequence is that fluctuations of the metric were imprinted. They are fundamentally quantum fluctuations stretched by inflation to macroscopic scales. One expects nearly scale-invariant Gaussian fluctuations and a spectral power-law index n close to unity, but slightly smaller. This expectation is also borne out by observations.

Cosmic Microwave Background (CMB)

The observation of the cosmic microwave background radiation by Penzias and Wilson in 1965 marked the true beginning of big-bang cosmology. This ubiquitous thermal radiation with a present-day temperature of 2.725 K proves that the universe was once hot. Photons decoupled from the cosmic plasma at a redshift of about 1100 when ordinary matter had "recombined" to form neutral atoms, a medium in which photons stream freely. From our perspective we are seeing a spherical shell of primordial matter around us where the photons originate, the "surface of last scattering." A different observer elsewhere in the universe sees a different spherical shell around his own location, so the surface of last scattering is different for each observer.

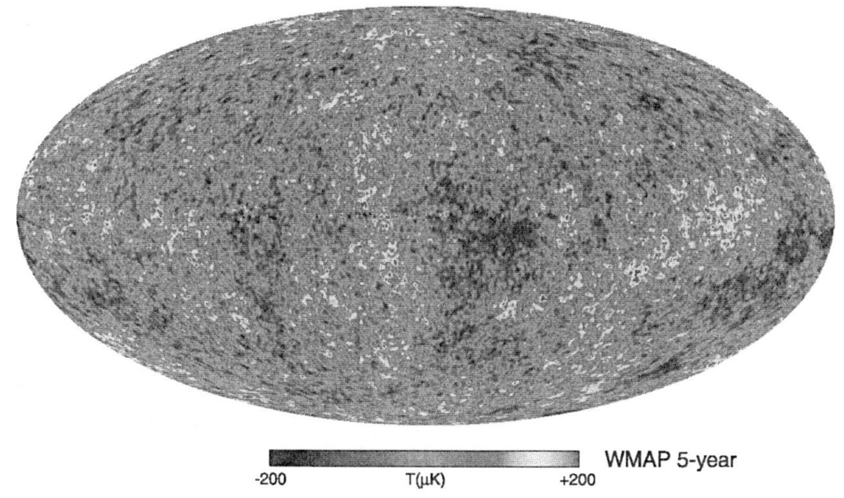

FIGURE 7. Sky map of the cosmic microwave background temperature fluctuations as measured by the WMAP satellite after 5 years of data taking. (WMAP Science Team [22].)

Within tiny experimental errors, the CMB shows an exact blackbody spectrum. The temperature is almost perfectly uniform over the sky. This is surprising because the different regions, for example in opposite directions from us, were never in causal contact. An early inflationary phase explains the uniformity in that these regions were in causal contact until they were driven apart by the early exponential expansion.

Upon closer scrutiny one observes a small dipole in the temperature sky map on the level of 10^{-3}. This shows our peculiar motion relative to the cosmic rest frame: we are seeing the same thermal radiation red-shifted in one direction and blue-shifted in the opposite direction. In other words, the CMB as a heat bath defines the cosmic reference frame where the universe is isotropic, i.e., the frame where the Robertson-Walker-Friedmann-Lemaître cosmology is formulated.

Upon even closer scrutiny, at the level of temperature differences of 10^{-5}, one finds temperature fluctuations that must exist because the universe is not perfectly homogeneous. The remarkable feat of observing these tiny fluctuations was first achieved by the COBE satellite in 1992. The most detailed modern CMB temperature map from the WMAP satellite's 5-year observations (2008) is shown in Fig. 7. What we are seeing are temperature fluctuations, corresponding to density fluctuations, of the small fraction of baryonic matter, whereas dark matter does not directly interact with photons.

This sky map provides little direct cosmological information. What is important once more are the statistical properties of the distribution. To this end one expands the sky map in spherical harmonics and displays the power spectrum as a function of multipole index ℓ as shown in Fig. 8. In contrast to the matter power spectrum discussed earlier, the multipole spectrum shows a distinct pattern of "acoustic peaks." The density fluctuations in the primordial soup of baryonic matter and photons lead to oscillations of this plasma that is an "elastic medium," in contrast to the collisionless dark matter. As in any elastic

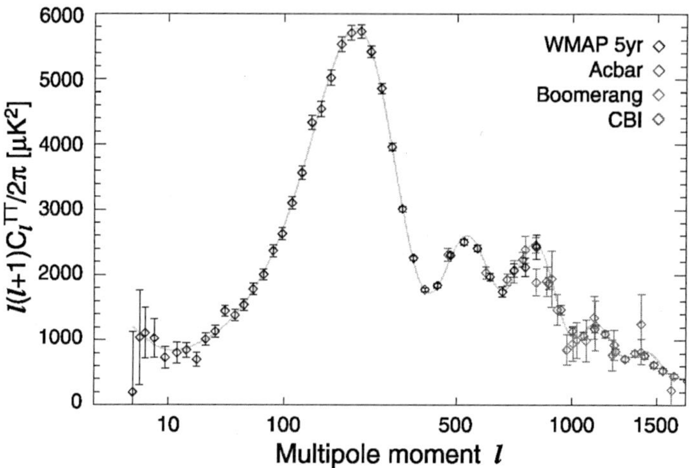

FIGURE 8. Angular power spectrum of the cosmic microwave background temperature fluctuations as measured by WMAP, ACBAR, Boomerang and CBI. (WMAP Science Team [23].)

medium, density fluctuations propagate as sound waves. Different modes oscillate with different frequencies, each reaching a different phase of oscillation since the big bang until the epoch of photon decoupling. The acoustic peaks in Fig. 8 are essentially a snapshot of the phases reached by the acoustic waves at photon decoupling.

The pattern of acoustic peaks has a lot of structure and therefore carries a lot of information about the parameters of the cosmological model. The sound waves in the early-universe plasma do not depend on the large-scale curvature of the universe that may exist today but was negligible at these early times. The sound waves provide an absolute length scale or "standard stick" that does not depend on the global cosmic parameters. On the other hand, upon propagating to us, we are seeing these scales as extending certain angles that do depend on the spatial curvature of the universe. In particular, the angular location of the first acoustic peak depends directly on the large-scale curvature of the universe. The measured result tells us with high precision that our universe is flat, the curvature radius exceeding about 30 Gpc, to be compared with a Hubble radius of about 3 Gpc. In the left panel of Fig. 4 we see the CMB best-fit region to be essentially aligned with the locus of flat models where $\Omega_M + \Omega_\Lambda = 1$. Moreover, it intersects with the SN Ia best-fit region for flat-universe parameters.

The measured curve of acoustic peaks carries so much information that it has been termed the "Cosmic Rosetta Stone." Another crucial parameter is the cosmic baryon density that provides the inertia for the acoustic waves. Roughly speaking, it is the relative height of the first and second acoustic peaks that give us Ω_B. The best-fit value in a common fit of all parameters is shown in Table 2 and tells us that most of the cosmic matter is dark. However, the density of luminous matter is only about a tenth of that, so actually most baryons are dark and presumably dispersed.

The baryon fraction also influences the relative abundances of light elements formed in the early universe, notably hydrogen, helium and deuterium. The tiny abundance of this latter isotope is a sensitive probe of the baryon density. One measures its cosmic abundance in the absorption spectra of distant quasars in intergalactic hydrogen clouds where one can differentiate between hydrogen and deuterium absorption lines by the tiny isotope shift. The baryon abundance inferred by this method is perfectly consistent with the CMB value, but its uncertainties are much larger so that this classic measurement no longer plays a crucial role for cosmological parameter fitting. Of course, one measures the baryon abundance in vastly different cosmic epochs so that the good agreement provides gratifying evidence for the overall consistency of the standard picture.

The CMB is slightly polarized and these polarizations have been measured. The sky map of polarizations, in principle, allows one to distinguish between density fluctuations and tensor disturbances (gravitational waves). The next big prize in CMB physics is to find the spectrum of these subdominant tensor modes because their detection would be an important further confirmation of early-universe inflation that would excite tensor modes besides density fluctuations. The PLANCK satellite, to be launched shortly, may have a first chance of observing this fundamental effect.

Acoustic oscillations of the primordial plasma manifest themselves primarily in the temperature fluctuations of the CMB, but should also imprint themselves in the matter power spectrum as tiny modulations. These "baryon acoustic oscillations" (BAO) have recently been found by the Sloan digital sky survey (SDSS). The cosmological best-fit parameters implied by the BAOs are shown in Fig. 4, intersecting with the regions from other observations and thus allowing one to perform a common fit.

In summary, we get a consistent picture of the universe, accommodating all crucial observations, if we begin with a simple Friedmann cosmology and allow for dark ingredients in the form of cold dark matter (that clusters on small scales) and dark energy (that only influences the overall Hubble expansion). Together with a nearly scale-invariant spectrum of density fluctuations as predicted by inflation and the theory of gravitational instability, the observed density fluctuations on all scales and the global properties of the universe are all accounted for. The few adjustable parameters of the model are very well determined by a common fit to all relevant observables.

DARK MATTER ON SMALL SCALES

Spiral Galaxies

If the global dynamics of the universe and a consistent picture of structure formation require large amounts of cold dark matter, its presence should also be evident in smaller systems. It is here, of course, where dark matter was first discovered. Spiral galaxies provide perhaps the most direct evidence. They consist of a central bulge and a thin disk. One may use the Doppler shift of spectral lines to obtain the orbital velocity of the disk as a function of radius ("rotation curve"). If the mass is indeed concentrated in the bulge, then the orbital velocity of the disk should show the Keplerian $1/\sqrt{r}$ decrease familiar from the solar system (left panel of Fig. 9).

However, one always finds that the orbital velocity stays essentially flat as a function of radius, implying much more gravitating mass interior to a given radius than implied by the luminous matter. This behavior became established in the 1970s, at first by the

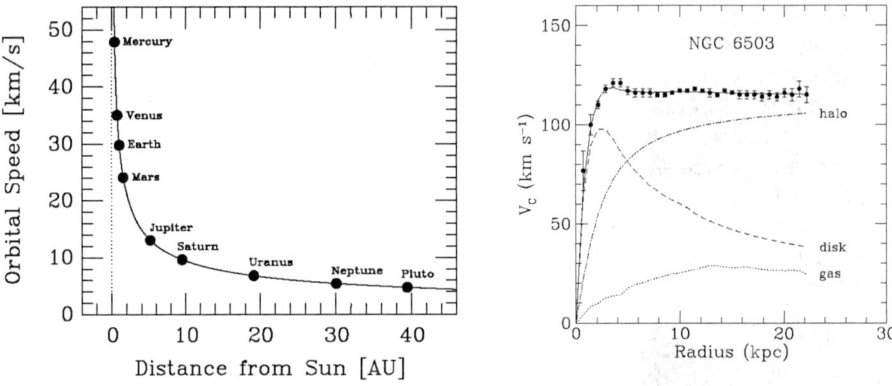

FIGURE 9. *Left:* "Rotation curve" of the solar system, showing the Keplerian $1/\sqrt{r}$ decrease of the planetary orbital velocities. *Right:* Rotation curve of the spiral galaxy NGC 6503 from radio observations of hydrogen in the disk [24]. The last measured point is at 12.8 disk scale-lengths. The dashed line is the rotation curve caused by the disk alone, the dot-dashed line from the dark matter halo alone.

optical observations of Vera Rubin. Spiral galaxies tend to have neutral hydrogen in the plane of their disks that reaches to much larger galactocentric radii than optical tracers. A typical case for such a rotation curve is shown in the right panel of Fig. 9. More than a thousand rotation curves have been studied and have allowed one to derive a "universal rotation curve" that depends only on a few empirical parameters [25].

Galaxy Clusters

Clusters of galaxies are the largest gravitationally bound systems in the universe. We know today several thousand clusters; they have typical radii of 1.5 Mpc and typical masses of $5 \times 10^{14} M_\odot$. Zwicky first noted in 1933 that these systems appear to contain large amounts of dark matter. He used the virial theorem which tells us that in a gravitationally bound system in equilibrium $2\langle E_{\text{kin}}\rangle = -\langle E_{\text{grav}}\rangle$ where $\langle E_{\text{kin}}\rangle = \frac{1}{2}m\langle v^2\rangle$ is the average kinetic energy of one of the bound objects of mass m and $\langle E_{\text{grav}}\rangle = -mG_{\text{N}}\langle M/r\rangle$ is the average gravitational potential energy caused by the other bodies. Measuring $\langle v^2\rangle$ from the Doppler shifts of the spectral lines and estimating the geometrical extent of the system gives one directly an estimate of its total mass M. As Zwicky stressed, this "virial mass" of the clusters far exceeds their luminous matter content, typically leading to a mass-to-light ratio of around 300.

With the beginning of x-ray astronomy it was recognized that clusters are powerful x-ray sources. They contain large amounts of hot gas, in virial equilibrium with the galaxies, that therefore is hot. The mass of this x-ray emitting matter far exceeds the luminous mass of the galaxies, yet is much less than the overall mass of the cluster. Besides the virial method, today detailed mass profiles can be measured using the gravitational light deflection of background galaxies. Galaxy clusters are so large that

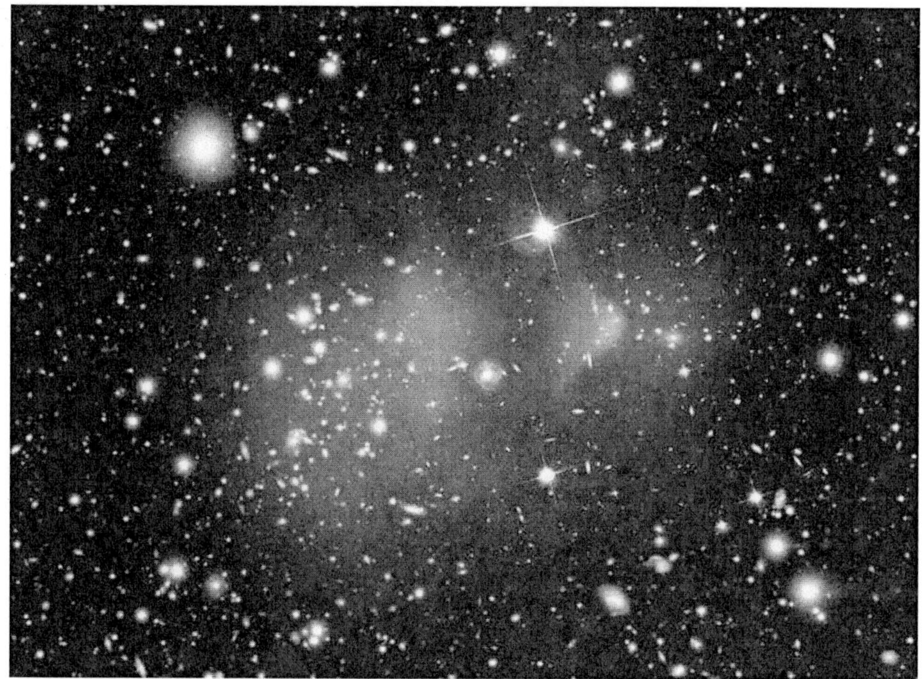

FIGURE 10. Bullet cluster (1E 0657-56). Optical image overlaid with a false-color x-ray image (red) and the mass density (blue) from gravitational lensing. (Credit: X-ray: NASA/CXC/CfA/M. Markevitch et al.; Optical: NASA/STScI; Magellan/U. Arizona/D. Clowe et al.; Lensing Map: NASA/STScI; ESO WFI; Magellan/U. Arizona/D. Clowe et al.)

over a Hubble time no significant mass exchange can have taken place with their environment so that their mass inventory should be typical for the universe. Indeed, one finds that the mass ratio of x-ray emitting gas to the total mass is similar to the ratio of $\Omega_B/(\Omega_B + \Omega_{CDM})$ implied by precision cosmological methods. Before the age of precision cosmology, this argument was turned around. The fraction of baryonic to total mass found in clusters together with the BBN-implied baryon density suggested that the total matter density could not be critical and that therefore it was very difficult to have a flat matter-dominated universe.

A spectacular case for dark matter in clusters was recently made by x-ray observations of the "bullet cluster" (Fig. 10). It consists of a small cluster (the "bullet") that has collided with a large one and penetrated it, emerging on the other side. The blue false-color indication of the mass distribution is found to be very different from the red false-color x-ray image. The x-ray emitting gas was stripped from the two clusters as one moved through the other, whereas the galaxies and dark matter, being collisionless, moved right through. This image has been taken as one of the most direct proofs for the existence of dark matter and its nature as a gas of "collisionless stuff."

WEAKLY INTERACTING PARTICLES

Neutrinos

If much of the gravitating matter in the universe is invisible and does not consist of baryons, then what could it be? Within the particle physics standard model, all stable matter consists of quarks of the first generation and electrons (Fig. 11). The simple periodic system of elementary particles leaves no room for weakly interacting dark matter except in the form of neutrinos. In spite of their weak interactions, neutrinos were in thermal equilibrium in the early universe until their interactions became too slow when the cosmic temperature fell below about 1 MeV. Since that time they streamed freely—we could see a neutrino sphere of last scattering if it were possible to detect neutrinos that today have energies as low as CMB photons.

The cosmic neutrino density is easy to predict if they were originally in thermal equilibrium. The only nontrivial point is that the disappearance of electrons and positrons at $T \lesssim m_e = 0.511$ MeV heats the photon gas relative to neutrinos which are already decoupled. One finds the standard result that the number density of neutrinos plus antineutrinos of one family is $n_\nu = \frac{3}{11} n_\gamma$, the latter being the present-day density of CMB photons of 410 cm^{-3} so that $n_\nu = 112$ cm^{-3}. The density contribution of one flavor is simply $\rho_\nu = \frac{3}{11} n_\gamma m_\nu$. Comparing with the cosmic critical density of Eq. (6) reveals

$$\Omega_\nu h^2 = \frac{\sum m_\nu}{94\ \text{eV}}. \tag{11}$$

With current cosmological parameters neutrinos would be all of dark matter if $\sum m_\nu \approx$ 11 eV. From oscillation experiments we know that the neutrino mass differences are negligible on this scale and from tritium beta decay experiments that the mass for a single flavor is below about 2.3 eV, so experimentally $\sum m_\nu \lesssim 7$ eV.

However, neutrinos can not be the bulk of dark matter. First, if neutrinos were to provide the dark matter in spiral galaxies, their mass would have to exceed about 20 eV or even 100 eV for dwarf galaxies. The phase space of neutrinos bound to a galaxy

	Quarks		Leptons	
	Charge +2/3	Charge −1/3	Charge −1	Charge 0
1. Family	Up u	Down d	Electron e	e-Neutrino ν_e
2. Family	Charm c	Strange s	Muon μ	μ-Neutrino ν_μ
3. Family	Top t	Bottom b	Tau τ	τ-Neutrino ν_τ
	Gravitation			
	Weak Interaction			
	Electromagnetic Interaction			
	Strong Interaction			

FIGURE 11. Three families of standard-model particles and their interactions.

is limited so that essentially the Pauli exclusion principle prevents too many low-mass fermions to accumulate ("Tremaine-Gunn limit").

From the perspective of structure formation, neutrinos form "hot dark matter" because they stay relativistic for a long time and their free streaming erases small-scale density fluctuations. Since neutrinos are some fraction of the total cosmic matter density, the matter power spectrum would be suppressed on small scales (large wave numbers). Indeed, $\sum m_\nu$ is an unavoidable cosmic fit parameter. On the present level of precision, cosmic data provide a best-fit value $\sum m_\nu = 0$ and an upper limit $\sum m_\nu \lesssim 0.6$ eV, implying $\Omega_\nu \lesssim 1.3\%$. The exact limit depends on the data sets used in the analysis. One problem is that the strongest limits come from the smallest scales where the matter power spectrum is in the nonlinear regime so that the data are more difficult to interpret. Whatever the exact limit, massive neutrinos can not provide the bulk of the dark matter. One may hope though that eventually precision cosmology will provide a measurement for $\sum m_\nu$ that can be compared with future experiments. Oscillation data imply $\sum m_\nu \gtrsim 50$ meV, a value that may be reachable by precision cosmology in the distant future.

Let us ignore oscillation experiments for the moment and assume that one of the neutrinos could have a large mass. When we show its contribution Ω_ν to the cosmic matter inventory as a function of mass, we find the "Lee-Weinberg-curve" of Fig. 12. The low-mass branch of course reproduces the previous result that the number density is $\frac{3}{11}$ that of photons and thus the mass contribution grows linearly with m_ν. Something new happens when the mass exceeds a few MeV, the temperature where neutrinos decouple from the thermal early-universe plasma. If $m_\nu \gg T$, their number density is suppressed by a Boltzmann factor $\exp(-m_\nu/T)$. In other words, as the cosmic T falls, neutrinos annihilate and disappear without being replenished by inverse reactions. This continues until they become so dilute that they no longer meet often enough to annihilate and a relic population "freezes out." Computing this "thermal relic population" is a straightforward exercise. For the given weak interaction strength of the standard model, Ω_ν steeply decreases with m_ν, providing the required dark matter density for $m_\nu \approx 10$ GeV. This would be safely in the regime of cold dark matter.

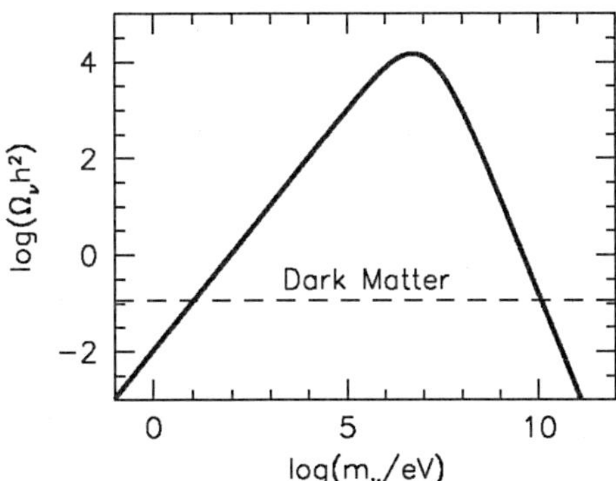

FIGURE 12. Cosmic neutrino mass density as a function of neutrino mass (Majorana case). The dashed line shows the required dark matter density.

Weakly Interacting Massive Particles (WIMPs)

The measured Z^0 decay width shows that there are exactly three neutrino families with masses up to $m_Z/2 = 46$ GeV. Even if a fourth generation with a larger mass existed, the abundance would be too low to provide for dark matter. Still, a hypothetical massive particle with approximately weak interactions, a generic weakly interacting massive particle (WIMP) provides a good dark matter candidate because it naturally survives as a thermal relic in sufficient numbers after annihilation freeze-out. If the interaction strength were somewhat weaker, the particle would freeze out earlier and a greater number would survive ("survival of the weakest"). Typically one thinks of these particles as Majorana fermions, i.e., neutrino-like particles that are their own antiparticles, and thus can annihilate with themselves.

Supersymmetric extensions of the standard model motivate precisely such particles. One postulates that every bosonic degree of freedom is matched by a supersymmetric fermionic one and vice versa. Normal and supersymmetric particles differ by a quantum number called R-parity which may be conserved so that the lightest supersymmetric particle (LSP) would be stable. If the LSP is the lightest "neutralino," i.e. the lightest mass eigenstate of a general superposition of the neutral spin-$\frac{1}{2}$ fermions photino, Zino, and Higgsino, then we have a perfect neutrino substitute. The interaction would be roughly, but not exactly, of weak strength. In detail the annihilation and scattering cross sections depend on specific assumptions and on the values of numerous parameters.

The main motivation for supersymmetry is the theoretical issue of stabilizing the electroweak scale of about 250 GeV against radiative corrections that would drive it up to the Planck mass of about 10^{19} GeV. For a typical gauge interaction strength, particles with masses around the electroweak scale have the right properties for cold dark matter ("WIMP miracle"). The Large Hadron Collider at CERN, taking up operation shortly, was designed to probe physics at this scale, notably to detect the Higgs particle and physics beyond the standard model that one may well expect at this scale such as supersymmetry. Therefore, if the ideas about the cosmological role of WIMPs are correct, the LHC may well discover these particles or indirect evidence for them.

Direct Search for WIMPs

In the mid-1980s it became clear that one can search for dark-matter WIMPs in the laboratory by a variety of methods. One always uses elastic WIMP collisions with the nuclei of a suitable target, for example a germanium crystal. Galactic WIMPs move with a typical virial velocity of around 300 km s^{-1}. If their mass is 10–100 GeV their energy transfer in such an elastic collision is of order 10 keV. Therefore, the task at hand is to identify such energy depositions in a macroscopic target sample. Three signatures can be used. (i) Scintillation light. (ii) Ionization, i.e., the liberation of free charges. (iii) Energy deposition as heat or phonons.

The direct search by such methods has turned into quite an industry. In Fig. 13 we show the three main techniques as a triangle, and list projects using them. The main problem with any such experiment is the extremely low expected signal rate. In detail it depends on the assumed WIMP properties and target material, but a typical number is below 1 event kg^{-1} day^{-1}, a counting-rate unit usually employed in this field. To reduce natural radioactive contaminations one must use extremely pure substances and to reduce cosmic-ray backgrounds requires underground locations, for example in deep mines ("underground physics"). Background can be suppressed by using two techniques simultaneously. In Fig. 13 the projects noted at a side of the triangle use the corresponding combination of techniques.

Most experiments have only reported limits on the interaction cross section of putative galactive WIMPs (Fig. 14). Intruigingly, the current experiments already bite deeply into the parameter space expected for supersymmetric particles and thus are in a position to actually detect supersymmetric dark matter if it exists.

One problem is how one would attribute a tentative signal unambiguously to galactic WIMPs rather than some unidentified radioactive background. One signature is the annual signal modulation which arises because the Earth moves around the Sun while the Sun orbits around the center of the galaxy. Therefore, the net speed of the Earth relative to the galactic dark matter halo varies. The DAMA experiment, using the NaI scintillation technique, has actually reported such a modulation for the past ten years [26]. Other experiments have now excluded this region when interpreted in terms of typical supersymmetric models. On the other hand, each detector uses a different material and their results do not directly compare. In fact, one way of identifying a WIMP is to see it with different detectors using different target materials. In this sense DAMA remains unconfirmed, but also is not strictly excluded as a dark matter signature. Ten years ago, the DAMA region in Fig. 14 was far below the sensitivity of any other experiment, whereas now it is vastly excluded if one compares the cross sections naively. This illustrates the enormous progress that has been made over the past decade.

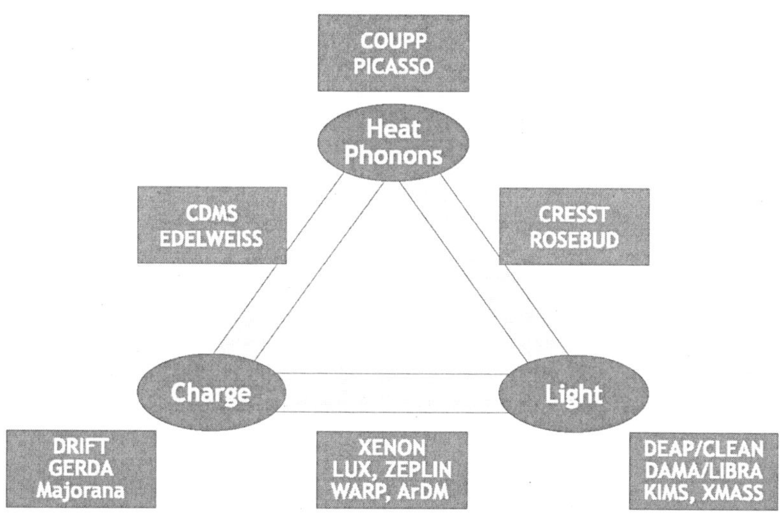

FIGURE 13. Possible signals for WIMPs and experiments using them singly or in combination to suppress backgrounds.

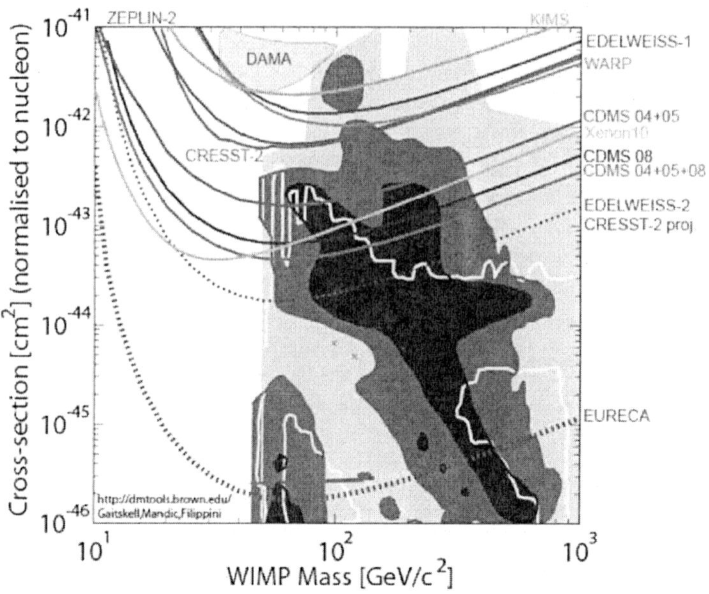

FIGURE 14. Experimental limits and foreseen sensitivities for WIMPs with spin-independent interactions. The theoretical benchmark is given by supersymmetric models.

The current level of sensitivity to scattering cross sections per nucleon is around 5×10^{-44} cm^2. The goal is to reach roughly the 10^{-45}–10^{-46} cm^2 level. On the path to this sensitivity there is a vast range of opportunity to find supersymmetric dark matter or other WIMPs.

Detecting WIMP Annihilation

In the WIMP paradigm it is usually assumed that these particles are self-conjugate spin-$\frac{1}{2}$ particles, i.e., they are their own antiparticles (Majorana fermions). Supersymmetric neutralinos have precisely this property. Therefore, one can perform the early-universe freeze-out calculation without having to worry about a possible asymmetry between WIMPs and anti-WIMPs. Of course, for ordinary matter it is precisely such an asymmetry which allows baryonic matter to exist; here the challenge is to explain how the universe developed this asymmetry (baryogenesis). One actually needs physics beyond the standard model to achieve the observed asymmetry. For WIMPs it is their very weak interaction that allows them to freeze out in sufficient numbers even without an asymmetry.

If WIMPs were produced and destroyed in the early universe by pair processes, the same goes on today, except with a much smaller rate. WIMPs froze out because they became too dilute to annihlate efficiently. However, whenever WIMPs concentrate in some location one may hope for a potentially detectable annihilation signal. Since WIMPs are thought to be heavy, their properties being likely connected to the electroweak scale close to the TeV range, there is enough energy in an annihilation event to produce all sorts of final states, including protons and antiprotons, electrons and positrons, high-energy γ-rays and neutrino pairs. The efficiency of different channels depends on detailed models.

Dark matter particles are naturally concentrated in galaxies where a typical density is about 10^5 times the cosmic average. The annihilation rate scales with the square of the particle density, so one gains 10 orders of magnitude relative to the cosmic average. Therefore, WIMP self-annihilation can lead to a contribution of cosmic-ray antiprotons, positrons and γ-rays, although the interpretation of possible signals depends on understanding the ordinary cosmic-ray background. The γ-ray excess observed by the EGRET satellite could be attributed to WIMP annihilation [27], although the accompanying antiproton flux seems too large. Once more one is confronted with the problem of how to unambiguously attribute a tentative signal to dark matter.

Therefore, "smoking-gun signatures" would be especially welcome that are difficult to ascribe to any other source. One example is WIMP annihilation into a single pair of γ-rays, producing a monochromatic line with energy corresponding to the WIMP mass. Hopes are high that the GLAST satellite, that was successfully launched on 11 June 2008, may detect such a signature. It explores a range of γ-ray energies in the few 100 GeV range that has not been explored yet.

In order to detect WIMPs in this way, one likely needs a significant "boost factor," meaning that dark matter is distributed so inhomogeneously that $\langle n_\chi^2 \rangle \gg \langle n_\chi \rangle^2$. Cold dark matter simulations indeed find that WIMPs tend to cluster on smaller and smaller scales in an almost fractal-like fashion. This effect may be a problem for CDM cosmology because the baryonic matter distribution does not show such behavior even though one would expect baryons and thus stars (luminous matter) to build up in these small-scale sub-galactic structures.

Another smoking-gun signature could come from high-energy neutrinos emitted by the Sun where WIMPs would build up. As they stream through the Sun, they sometimes collide with nuclei and get gravitationally trapped, sinking to the solar center where they concentrate. Only final-state neutrinos could escape after annihilation. The high-energy neutrino telescopes IceCube that is built at the South Pole and ANTARES being built in the Mediterranean could eventually detect this flux. In terms of supersymmetric models, this method is complementary to direct laboratory searches. At present only limits exist that begin to bite into the supersymmetric parameter space.

If the dark matter indeed consists of WIMPs with properties as discussed here, the upcoming 5–10 years could prove crucial. With the LHC turning on, GLAST having been launched, IceCube reaching the cubic-kilometer size, and direct searches gearing up for larger size and greater sensitivity, one may hope that one or more methods will lead to a clear detection.

AXIONS

The search for WIMPs has strong synergy with other experimental activities, notably the LHC, neutrino and γ-ray astronomy, and the development of low-background detectors. Moreover, the supersymmetric motivation for neutralinos is compelling to many particle physicists. Still, we should not blind ourselves to the fact that the nature of dark matter is completely unknown and one should keep an open mind to other possibilities of which there exist many. However, few are both well-motivated and at the same time offer systematic opportunities to search for them. One alternative to supersymmetric particles that is both well motivated and searchable are axions. They form a particle dark-matter candidate sui generis in that they are very weakly interacting very low-mass bosons and yet a candidate for cold dark matter.

Axions are motivated by the Peccei-Quinn solution of the "strong CP problem." The nontrivial vacuum structure of QCD produces CP violating effects that are represented by a parameter $0 \leq \bar{\Theta} \leq 2\pi$. The experimental limits on the neutron electric dipole moment (a CP-violating quantity) reveal $\bar{\Theta} \lesssim 10^{-10}$, although naively it should be of order unity. The Peccei-Quinn mechanism holds that $\bar{\Theta}$ should be re-interpreted as a physical field $a(x)$ in the form $\bar{\Theta} \to a(x)/f_a$, where f_a is an energy scale, the Peccei-Quinn scale or axion decay constant. The CP-violating Lagrangian produces a potential which drives the axion field to the CP-conserving position corresponding to $\bar{\Theta} = 0$ ("dynamical symmetry restoration").

The excitations of the new field are axions. Their properties are essentially fixed by the value of f_a. Axions are closely related to neutral pions: they mix with each other with an amplitude of about f_π/f_a where $f_\pi = 93$ MeV is the pion decay constant. The axion mass and interactions follow roughly by scaling the corresponding π^0 properties with f_π/f_a. The axion interactions are inversely proportional to f_a and thus can be arbitrarily small ("invisible axions"). Strong laboratory and stellar energy-loss limits suggest that axions, if they exist, must be very light ($m_a \lesssim 10^{-2}$ eV) and very weakly interacting (Fig. 15).

In concrete models the axion field is interpreted as the phase of a new Higgs field $\Phi(x)$ after spontaneous breakdown of a new chiral U(1) symmetry, i.e., it is a Nambu-Goldstone boson. When the temperature in the early universe falls below f_a, the symmetry breaks down and $\Phi(x)$ finds a minimum somewhere in the rim of the Mexican hat, selecting one value for the axion field and thus for $\bar{\Theta}$. Later at a temperature $T = \Lambda_{\rm QCD} \approx 200$ MeV the QCD phase transition occurs, switching on the potential that drives the axion field to the CP-conserving minimum. In other words, at the QCD transition the Peccei-Quinn symmetry is explicitly broken, the Mexican hat tilts, and the axion field rolls towards the CP-conserving minimum.

In this way the initial "misalignment angle" $\bar{\Theta}_i$ sets the axion field into motion and thus excites coherent oscillations. They correspond to an axionic mass density of the present-day universe of about [11]

$$\Omega_a h^2 \approx 0.7 \left(\frac{f_a}{10^{12}\,\mathrm{GeV}} \right)^{7/6} \left(\frac{\bar{\Theta}_i}{\pi} \right)^2 . \qquad (12)$$

If inflation occurs before the PQ transition, the axion field will start with a different $\bar{\Theta}_i$ in each region which is causally connected at $T \approx f_a$. Then one has to average over all regions to obtain the present-day axion density and finds

$$\Omega_a h^2 \approx 0.3 \left(\frac{f_a}{10^{12}\,\mathrm{GeV}} \right)^{7/6} \tag{13}$$

Because axions are the Nambu-Goldstone mode of a spontaneously broken U(1) symmetry, cosmic axion strings form by the Kibble mechanism. This and other contributions from axion-field inhomogeneities are comparable to the homogeneous mode from misalignment, even though significant modification factors can occur. Therefore, the axion mass required for dark matter is rather uncertain as indicated in Fig. 15.

Axions produced by strings or the misalignment mechanism were never in thermal equilibrium. The field modes are highly occupied, forming a Bose-Einstein condensate. Axions are nonrelativistic almost from the start and thus form cold dark matter, in spite

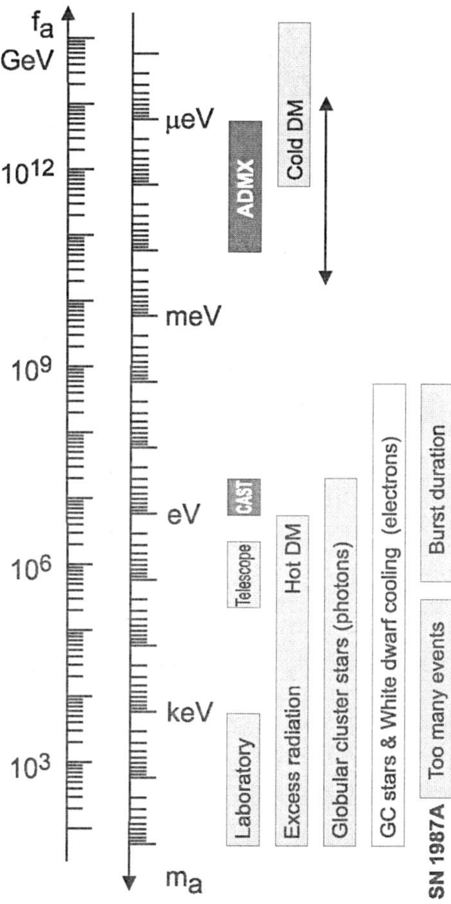

FIGURE 15. Limits on the axion decay constant and axion mass from different sources.

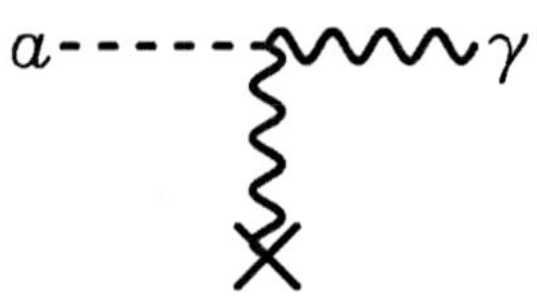

FIGURE 16. Primakoff conversion of axions into photons in an external electromagnetic field.

of their small mass. If axion interactions were sufficiently strong ($f_a \lesssim 10^8$ GeV) they would have reached thermal equilibrium, leading to a thermal axion background in analogy to neutrinos. Axions, like neutrinos, contribute hot dark matter if their mass is in the eV range, but they are cold dark matter for much smaller masses, whereas WIMPs are cold dark matter for much larger masses. WIMPs are thermal relics whereas cold dark matter axions are nonthermal relics.

If axions are the galactic dark matter one can search for them in the laboratory. The detection principle is analogous to the Primakoff effect for neutral pions which can convert into photons in an external electromagnetic field due to their two-photon vertex (Fig. 16). Dark-matter axions would have a mass in the μeV–meV range. Their velocity dispersion is of order the galactic virial velocity of around $10^{-3}c$ so that their kinetic energy is exceedingly small relative to their rest mass. Noting that a frequency of 1 GHz corresponds to 4 μeV the Primakoff conversion produces microwaves. Because the galactic axions are nonrelativistic while the resulting photons are massless the conversion involves a huge momentum mismatch which can be overcome by looking for the appearance of excitations of a microwave cavity.

After several pilot experiments, a full scale axion search, the ADMX experiment in Livermore, has now begun and is expected to cover the search range indicated in Fig. 15, i.e., two decades in f_a or m_a in a well-motivated regime of parameters. The time scale for this search is about one decade. ADMX is the first axion search experiment that should definitely find axions if they are the dark matter and have masses in the search range.

SUMMARY

As astronomy and cosmology investigate the world at larger and larger scales, and as elementary particle physics probes the microscopic structure of the world and its forces at smaller and smaller scales, we realize that "inner space" and "outer space" are closely related and the two directions in many ways lead to similar questions.

The different components of the dark universe, and even the creation of the ordinary matter-antimatter asymmetry, cry out for physics beyond the standard model. Some of the open questions, notably the nature of dark matter, may boil down to the experimental challenge of detecting dark-matter particles in the laboratory. For the wide-spread hypothesis that supersymmetric particles are responsible, the next 5–10 years may become crucial with the LHC perhaps finding evidence for supersymmetry and with many direct and indirect search strategies reaching critical size. Likewise, the experimental search for axion dark matter, a completely different type of candidate, has finally reached the critical size to corner this hypothesis.

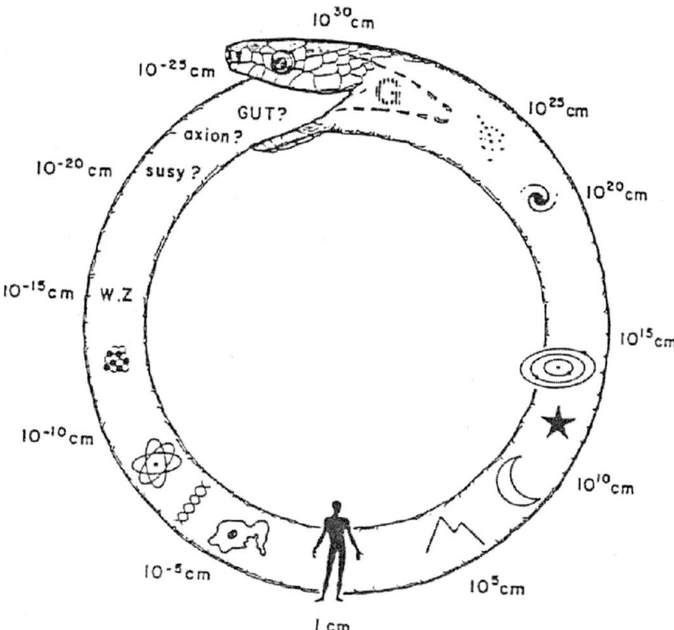

FIGURE 17. Uruborus—the mystical snake eating its own tail—symbolizing the connection between large and small, between "inner space" and "outer space."

While the dark-matter problem may "simply" consist of the experimental challenge to detect very weakly interacting new particles, the dark-energy problem looks far more profound. The accelerated expansion of the universe is best fit, at current evidence, by a simple cosmological constant. If a nontrivial dynamical evolution were eventually discovered one would have a handle for a physical interpretation, perhaps in terms of new scalar fields. As long as dark energy is just a single measured number, nature does not provide us with much information about the underlying physics. So, if dark matter looks like an experimental challenge, dark energy looks like one for fundamental theory. Answers may depend on a profound understanding of a quantum theory of gravity, perhaps ultimately provided by string theory or some other novel approach.

The dark universe provides challenges to theory and experiment, but above all it provides us with windows of opportunity to test new ideas and to make exciting new discoveries. It may be the upcoming decade where light is shed on the dark universe.

ACKNOWLEDGMENTS

This work was partly supported by the Deutsche Forschungsgemeinschaft (grant TR-27 "Neutrinos and Beyond"), by the Cluster of Excellence "Origin and Structure of the Universe" and by the European Union (contract No. RII3-CT-2004-506222).

REFERENCES

1. F. Zwicky, "Die Rotverschiebung von extragalaktischen Nebeln" ("Spectral displacement of extra galactic nebulae"), Helv. Phys. Acta **6**, 110 (1933).
2. R. Cowsik and J. McClelland, "Gravity of neutrinos of nonzero mass in astrophysics," Astrophys. J. **180**, 7 (1973).
3. P. Gondolo, "Non-baryonic dark matter," NATO Sci. Ser. II **187**, 279 (2005) [arXiv:astro-ph/0403064].
4. M. Trodden and S. M. Carroll, "TASI lectures: Introduction to cosmology," Proc. Theoretical Advanced Study Institute in Elementary Particle Physics (TASI 2003): Recent Trends in String Theory, Boulder, Colorado, 1–27 June 2003, edited by J. M. Maldacena (Hackensack, World Scientific, 2005) [arXiv:astro-ph/0401547].
5. G. Bertone, D. Hooper and J. Silk, "Particle dark matter: Evidence, candidates and constraints," Phys. Rept. **405**, 279 (2005) [arXiv:hep-ph/0404175].
6. J. R. Primack, D. Seckel and B. Sadoulet, "Detection of cosmic dark matter," Ann. Rev. Nucl. Part. Sci. **38**, 751 (1988).
7. G. Jungman, M. Kamionkowski and K. Griest, "Supersymmetric dark matter," Phys. Rept. **267**, 195 (1996) [arXiv:hep-ph/9506380].
8. P. J. E. Peebles and B. Ratra, "The cosmological constant and dark energy," Rev. Mod. Phys. **75**, 559 (2003) [arXiv:astro-ph/0207347].
9. E. J. Copeland, M. Sami and S. Tsujikawa, "Dynamics of dark energy," Int. J. Mod. Phys. D **15**, 1753 (2006) [arXiv:hep-th/0603057].
10. B. Ratra and M. S. Vogeley, "Resource Letter BE-1: The beginning and evolution of the universe," Publ. Astron. Soc. Pac. **120**, 235 (2008) [arXiv:0706.1565]. See also http://www.physics.drexel.edu/~vogeley/universe
11. C. Amsler *et al.* [Particle Data Group], "Review of particle physics," Phys. Lett. B **667**, 1 (2008). See also http://pdg.lbl.gov
12. E. W. Kolb and M. S. Turner, *The Early Universe* (Addison-Wesley, Redwood City, 1988).
13. A. Linde, *Particle Physics and Inflationary Cosmology* (CRC Press, 1990)
14. G. Börner, *The Early Universe*, 4th edition, 2nd corrected printing (Springer-Verlag, Berlin, 2004).
15. S. Dodelson, *Modern Cosmology* (Academic Press, San Diego, 2003).
16. V. Mukhanov, *Physical Foundations of Cosmology* (Cambridge University Press 2005).
17. S. Weinberg, *Cosmology* (Oxford University Press 2008).
18. M. Kowalski *et al.*, "Improved cosmological constraints from new, old and combined supernova datasets," arXiv:0804.4142 [astro-ph].
19. D. Rubin *et al.*, "Looking beyond Lambda with the Union Supernova Compilation," arXiv:0807.1108 [astro-ph].
20. V. Springel, C. S. Frenk and S. D. M. White, "The large-scale structure of the Universe," Nature **440**, 1137 (2006) [arXiv:astro-ph/0604561].
21. M. Tegmark *et al.* [SDSS Collaboration], "The 3D power spectrum of galaxies from the SDSS," Astrophys. J. **606**, 702 (2004) [arXiv:astro-ph/0310725].
22. G. Hinshaw *et al.* [WMAP Collaboration], "Five-Year Wilkinson Microwave Anisotropy Probe (WMAP) observations: Data processing, sky maps, and basic results," arXiv:0803.0732 [astro-ph].
23. M. R. Nolta *et al.* [WMAP Collaboration], "Five-Year Wilkinson Microwave Anisotropy Probe (WMAP) observations: Angular power spectra," arXiv:0803.0593 [astro-ph].
24. K. G. Begeman, A. H. Broeils and R. H. Sanders, "Extended rotation curves of spiral galaxies: Dark haloes and modified dynamics," Mon. Not. Roy. Astron. Soc. **249**, 523 (1991).
25. M. Persic, P. Salucci and F. Stel, "The Universal rotation curve of spiral galaxies: 1. The Dark matter connection," Mon. Not. Roy. Astron. Soc. **281**, 27 (1996) [arXiv:astro-ph/9506004].
26. R. Bernabei *et al.* [DAMA Collaboration], "First results from DAMA/LIBRA and the combined results with DAMA/NaI," arXiv:0804.2741 [astro-ph].
27. W. de Boer, C. Sander, V. Zhukov, A. V. Gladyshev and D. I. Kazakov, "EGRET excess of diffuse galactic gamma rays as tracer of dark matter," Astron. Astrophys. **444**, 51 (2005) [arXiv:astro-ph/0508617].

A Primer on Resonances in Quantum Mechanics

Oscar Rosas-Ortiz*, Nicolás Fernández-García* and Sara Cruz y Cruz*†

*Departamento de Física, Cinvestav, AP 14-740, 07000 México DF, Mexico
†Sección de Estudios de Posgrado e Investigación, UPIITA-IPN, Av IPN 2508, CP 07340, México DF, Mexico

Abstract. After a pedagogical introduction to the concept of resonance in classical and quantum mechanics, some interesting applications are discussed. The subject includes resonances occurring as one of the effects of radiative reaction, the resonances involved in the refraction of electromagnetic waves by a medium with a complex refractive index, and quantum decaying systems described in terms of resonant states of the energy (Gamow-Siegert functions). Some useful mathematical approaches like the Fourier transform, the complex scaling method and the Darboux transformation are also reviewed.

INTRODUCTION

Solutions of the Schrödinger equation associated to complex eigenvalues $\varepsilon = E - i\Gamma/2$ and satisfying purely outgoing conditions are known as Gamow-Siegert functions [1, 2]. These solutions represent a special case of scattering states for which the 'capture' of the incident wave produces delays in the scattered wave. The 'time of capture' can be connected with the lifetime of a decaying system (resonance state) which is composed by the scatterer and the incident wave. Then, it is usual to take $\text{Re}(\varepsilon)$ as the binding energy of the composite while $\text{Im}(\varepsilon)$ corresponds to the inverse of its lifetime. The Gamow-Siegert functions are not admissible as physical solutions into the mathematical structure of quantum mechanics since, in contrast with conventional scattering wave-functions, they are not finite at $r \to \infty$. Thus, such a kind of functions is acceptable in quantum mechanics only as a convenient model to solve scattering equations. However, because of the resonance states relevance, some approaches extend the formalism of quantum theory so that they can be defined in a precise form [3, 4, 5, 6, 7, 8].

The concept of resonance arises from the study of oscillating systems in classical mechanics and extends its applications to physical theories like electromagnetism, optics, acoustics, and quantum mechanics, among others. In this context, resonance may be defined as the excitation of a system by matching the frequency of an applied force to a characteristic frequency of the system. Among the big quantity of examples of resonance in daily life one can include the motion of a child in a swing or the tuning of a radio or a television receiver. In the former case you must push the swing from time to time to maintain constant the amplitude of the oscillation. In case you want to increase the amplitude you should push 'with the natural motion' of the swing. That is, the acting of the force you are applying on the swing should be in 'resonance' with the swing motion. On the other hand, among the extremely large number of electromagnetic signals in space,

CP1077, *Advanced Summer School in Physics 2008, Frontiers in Contemporary Physics—EAV08*
edited by L. M. Montaño Zetina, G. Torres Vega, M. García Rocha, L. F. Rojas Ochoa, and R. López Fernández

your radio responds only to that one for which it is tuned. In other words, the set has to be in resonance with a specific electromagnetic wave to permit subsequent amplification to an audible level. In this paper we present some basics of the resonance phenomenon. We intend to provide a strong primer introduction to the subject rather than a complete treatment. In the next sections we shall discuss classical models of vibrating systems giving rise to resonance states of the energy. Then we shall review some results arising from the Fourier transform widely used in optics and quantum mechanics. This material will be useful in the discussions on the effects of radiative reaction which are of great importance in the study of atomic systems. We leave for the second part of these notes the discussion on the resonances in quantum decaying systems and their similitudes with the behavior of optical devices including a complex refractive index. Then the complex scaling method arising in theories like physical chemistry is briefly reviewed to finish with a novel application of the ancient Darboux transformation in which the transformation function is a quantum resonant state of the energy. At the very end of the paper some lines are included as conclusions.

VIBRATION, WAVES AND RESONANCES

Mechanical Models

Ideal vibrating (or oscillating) systems undergo the same motion over and over again. A very simple model consists of a mass m at the end of a spring which can slide back and forth without friction. The time taken to make a complete vibration is the *period* of oscillation while the *frequency* is the number of vibration cycles completed by the system in unit time. The motion is governed by the acceleration of the vibrating mass

$$\frac{d^2x}{dt^2} = -\left(\frac{k}{m}\right)x \equiv -w_0^2 x \tag{1}$$

where $w_0 := \sqrt{k/m}$ is the *natural angular frequency* of the system. In other words, a general displacement of the mass follows the rule

$$x = A\cos w_0 t + B\sin w_0 t \tag{2}$$

A and B being two arbitrary constants. To simplify our analysis we shall consider a particular solution by taking $A = a\cos\theta$ and $B = -a\sin\theta$, therefore we can write

$$x = a\cos(w_0 t + \theta). \tag{3}$$

At $t_n = \frac{(2n+1)\pi - 2\theta}{2w_0}$, $n = 0, 1, 2\ldots$, the kinetic energy $T = \frac{1}{2}m(dx/dt)^2$ reaches its maximum value $T_{\max} = mw_0^2a^2/2$ while x passes through zero. On the other hand, the kinetic energy is zero and the displacement of the mass is maximum ($x = a$ is the *amplitude* of the oscillation) at $t_m = \frac{m\pi - \theta}{w_0}$, $m = 0, 1, 2, \ldots$ This variation of T is just opposite of that of the potential energy $V = kx^2/2$. As a consequence, the total stored energy E is a constant of motion which is proportional to the square of the amplitude (twice the amplitude means an oscillation which has four times the energy):

$$E = T + V = \frac{1}{2} m w_0^2 a^2. \tag{4}$$

Systems exhibiting such behavior are known as *harmonic oscillators*. There are plenty of examples: a weight on a spring, a pendulum with small swing, acoustical devices producing sound, the oscillations of charge flowing back and forth in an electrical circuit, the 'vibrations' of electrons in an atom producing light waves, the electrical and magnetical components of electromagnetic waves, and so on.

Steady-state oscillations

In actual vibrating systems there is some loss of energy due to friction forces. In other words, the amplitude of their oscillations is a decreasing function of time (the vibration *damps down*) and we say the system is *damped*. This situation occurs, for example, when the oscillator is immersed in a viscous medium like air, oil or water. In a first approach the friction force is proportional to the velocity $F_f = -\alpha \frac{dx}{dt}$, with α a *damping constant* expressed in units of mass times frequency. Hence, an external energy must be supplied into the system to avoid the damping down of oscillations. In general, vibrations can be driven by a repetitive force $F(t)$ acting on the oscillator. So long as $F(t)$ is acting there is an amount of work done to maintain the stored energy (i.e., to keep constant the amplitude). Next we shall discuss the forced oscillator with damping for a natural frequency w_0 and a damping constant α given.

Let us consider an oscillating force defined as the real part of $F(t) = Fe^{iwt} \equiv F_0 e^{i(wt+\eta)}$. Our problem is to solve the equation

$$\frac{d^2x}{dt^2} + \gamma \frac{dx}{dt} + w_0^2 x = \text{Re}\left(\frac{Fe^{iwt}}{m}\right), \qquad \gamma := \frac{\alpha}{m}. \tag{5}$$

Here the new damping constant γ is expressed in units of frequency. The ansatz $x = \text{Re}(ze^{iwt})$ reduces (5) to a factorizable expression of z, from which we get

$$z = \frac{F\Omega}{m}, \qquad \Omega = \frac{1}{w_0^2 - w^2 + i\gamma w}. \tag{6}$$

We realize that z is proportional to the complex function Ω, depending on the driving force's frequency w and parameterized by the natural frequency w_0 and the damping constant γ. In polar form $\Omega = |\Omega| e^{i\phi}$, the involved phase angle ϕ is easily calculated by noticing that $\Omega^{-1} = e^{-i\phi}/|\Omega| = w_0^2 - w^2 + i\gamma w$, so we get

$$\tan\phi = -\frac{\gamma w}{w_0^2 - w^2}. \tag{7}$$

Let us construct a single valued phase angle ϕ for finite values of w_0 and γ. Notice that $w < w_0$ leads to $\tan\phi < 0$ while $w \to w_0^-$ implies $\tan\phi \to -\infty$. Thereby we can set

$\phi(w=0)=0$ and $\phi(w_0)=-\pi/2$ to get $\phi\in[-\pi/2,0]$ for $w\leq w_0$. Now, since $w>w_0$ produces $\tan\phi>0$, we use $\tan(-\phi)=-\tan\phi$ to extend the above defined domain $\phi\in(-\pi,0]$, no matter the value of the angular frequency w. Bearing these results in mind we calculate the real part of z (see equation 6) to get the physical solution

$$x=x_0\cos(wt+\eta+\phi), \qquad x_0:=\frac{F_0|\Omega|}{m}. \tag{8}$$

Notice that the mass oscillation is not in phase with the driving force but is shifted by ϕ. Moreover, $\gamma\to 0$ produces $\phi\to 0$, so that this phase shift is a measure of the damping. Since the phase angle is always negative or zero ($-\pi/2<\phi\leq 0$), equation (8) also means that the displacement x lags behind the force $F(t)$ by an amount ϕ. On the other hand, the amplitude x_0 results from the quotient F_0/m scaled up by $|\Omega|$. Thus, such a scale factor gives us a measure of the response of the oscillator to the action of the driving force. The total energy (4), with $a=x_0$, is then a function of the angular frequency:

$$E(w)=\frac{(w_0F_0)^2}{2m}|\Omega|^2\equiv\frac{(w_0F_0)^2}{2m}\left[\frac{1}{(w_0^2-w^2)^2+(\gamma w)^2}\right]. \tag{9}$$

Equations (7) and (9) comprise the complete solution to the problem. The last one, in particular, represents the *spectral energy distribution* of the forced oscillator with damping we are dealing with. It is useful, however, to simplify further under the assumption that $\gamma<<1$. For values of w closer to that of w_0 the energy approaches its maximum value $2mE(w_0)\approx(F_0/\gamma)^2$ while $E(w\to+\infty)$ goes to zero as w^{-4}. In other words, $E(w)$ shows rapid variations only near w_0. It is then reasonable to substitute

$$w_0^2-w^2=(w_0-w)(w_0+w)\approx(w_0-w)2w_0 \tag{10}$$

in the expressions of the energy and the phase shift to get

$$E(w\to w_0)\approx\frac{1}{2m}\left(\frac{F_0}{\gamma}\right)^2\omega(w,w_0,\gamma), \qquad \tan\phi\approx\frac{\gamma}{2(w-w_0)} \tag{11}$$

with

$$\omega(w,w_0,\gamma):=\frac{(\gamma/2)^2}{(w_0-w)^2+(\gamma/2)^2}. \tag{12}$$

Equation (12) describes a bell-shaped curve known as the *Cauchy* (mathematics), *Lorentz* (statistical physics) or *Fock-Breit-Wigner* (nuclear and particle physics) distribution. It is centered at $w=w_0$ (the *location parameter*), with a half-width at half-maximum equal to $\gamma/2$ (the *scale parameter*) and amplitude (height) equal to 1. That is, the damping constant γ defines the width of the spectral line between the half-maximum points $w_0-w=\pm\gamma/2$. Fig. 1 shows the behavior of the curve ω for different values of the damping constant (*spectral width*) γ.

These last results show that the supplying of energy to the damping oscillator is most efficient if the vibrations are sustained at a frequency $w=w_0$. In such a case it is said

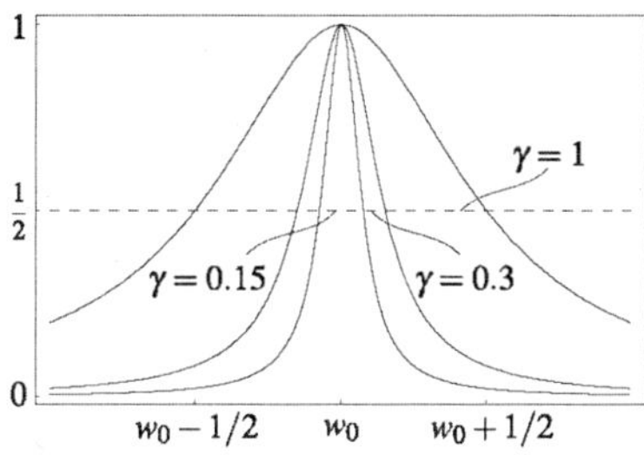

FIGURE 1. The Fock-Breit-Wigner (Lorentz-Cauchy) distribution ω for different values of the width at half maximum γ.

that the driving force is in *resonance* with the oscillator and w_0 is called the *resonance frequency*. Besides the above discussed spring-mass system, the motion of a child in a swing is another simple example giving rise to the same profile. To keep the child+swing system oscillating at constant amplitude you must push it from time to time. To increase the amplitude you should push 'with the motion': the oscillator vibrates most strongly when the frequency of the driving force is equal to the frequency of the free vibration of the system. On the other hand, if you push against the motion, the oscillator do work on you and the vibration can be brought to a stop. The above cases include an external force steading the oscillations of the system. This is why equations (7–9) are known as the *steady-state* solutions of the problem.

We can calculate the amount of work W_{work} which is done by the driving force. This can be measured in terms of the *power* P, which is the work done by the force per unit time:

$$P = \frac{d}{dt} W_{\text{work}} = F(t) \frac{dx}{dt} = \frac{dE}{dt} + m\gamma \left(\frac{dx}{dt} \right)^2. \tag{13}$$

The *average* power $\langle P \rangle$ corresponds to the *mean* of P over many cycles. To calculate it we first notice that $\langle dE/dt \rangle = 0$. That is, the energy E does not change over a period of time much larger than the period of oscillation. Now, since the square of any sinusoidal function has an average equal to $1/2$, the last term in (13) has an average which is proportional to the square of the frequency times the amplitude of the oscillation. From (8) we get

$$\langle P \rangle = \frac{m\gamma w^2 x_0^2}{2}. \tag{14}$$

It is then clear that the driving force does a great deal of work to cancel the action of the friction force. In a similar form we obtain the average of the stored energy:

$$\langle E \rangle = \frac{m x_0^2 (w^2 + w_0^2)}{4}. \tag{15}$$

Remark that the mean of E does not depend on the friction but on the angular frequency of the driving force. If w is close to the resonance frequency w_0, then $\langle E \rangle$ goes to the ideal oscillator's energy (4), scaled by $(x_0/a)^2$. Moreover, the same result is obtained no matter the magnitude of the driving force, since it does not play any role in (15).

Transient oscillations

Suppose a situation in which the driving force is turned off at a given time $t = t_0$. This means that no work is done to sustain the oscillations so there is no supplied energy to preserve the motion any longer. This system can be studied by solving (5) with $F = 0$. After introducing the ansatz $x = \mathrm{Re}(ze^{iwt})$ we get a quadratic equation for w, the solution of which reads

$$w_\pm = i\gamma/2 \pm \vartheta, \qquad \vartheta := \sqrt{w_0^2 - (\gamma/2)^2}. \tag{16}$$

If $\gamma < w_0$ then $\vartheta \in \mathbf{R}$ and any of these two roots produces the desired solution:

$$x = \mathrm{Re}(ze^{iwt}) = |z|e^{-\frac{\gamma}{2}t}\cos(\vartheta t + z_0), \quad t \geq t_0. \tag{17}$$

First, notice that the energy is not a constant of motion but decreases in exponential form $E \propto |z|^2 e^{-\gamma t}$. The damping constant γ is then a measure of the lifetime of the oscillation because at the time $\tau = 1/\gamma$, the energy is reduced to approximately the 36% ($E \rightarrow E/e$) while the amplitude goes to the 60% of its initial value ($|z| \rightarrow |z|/\sqrt{e}$). Thus, the smaller the value of γ the larger the lifetime τ of the oscillation. In this way, for values of γ such that $w_0 >> \gamma/2$, the discriminant in (16) becomes $\vartheta \approx w_0$. Thereby, the system exhibits an oscillation of frequency close to the resonance frequency w_0. This means that large lifetimes are intimately connected with resonances for small values of the damping constant.

As we can see, the resonance phenomenon is a characteristic of vibrating systems even in absence of forces steading the oscillations. Solutions like (17) are known as *transient* oscillations because there is no force present which can ensure their prevalentness. They are useful to describe mechanical oscillators for which the driven force has been turned off at the time $t = t_0$ or, more general, decaying systems like the electric field emitted by an atom. In general, 'resonance' is the tendency of a vibrating system to oscillate at maximum amplitude under certain frequencies w_n, $n = 0, 1, 2, \ldots$ At these resonance frequencies even small driving forces produce large amplitude vibrations. The phenomenon occurs in all type of oscillators, from mechanical and electromagnetic systems to quantum probability waves. A resonant oscillator can produce waves oscillating at specific frequencies. Even more, this can be used to pick out a specific frequency from an arbitrary vibration containing many frequencies.

Fourier Optics Models

In this section we shall review some interesting results arising from the Fourier transform. This mathematical algorithm is useful in studying the properties of optical devices, the effects of radiative reaction on the motion of charged particles and the energy spectra of quantum systems as well. Let $\{e^{ikx}\}$ be a set of plane waves orthonormalized as follows

$$(e^{i\kappa x}, e^{ikx}) = \lim_{a\to+\infty} \int_{-a}^{a} e^{i(k-\kappa)x}dx = \lim_{a\to+\infty} 2\frac{\sin(k-\kappa)a}{k-\kappa} \equiv 2\pi\delta(k-\kappa) \tag{18}$$

with $\delta(x-x_0)$ the Dirac's delta distribution. This 'function' arises in many fields of study and research as representing a sharp impulse applied at x_0 to the system one is dealing with. The response of the system is then the subject of study and is known as the *impulse response* in electrical engineering, the *spread function* in optics or the *Green's function* in mathematical-physics. Among its other peculiar properties, the Dirac function is defined in such a way that it can sift out a single ordinate in the form

$$f(x_0) = \int_{-\infty}^{\infty} \delta(x-x_0) f(x) dx. \tag{19}$$

In general, a one-dimensional function $\varphi(x)$ can be expressed as the linear combination

$$\varphi(x) = \frac{1}{2\pi} \int_{-\infty}^{+\infty} \widetilde{\varphi}(k) e^{-ikx} dk \tag{20}$$

where the coefficient of the expansion $\widetilde{\varphi}(k)$ is given by the following inner product

$$\begin{aligned} (e^{-ikx}, \varphi) &= \int_{-\infty}^{+\infty} \varphi(x) e^{ikx} dx = \frac{1}{2\pi} \int_{-\infty}^{+\infty} \left[\int_{-\infty}^{+\infty} e^{i(k-\kappa)x} dx \right] \widetilde{\varphi}(\kappa) d\kappa \\ &= \int_{-\infty}^{+\infty} \delta(k-\kappa) \widetilde{\varphi}(\kappa) d\kappa = \widetilde{\varphi}(k). \end{aligned} \tag{21}$$

If (20) is interpreted as the Fourier series of $\varphi(x)$, then the continuous index k plays the role of an *angular spatial frequency*. The coefficient $\widetilde{\varphi}(k)$, in turn, is called the *Fourier transform* of $\varphi(x)$ and corresponds to the amplitude of the *spatial frequency spectrum* of $\varphi(x)$ between k and $k+dk$. It is also remarkable that functions (20) and (21) are connected via the Parseval formula

$$\int_{-\infty}^{+\infty} |\varphi(x)|^2 dx = \int_{-\infty}^{+\infty} |\widetilde{\varphi}(k)|^2 dk. \tag{22}$$

This expression often represents a conservation principle. For instance, in quantum mechanics it is a conservation of probability [9]. In optics, it represents the fact that all the light passing through a diffraction aperture eventually appears distributed throughout the diffraction pattern [10]. On the other hand, since x and k represent arbitrary (canonical conjugate) variables, if φ were a function of time rather than space we would replace x by t and then k by the angular temporal frequency w to get

$$\varphi(t) = \frac{1}{2\pi} \int_{-\infty}^{+\infty} \widetilde{\varphi}(w) e^{-iwt} dw, \qquad \widetilde{\varphi}(w) = \int_{-\infty}^{+\infty} \varphi(t) e^{iwt} dt \tag{23}$$

and

$$\int_{-\infty}^{+\infty} |\varphi(t)|^2 dt = \int_{-\infty}^{+\infty} |\widetilde{\varphi}(w)|^2 dw. \tag{24}$$

Now, let us take a time depending wave $\varphi(t)$, defined at $x=0$ by

$$\varphi(t) = \varphi_0 \Theta(t) e^{-\frac{\gamma}{2}t} \cos w_0 t, \qquad \Theta(t) = \begin{cases} 1 & t > 0 \\ 0 & t < 0 \end{cases} \tag{25}$$

Function (25) is a transient oscillation as it has been defined in the above sections. From our experience with the previous cases we know that it is profitable to represent $\varphi(t)$ in terms of a complex function. In this case $\varphi = \mathrm{Re}(Z)$, with

$$Z(t) = A(t)e^{-iw_0t}, \qquad A(t) = \varphi_0\Theta(t)\exp\left(-\frac{\gamma}{2}t\right). \tag{26}$$

Observe that $|Z(t)|^2 = |A(t)|^2$. Then the Parseval formula (24) gives

$$\int_{-\infty}^{+\infty} |A(t)|^2 dt = \int_{-\infty}^{+\infty} |\widetilde{A}(w)|^2 dw. \tag{27}$$

Since the stored energy at the time t is proportional to $|A(t)|^2$, both integrals in equation (27) give the total energy W of the wave as it is propagating throughout $x = 0$. Thereby the power involved in the oscillation as a function of time is given by $P(t) = dW/dt \propto |A(t)|^2$. In the same manner $I_w = dW/dw \propto |\widetilde{A}(w)|^2$ is the energy per unit frequency interval. Now, from (26) we get $A(t) = Z(t)e^{iw_0t}$, so that

$$A(t) = \frac{1}{2\pi}\int_{-\infty}^{+\infty} \widetilde{Z}(w)e^{-(w-w_0)t}dw \equiv \frac{1}{2\pi}\int_{-\infty}^{+\infty} a(\varepsilon)e^{-i\varepsilon t}d\varepsilon \tag{28}$$

where $\varepsilon := w - w_0$ and $a(\varepsilon) := \widetilde{Z}(\varepsilon + w_0)$. This last term is given by

$$a(\varepsilon) = \int_{-\infty}^{+\infty} A(t)e^{i\varepsilon t}dt = \varphi_0\int_0^{+\infty} e^{(i\varepsilon-\frac{\gamma}{2})t}dt = \frac{\varphi_0}{\frac{\gamma}{2} - i(w - w_0)}. \tag{29}$$

Then, we have

$$I_w = \left(\frac{2\varphi_0}{\gamma}\right)^2 \frac{(\gamma/2)^2}{(w - w_0)^2 + (\gamma/2)^2}. \tag{30}$$

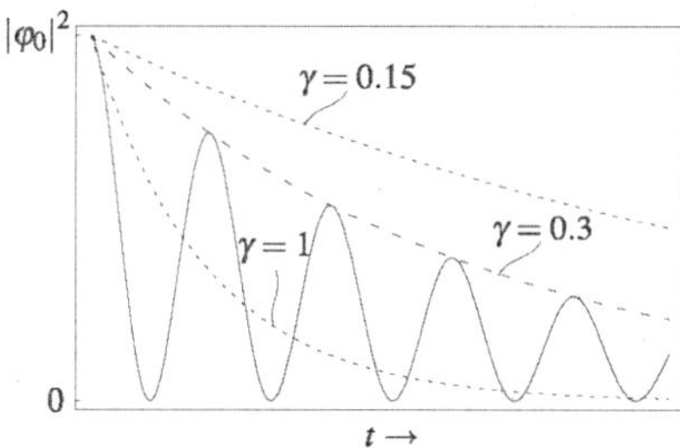

FIGURE 2. The power $P(t) \propto |\varphi_0|^2e^{-\gamma t}$ involved with the transient oscillation (25) for the same values of γ as those given in Figure 1. Remark that the area under the dotted curves is larger for smaller values of the line breadth γ.

That is, the Fock-Breit-Wigner (FWB) function (30) is the spectral energy distribution of the transient oscillation (25). In the previous section we learned that the inverse of the damping constant γ measures the oscillation lifetime. The same rule holds for the power $|A(t)|^2 = |\varphi_0|^2 e^{-\gamma t}$, as it is shown in Fig. 2. Let us investigate the extremal case of infinite lifetimes. Thus we calculate I_w in the limit $\gamma \to 0$. The result reads

$$I_w \to 2\pi \left(\frac{\varphi_0^2}{\gamma}\right) \delta(w - w_0), \qquad \gamma \to 0 \tag{31}$$

where we have used

$$\delta(x) = \frac{1}{\pi} \lim_{\varepsilon \to 0} \left(\frac{\varepsilon}{x^2 + \varepsilon^2}\right). \tag{32}$$

Infinite lifetimes ($1/\gamma = \tau \to +\infty$) correspond to spectral energy distributions of infinitesimal width ($\gamma \to 0$) and very high height (φ_0^2/γ). As a consequence, the transient oscillation (25) has a definite frequency ($w \to w_0$) along the time. The same conclusion is obtained for steady-state oscillations by taking $F_0^2 = 8m\varphi_0^2$ in equations (11–12). In the next sections we shall see that this lifetime$\leftrightarrow$width relationship links bound and decaying energy states in quantum mechanics.

As a very simple application of the above results let the plane wave (25) be the electric field emitted by an atom. Then W corresponds to the total energy radiated per unit area perpendicular to the direction of propagation. Equation (30) in turn, relates in a quantitative way the behavior of the power radiated as a function of the time to the frequency spectrum of the energy radiated. To give a more involved example let us consider a nonrelativistic charged particle of mass m_e and charge q_e, acted on by an external force $\vec{F}$. The particle emits radiation since it is accelerated. To account for this radiative energy loss and its effect on the motion of the particle it is necessary to add a *radiative reaction force* $\vec{F}_{\text{rad}}$ in the equation of motion to get

$$\vec{F} + \vec{F}_{\text{rad}} \equiv \vec{F} + \left(\frac{2q_e^2}{3c^3}\right) \frac{d^3}{dt^3}\vec{r} = m_e \frac{d^2}{dt^2}\vec{r} \tag{33}$$

where c is the speed of light. This last expression is known as the *Abraham-Lorentz* equation and is useful only in the domain where the reactive term is a small correction, since the third order derivative term does not fulfill the requirements for a dynamical equation (see e.g., reference [11], Ch 16). Bearing this condition in mind let us investigate the effect of an external force of the form $\vec{F} = -m_e w_0^2 \vec{r}$. The Abraham-Lorentz equation is written

$$\left(\frac{2q_e^2}{3m_e c^3}\right) \frac{d^3}{dt^3}\vec{r} = \frac{d^2}{dt^2}\vec{r} + w_0^2 \vec{r}. \tag{34}$$

For small values of the third order term one has $\frac{d^2}{dt^2}\vec{r} \approx -w_0^2\vec{r}$. That is, the particle oscillates like a mass at the end of a spring with frequency w_0. Hence $\frac{d^3}{dt^3}\vec{r} \approx -w_0^2 \frac{d}{dt}\vec{r}$, so that the problem is reduced to the transient equation

$$\frac{d^2}{dt^2}\vec{r} + \gamma \frac{d}{dt}\vec{r} + w_0^2 \vec{r} = 0, \qquad \gamma = \frac{2q_e^2 w_0^2}{3m_e c^3}, \tag{35}$$

the solution of which has the form (25-26) with $\varphi(t)$ replaced by $\vec{r}(t)$ and φ_0 by a constant vector $\vec{r}_0$. To get an idea of the order of our approach let us evaluate the quotient

γ/w_0^2. A simple calculation gives the following constant

$$\frac{\gamma}{w_0^2} \approx 0.624 \times 10^{-23} s. \tag{36}$$

Thus the condition $\gamma << 1s^{-1}$ defines the appropriate values of the frequency w_0. For instance, let γ take the value $10^{-3}s^{-1}$. The value of w_0 is then of the order of an infrared frequency $w_0 \sim 10^{10}s^{-1}$. But if $\gamma \sim 10^{-11}s^{-1}$ then $w_0 \sim 10^6 s^{-1}$, that is, the particle will oscillate at a radio wave frequency.

Under the limits of our approach the radiative reaction force $\vec{F}_{\text{rad}}$ plays the role of a friction force which damps the oscillations of the electric field. The *resonant line shape* defined by the Fock-Breit-Wigner function (30) is broadened and shifted in frequency due to the reactive effects of radiation. That is, because the decaying of the power radiated $P(t) \propto |\varphi_0|^2 e^{-\gamma t}$, the emitted radiation corresponds to a pulse (wave train) with effective length $\lambda \approx c/\gamma$ and covering an interval of frequencies equal to γ rather than being monochromatic. The infinitesimal finiteness of the width in the spectral energy distribution (31) is then justified by the 'radiation friction'. In the language of radiation the damping constant γ is known as the *line breadth*.

Finally, it is well known that the effects of radiative reaction are of great importance in the detailed behavior of atomic systems. It is then remarkable that the simple plausibility arguments discussed above led to the qualitative features derived from the formalism of quantum electrodynamics. By proceeding in a similar manner, it is also possible to verify that the scattering and absorption of radiation by an oscillator are also described in terms of FBW-like distributions appearing in the scattering cross section. The reader is invited to review the approach in classical references like [11, 12].

Fock's Energy Distribution Model

Let $\{\phi_E(x)\}$ be a set of eigenfunctions belonging to energy eigenvalues E in the continuous spectrum of a given one-dimensional Hamiltonian H. The vectors are orthonormalized as follows

$$\int_{-\infty}^{\infty} \overline{\phi}_E(x)\phi_{E'}(x)dx = \delta(E' - E). \tag{37}$$

Notice we have taken for granted that the continuous spectrum is not degenerated, otherwise equation (37) requires some modifications. Let us assume that a wave function $\psi_0(x)$ can be expanded in a series of these functions, that is:

$$\psi_0(x) = \int_{-\infty}^{\infty} C(E)\phi_E(x)dE, \qquad C(E) = \int_{-\infty}^{+\infty} \overline{\phi}_E(x)\psi_0(x)dx. \tag{38}$$

The inner product of ψ_0 with itself leads to the Parseval relation

$$W := (\psi_0, \psi_0) = \int_{-\infty}^{+\infty} |\psi_0(x)|^2 dx = \int_{-\infty}^{+\infty} |C(E)|^2 dE. \tag{39}$$

In the previous sections we learned that W allows the definition of the energy distribution $\omega(E)$. In this case we have

$$\omega(E) := \frac{dW}{dE} = |C(E)|^2. \tag{40}$$

At an arbitrary time $t > 0$ the state of the system reads

$$\psi_t(x) = \int_{-\infty}^{\infty} C(E)\phi_E(x)e^{-iEt/\hbar}dE. \tag{41}$$

The *transition amplitude* $T(t \geq 0)$ from ψ_0 to ψ_t is given by the inner product

$$T(t \geq 0) \equiv (\psi_0, \psi_t) = \int_{-\infty}^{+\infty} \overline{\psi}_0(x)\psi_t(x)dx = \int_{-\infty}^{+\infty} \omega(E)e^{-iEt/\hbar}dE. \tag{42}$$

This function rules the transition probability $|T(t)|^2$ from $\psi_0(x)$ to $\psi_t(x)$ by relating the wave function at two different times: t_0 and $t \geq t_0$. It is known as the *propagator* in quantum mechanics and can be identified as a (spatial) Green's function for the time-dependent Schrödinger equation (see next section). From (42), it is clear that T can be investigated in terms of spatial coordinates x or as a function of the energy distribution. Next, following the Fock's arguments [9], we shall analyze the transition probability for a decaying system by assuming that $\omega(E)$ is given. Let $\omega(E)$ be the Fock-Breit-Wigner distribution

$$\omega(a) = \frac{1}{\pi}\left[\frac{(\Gamma/2)^2}{a^2 + (\Gamma/2)^2}\right] = \frac{1}{\pi}\left[\frac{(\Gamma/2)^2}{(a+i\Gamma/2)(a-i\Gamma/2)}\right], \quad a := E - E_0. \tag{43}$$

Then equation (42) reads

$$T(t \geq 0) = \frac{1}{\pi}\left(\frac{\Gamma}{2}\right)^2 e^{-iE_0t/\hbar}\int_{-\infty}^{+\infty}\frac{e^{-iat/\hbar}}{(a+i\Gamma/2)(a-i\Gamma/2)}. \tag{44}$$

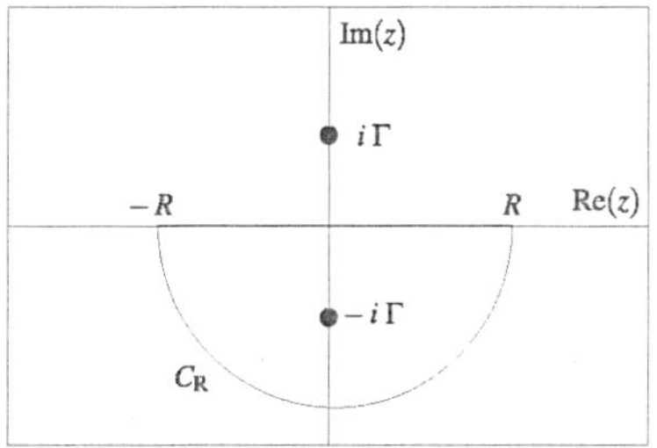

FIGURE 3. Contour of integration in the complex E-plane.

We start with the observation that function (43) has two isolated singularities at $a = \pm i\Gamma/2$ (none of them lies on the real axis!). When $R > 0$, the point $-i\Gamma/2$ lies in the interior of the semicircular region which is depicted in Fig 3. At this stage, it is convenient to introduce the function

$$f(z) = \frac{g(z)}{z + i\Gamma/2}, \qquad g(z) = \frac{e^{-izt/\hbar}}{z - i\Gamma/2}. \tag{45}$$

Integrating $f(z)$ counterclockwise around the boundary of the semicircular region the *Cauchy integral formula* (see e.g., [13]) is written

$$\int_{R}^{-R} f(a)da + \int_{C_R} f(z)dz = 2\pi i g(-i\Gamma/2). \tag{46}$$

This last expression is valid for all values of $R > 0$. Since the value of the integral on the right in (46) tends to 0 as $R \to +\infty$, we finally arrive at the desired result:

$$\int_{-\infty}^{+\infty} f(a)da = \frac{2\pi}{\Gamma} \exp\left(-\frac{\Gamma t}{2\hbar}\right). \tag{47}$$

In this way the transition rate $T(t)$ acquires the form of a transient oscillation

$$T(t \geq 0) = \frac{\Gamma}{2} \exp(-i\varepsilon t/\hbar), \qquad \varepsilon := E_0 - i\Gamma/2. \tag{48}$$

It is common to find $T(t)$ as free of the factor $\Gamma/2$ (see e.g., [9], pp 159). This factor arises here because (43) has been written to be consistent with the previous expressions of a FWB distribution. The factor is easily removed if we take $\Gamma/2$ rather than $(\Gamma/2)^2$ in the numerator of $\omega(a)$. Now, we know that transient oscillations involve lifetimes and the present case is not an exception. Equation (48) means that the transition from $\psi_0(x)$ to $\psi_t(x)$ is an exponential decreasing function of the time. Since this rate of change is symmetrical, it also gives information about the rate of decaying of the initial wave. Thus, the probability that the system has not yet decayed at time t is given by

$$|T(t)|^2 = \left(\frac{\Gamma}{2}\right)^2 \exp(-\Gamma t/\hbar), \quad t \geq 0. \tag{49}$$

We note that the state of the undecayed system ψ_0 does not change but decays suddenly. That is, the time t in (49) is counted off starting from the latest instant when the system has not decayed. The above description was established by Fock in his famous book on quantum mechanics [9] and this is why functions like (43) bears his name. It is remarkable that the first Russian edition is dated on August 1931.

Finally, remark we have written (48) in terms of the complex number $\varepsilon = E_0 - i\Gamma/2$. The reason is not merely aesthetic because, up to a constant factor, $T(t \geq 0)$ is the Fourier transform[1] of the expansion coefficient $C(E)$:

$$C(E) := \frac{\Gamma/2}{\sqrt{\pi}(E - E_0 + i\Gamma/2)} = \frac{\Gamma/2}{\sqrt{\pi}(E - \varepsilon)} \tag{50}$$

where we have used (43). The relevance of this result will be clarified in the sequel.

---

[1] The Fourier transforms in this case correspond to the equations (23), with the energy E and the angular frequency w related by the Einstein's expression $E = \hbar w$.

QUANTA, TUNNELING AND RESONANCES

Let us consider the motion of a particle of mass m constrained to move on the straight-line in a given potential $U(x)$. Its *time-dependent Schrödinger equation* is

$$H\psi(x,t) := \left[-\frac{\hbar^2}{2m}\frac{\partial^2}{\partial x^2} + U(x) \right] \psi(x,t) = i\hbar \frac{\partial}{\partial t}\psi(x,t). \tag{51}$$

Let us assume the wave-function $\psi(x,t)$ is separable, that is $\psi(x,t) = \varphi(x)\theta(t)$. A simple calculation leads to $\theta(t) = \exp(-iEt/\hbar)$, with E a constant, and $\varphi(x)$ a function fulfilling

$$\left[-\frac{\hbar^2}{2m}\frac{\partial^2}{\partial x^2} + U(x) \right] \varphi(x) \equiv H\varphi(x) = E\varphi(x). \tag{52}$$

This time-independent Schrödinger equation (plainly the Schrödinger equation) defines a set of *eigenvalues* E and *eigenfunctions* of the Hamiltonian operator H which, in turn, represents the *observable* of the energy. Now, to get some intuition about the separableness of the wave-function let us take the Fourier transform of its temporal term

$$\widetilde{\theta}(E) = \lim_{a \to +\infty} \int_{-a}^{a} e^{i(E-E')t/\hbar} dt = 2\pi\delta(E - E') \tag{53}$$

where we have used (18). This last result means that the energy distribution is of infinitesimal width and very high height. In other words, the system has a definite energy $E = E'$ along the time. Systems exhibiting this kind of behavior are known as *stationary* and it is said that they are *conservative*. Since the Hamiltonian operator H does not depend on t and because for any analytic function f of H one has

$$f(H)\varphi(x) = f(E)\varphi(x), \tag{54}$$

our separability ansatz $\psi \to \varphi\theta$ can be written

$$\psi(x,t) = \exp\left(-\frac{i}{\hbar}Ht \right) \varphi(x) = e^{-iEt/\hbar}\varphi(x). \tag{55}$$

According with the Born's interpretation, the wave function $\varphi(x)$ defines the probability density $\rho(x) = |\varphi(x)|^2$ of finding the quantum particle between x and $x + dx$. Thereby, the sum of all probabilities (i.e., the probability of finding the particle anywhere in the straight-line at all) is unity:

$$\int_{-\infty}^{+\infty} \rho(x)dx = \int_{-\infty}^{+\infty} \overline{\varphi}(x)\varphi(x)dx \equiv \int_{-\infty}^{+\infty} |\varphi(x)|^2 dx = 1. \tag{56}$$

The above equation represents the *normalization condition* fulfilled by the solutions $\varphi(x)$ to be physically acceptable. Hence, they are elements of a vector space $\mathscr{H}$ consisting of *square-integrable* functions and denoted as $\mathscr{H} = L^2(\mathbf{R}, \mu)$, with μ the Lebesgue measure (for simplicity in notation we shall omit μ by writing $L^2(\mathbf{R})$). As an example,

$\mathscr{H}$ can be the space spanned by the Hermite polynomials $H_n(x)$, weighted by the factor $\mu(x) = e^{-x^2/2}$ and defined as follows

$$H_n(x) = (-1)^n e^{x^2/2} \frac{d^n}{dx^n} e^{-x^2/2}. \tag{57}$$

In quantum mechanics, observables are represented by the so-called Hermitian operators in the Hilbert space $\mathscr{H}$. A differential operator A defined on $L^2(\mathbf{R})$ is said to be *Hermitian* if, whenever Af and Ag are defined for $f,g \in L^2(\mathbf{R})$ and belong to $L^2(\mathbf{R})$, then

$$(Af,g) = (f,Ag) = \int_{-\infty}^{+\infty} \overline{f}(x) Ag(x) \mu(x) dx. \tag{58}$$

In particular, if $f = g$ the above definition means that the action of the Hermitian operator A on $f \in L^2(\mathbf{R})$ is symmetrical. If $Af(x) = \alpha f(x)$, we have $(Af,f) = \overline{\alpha}$ and $(f,Af) = \alpha$, so that $\overline{\alpha} = \alpha \Rightarrow \alpha \in \mathbf{R}$. In other words, the eigenvalues of a Hermitian operator acting on $L^2(\mathbf{R})$ are real numbers. It is important to stress, however, that this rule is not true in the opposite direction. In general, as we are going to see in the next sections, there is a wide family of operators A_λ sharing the same set of real eigenvalues $\{\alpha\}$. (The family includes some non-Hermitian operators!) Moreover, notice that the rule is not necessarily true if $f(x) \notin L^2(\mathbf{R})$. In general, the set of solutions of $Af(x) = \alpha f(x)$ is wider than $L^2(\mathbf{R})$. That is, the complete set of mathematical solutions $\varphi(x)$ embraces functions such that its absolute value $|\varphi(x)|$ diverges even for real eigenvalues. A plain example is given by the solutions representing scattering states because they do not fulfill (56). In such a case one introduces another kind of normalization like that defined in (37), with a similar notion of the Born's probability as we have seen in the previous section. Normalization (56) is then a very restrictive condition picking out the appropriate physical solutions among the mathematical ones. This is why the Schrödinger equation (51) is "physically solvable" for a very narrow set of potentials.

If E is real, the time-dependent factor in (55) is purely oscillatory (a phase) and the time displacement $\psi(x,t)$ gives the same 'prediction' (probability density) as $\varphi(x)$. As a result, both of these vectors lead to the same expectation values of the involved observables

$$\langle A \rangle := \int_{-\infty}^{+\infty} \overline{\psi}(x,t) A \psi(x,t) dx = \int_{-\infty}^{+\infty} \overline{\varphi}(x) A \varphi(x) = \alpha. \tag{59}$$

In particular, $\langle H \rangle = E$ shows that the eigenvalue E is also the expectation value of the energy. Notice that stationary states are states of well-defined energy, E being the definite value of its energy and not only its expectation value (see equation 53). That is, any determination of the energy of the particle always yields the particular value E. Again, as an example, let us consider a scattering state. Far away from the influence of the scatterer, it is represented by a plane wave like

$$\psi(x,t) = \exp\left(-\frac{i}{\hbar} Et\right) \exp\left(-\frac{i}{\hbar} xp\right) \tag{60}$$

where p is the linear momentum of the particle. Function (60) represents a state having a definite value for its energy. However, there is no certainty neither on the position of the particle nor in the transit time of the particle at a given position. In general, the energy distribution will not be a continuous function. It could include a set of isolated points (discrete energy levels) and/or continuous portions showing a set of very narrow and high peaks (resonance levels). The former correspond to infinite lifetime states (the observed discrete energy levels of atoms are good examples) while the lifetime of the second ones will depend on the involved interactions.

Quasi-stationary States and Optical Potentials

It is also possible to define a *probability current density* j

$$j = \frac{\hbar}{2mi}\left[\overline{\psi}\left(\frac{d}{dx}\psi\right) - \left(\frac{d}{dx}\overline{\psi}\right)\psi\right] \tag{61}$$

which, together with the probability density $\rho = |\psi|^2$ satisfies a continuity equation

$$\frac{d\rho}{dt} + \frac{dj}{dx} = 0 \tag{62}$$

exactly as in the case of conservation of charge in electrodynamics. Observe that stationary states fulfill $d\rho/dt = 0$, so that $\rho \neq \rho(t)$ and $j = 0$. What about decaying systems for which the transition amplitude $T(t \geq 0)$ involves a complex number $\varepsilon = E_0 - i\Gamma/2$ like that found in equation (48)? Let us assume a complex eigenvalue of the energy $H\varphi_\varepsilon = \varepsilon\varphi_\varepsilon$ is admissible in (55). Then we have $\rho(x,t) = \rho_\varepsilon(x)e^{-\Gamma t/\hbar}$ and $j \neq 0$. That is, complex energies are included at the cost of adding a non-trivial value of the probability current density j. A conventional way to solve this 'problem' is to consider a complex potential $U = U_R + iU_I$. Then equation (62) acquires the form

$$\frac{d\rho}{dt} + \frac{dj}{dx} = \frac{2}{\hbar}U_I(x)\rho(x), \tag{63}$$

the integration of which can be identified with the variation of the number of particles

$$\frac{d}{dt}N = \frac{2}{\hbar}\int_{-\infty}^{+\infty} U_I(x)\rho(x)dx. \tag{64}$$

Let $U_I(x) = U_0$ be a constant. If $U_0 > 0$, there is an increment of the number of particles ($dN/dt > 0$) and viceversa, $U_0 < 0$ leads to a decreasing number of particles ($dN/dt < 0$). In the former case the imaginary part of the potential works like a source of particles while the second one shows U_I as a sink of particles. The introduction of this potential into the Schrödinger equation gives

$$\left[-\frac{\hbar^2}{2m}\frac{\partial^2}{\partial x^2} + U_R(x) + iU_0\right]\varphi_\varepsilon(x) = \left(E_0 - i\frac{\Gamma}{2}\right)\varphi_\varepsilon(x). \tag{65}$$

Then, the identification $U_0 = -\Gamma/2$ reduces the solving of this last equation to a stationary problem. The exponential decreasing probability density $\rho_\varepsilon(x)e^{-\Gamma t/\hbar}$ is then justified by the presence of a sink-like potential $U_I(x) = U_0 < 0$. However, this solution requires the introduction of a non-Hermitian Hamiltonian because the involved potential is complex. Although such a Hamiltonian is not an observable in the sense defined in the above section, notice that $U_0 = -i\Gamma/2$ is a kind of damping constant. The lifetime of the probability $\rho(x,t)$ is defined by the inverse of Γ and, according with the derivations of the previous section, the energy distribution shows a bell-shaped peak at $E = E_0$. In other words, the complex eigenvalue $\varepsilon = E_0 - i\Gamma/2$ is a pole of $\omega(E)$ and represents a resonance of the system. The modeling of decay by the use of complex potentials with constant imaginary part is know as the *optical model* in nuclear physics. The reason for such a name is clarified in the next section.

Complex Refractive index in Optics

It is well known that the spectrum of *electromagnetic energy* includes radio waves, infrared radiation, the visible spectrum of colors red through violet, ultraviolet radiation, x-rays and gamma radiation. All of them are different forms of light and are usually described as *electromagnetic waves*. The physical theory treating the *propagation* of light is due mainly to the work of James Clerk Maxwell. The interaction of light and matter or the absorption and emission of light, on the other hand, is described by the quantum theory. Hence, a consistent theoretical explanation of all optical phenomena is furnished jointly by Maxwell's electromagnetic theory and the quantum theory. In particular, the speed of light $c = 299,792,456.2 \pm 1.1 m/s$ is a part of the *wave equation* fulfilled by the electric field $\vec{E}$ and the magnetic field $\vec{H}$:

$$\nabla^2 \vec{A} = \frac{1}{c^2}\frac{\partial^2 \vec{A}}{\partial t^2}, \qquad \vec{A} = \vec{E}, \vec{H}. \tag{66}$$

The above expression arises from the Maxwell's equations in empty space with $c = (\mu_0 \varepsilon_0^*)^{-1/2}$. The constant μ_0 is known as the *permeability of the vacuum* and the constant ε_0^* is called the *permitivity of the vacuum* [11]. In isotropic nonconducting media these constants are replaced by the corresponding constants for the medium, namely μ and ε^*. Consequently, the speed of propagation v of the electromagnetic fields in a medium is given by $v = (\mu\varepsilon^*)^{-1/2}$. The *index of refraction* n is defined as the ratio of the speed of light in vacuum to its speed in the medium: $n = c/v$. Most transparent optical media are nonmagnetic so that $\mu/\mu_0 = 1$, in which case the index of refraction should be equal to the square root of the relative permitivity $n = (\varepsilon^*/\varepsilon_0^*)^{1/2} \equiv K^{1/2}$. In a nonconducting, isotropic medium, the electrons are permanently bound to the atoms comprising the medium and there is no preferential direction. This is what is meant by a simple isotropic dielectric such as glass [14]. Now, consider a (one-dimensional) plane harmonic wave incident upon a plane boundary separating two different optical media. In agreement with the phenomena of reflection and refraction of light ruled by the Huygens's principle, there will be a reflected wave and a transmitted wave (see e.g. [10]). Let the first medium be the empty space and the second one having a complex index of refraction

$$\mathcal{N} = n + i n_I. \tag{67}$$

Then, the wavenumber of the refracted wave is complex:

$$\mathscr{K} = k + i\alpha \tag{68}$$

and we have

$$\vec{E} = \vec{E}_0 \, e^{i(k_0 x - wt)} \qquad \text{(incident wave)}$$
$$\vec{E}' = \vec{E}'_0 \, e^{i(k'_0 x - wt)} \qquad \text{(reflected wave)}$$
$$\vec{E}'' = \vec{E}''_0 \, e^{i(\mathscr{K} x - wt)} = \vec{E}''_0 \, e^{-\alpha x} e^{i(kx - wt)} \qquad \text{(refracted wave)}$$

If $\alpha > 0$ the factor $e^{-\alpha x}$ indicates that the amplitude of the wave decreases exponentially with the distance. That is, the energy of the wave is absorbed by the medium and varies with distance as $e^{-2\alpha x}$. Hence 2α is the *coefficient of absorption* of the medium. The imaginary part n_I of $\mathscr{N}$, in turn, is known as the *extinction index*. In general, it can be shown that the corresponding polarization is similar to the amplitude formula for a driven harmonic oscillator [14]. Thus, an optical resonance phenomenon will occur for light frequencies in the neighborhood of the resonance frequency $w_0 = (K/m)^{1/2}$. The relevant aspect of these results is that a complex index of refraction $\mathscr{N}$ leads to a complex wavenumber $\mathscr{K}$. That is, the properties of the medium (in this case an absorbing medium) induce a specific behavior of the electromagnetic waves (in this case, the exponential decreasing of the amplitude). This is the reason why the complex potential discussed in the previous section is named the 'optical potential'.

Quantum Tunneling and Resonances

In quantum mechanics the complex energies were studied for the first time in a paper by Gamow concerning the alpha decay (1928) [1]. In a simple picture, a given nucleus is composed in part by alpha particles (^4_2He nuclei) which interact with the rest of the nucleus via an attractive well (obeying the presence of nuclear forces) plus a potential barrier (due, in part, to repulsive electrostatic forces). The former interaction constrains the particles to be bounded while the second holds them inside the nucleus. The alpha particles have a small (non–zero) probability of tunneling to the other side of the barrier instead of remaining confined to the interior of the well. Outside the potential region, they have a finite lifetime. Thus, alpha particles in a nucleus should be represented by *quasi–stationary* states. For such states, if at time $t = 0$ the probability of finding the particle inside the well is unity, in subsequent moments the probability will be a slowly decreasing function of time (see e.g. Sections 7 and 8 of reference [9]). In his paper of 1928, Gamow studied the escape of alpha particles from the nucleus via the tunnel effect. In order to describe eigenfunctions with exponentially decaying time evolution, Gamow introduced energy eigenfunctions ψ_G belonging to complex eigenvalues $Z_G = E_G - i\Gamma_G$, $\Gamma_G > 0$. The real part of the eigenvalue was identified with the energy of the system and the imaginary part was associated with the inverse of the lifetime. Such 'decaying states' were the first application of quantum theory to nuclear physics.

Three years later, in 1931, Fock showed that the law of decay of a quasi–stationary state depends only on the energy distribution function $\omega(E)$ which, in turn, is meromorphic [9]. According to Fock, the analytical expression of $\omega(E)$ is rather simple and has only two poles $E = E_0 \pm i\Gamma$, $\Gamma > 0$ (see our equation (43) and equation (8.13) of [9]). A close result was derived by Breit and Wigner in 1936. They studied the cross section of slow neutrons and found that the related energy distribution reaches its maximum at E_R with a half–maximum width Γ_R. A resonance is supposed to take place at E_R and to have "half–value breath" Γ_R [2]. It was in 1939 that Siegert introduced the concept of a purely outgoing wave belonging to the complex eigenvalue $\varepsilon = E - i\Gamma/2$ as an appropriate tool in the studying of resonances [15]. This complex eigenvalue also corresponds to a first–order pole of the S matrix [16] (for more details see e.g. [17]). However, as the Hamiltonian is a Hermitian operator, then (in the Hilbert space $\mathscr{H}$) there can be no eigenstate having a strict complex exponential dependence on time. In other words, decaying states are an approximation within the conventional quantum mechanics framework. This fact is usually taken to motivate the study of the rigged (equipped) Hilbert space $\overline{\mathscr{H}}$ [3, 4, 5] (For a recent review see [6]). The mathematical structure of $\overline{\mathscr{H}}$ lies on the nuclear spectral theorem introduced by Dirac in a heuristic form [18] and studied in formal rigor by Maurin [19] and Gelfand and Vilenkin [20].

In general, solutions of the Schrödinger equation associated to complex eigenvalues and fulfilling purely outgoing conditions are known as *Gamow-Siegert functions.* If u_ε is a function solving $Hu_\varepsilon = \varepsilon u_\varepsilon$, the appropriate boundary condition may be written

$$\lim_{x\to\pm\infty} \left(u'_\varepsilon \mp iku_\varepsilon\right) = \lim_{x\to\pm\infty} \{(-\beta \mp ik)\, u_\varepsilon\} = 0, \tag{69}$$

with β defined as the derivative of the logarithm of u_ε:

$$\beta := -\frac{d}{dx} \ln u_\varepsilon. \tag{70}$$

Now, let us consider a one-dimensional short-range potential $U(x)$, characterized by a cutoff parameter $\zeta > 0$. The general solution of (52) can be written in terms of ingoing and outgoing waves:

$$u_< := u_\varepsilon(x < -\zeta) = Ie^{ikx} + Le^{-ikx}, \qquad u_> := u_\varepsilon(x > \zeta) = Ne^{-ikx} + Se^{ikx} \tag{71}$$

where the coefficients I, L, N, S, depend on the potential parameters and the incoming energy $k^2 = 2m\varepsilon/\hbar^2$ (the kinetic parameter k is in general a complex number $k = k_R + ik_I$), they are usually fixed by imposing the continuity conditions for u and du/dx at the points $x = \pm\zeta$. Among these solutions, we are interested in those which are purely outgoing waves. Thus, the second term in each of the functions (71) must dominate over the first one. For such states, equation (61) takes the form:

$$j_< = -v|u_<|^2, \qquad j_> = v|u_>|^2, \qquad v := \frac{\hbar}{2m}(k + \overline{k}) = \frac{\hbar k_R}{m}. \tag{72}$$

This last equation introduces the *flux velocity* v. If ε is a real number $\varepsilon = E$, then k is either pure imaginary or real according to E negative or positive. If we assume that the potential admits negative energies, we get $k_\pm = \pm i\sqrt{|2mE/\hbar^2|}$ and (72) vanishes (the flux velocity $v \propto k_R$ is zero outside the interaction zone). Notice that the solutions $u_E^{(+)}$, connected with k_+, are bounded so that they are in $L^2(\mathbf{R})$. That is, they are the physical solutions φ associated with a discrete set of eigenvalues $2mE_n/\hbar^2 = k_{n,+}^2$ solving the continuity equations for u and du/dx at $x = \pm\zeta$. On the other hand, *antibound states* $u_E^{(-)}$ increase exponentially as $|x| \to +\infty$. To exhaust the cases of a real eigenvalue ε, let us take now $2mE/\hbar^2 = \kappa^2 > 0$. The outgoing condition (69) drops the interference term in the density

$$\rho(x;t) = |N|^2 + |S|^2 + 2|\overline{N}S|\cos(2\kappa x + \operatorname{Arg} S/N), \qquad x > \zeta, \tag{73}$$

so that the integral of $\rho = |S|^2$ is not finite neither in space nor in time (similar expressions hold for $x < -\zeta$). Remark that flux velocity is not zero outside the interaction zone. Thereby, $E > 0$ provides outgoing waves at the cost of a net outflow $j \neq 0$. To get solutions which are more appropriate for this nontrivial j, we shall consider complex eigenvalues ε. Let us write

$$\varepsilon = E - \frac{i}{2}\Gamma, \qquad \varepsilon_R \propto k_R^2 - k_I^2, \qquad \varepsilon_I \propto 2k_R k_I \tag{74}$$

where $2m\varepsilon/\hbar^2 = (k_R + ik_I)^2$. According to (69), the boundary condition for β reads now

$$\lim_{x\to\pm\infty} \{-\beta \pm (k_I - ik_R)\} = 0 \tag{75}$$

so that the flux velocity is $v_+ \propto k_R$ for $x > \zeta$ and $v_- \propto -k_R$ for $x < -\zeta$. Hence, the "correct" direction in which the outgoing waves move is given by $k_R > 0$. In this case, the density

$$\rho(x;t) \equiv |u(x,t)|^2 = e^{-\Gamma t/\hbar}|u(x)|^2, \qquad \lim_{x\to\pm\infty} \rho(x;t) \propto e^{-\Gamma(t-x/v_\pm)/\hbar} \tag{76}$$

can be damped by taking $\Gamma > 0$. Thereby, $k_I \neq 0$ and $k_R \neq 0$ have opposite signs. Since $k_R > 0$ has been previously fixed, we have $k_I < 0$. Then, purely outgoing, exponentially increasing functions (resonant states) are defined by points in the fourth quadrant of the complex k-plane. In general, it can be shown that the *transmission amplitude* S in (71) is a meromorphic function of k, with poles restricted to the positive imaginary axis (bound states) and the lower half-plane (resonances) [21]. Let k_n be a pole of S in the fourth quadrant of the k-plane, then $-\overline{k}_n$ is also a pole while $\overline{k}_n$ and $-k_n$ are zeros of S (see Figure 4, left). On the other hand, if S is studied as a function of ε, a Riemann surface of $\varepsilon^{1/2} = k$ is obtained by replacing the ε-plane with a surface made up of two sheets R_0 and R_1, each cut along the positive real axis and with R_1 placed in front of R_0 (see e.g. [13], pp 337). As the point ε starts from the upper edge of the slit in R_0 and describes a continuous circuit around the origin in the counterclockwise direction (Figure 4, right), the angle increases from 0 to 2π. The point then passes from the sheet R_0 to the sheet R_1, where the angle increases from 2π to 4π. The point then passes back to the sheet R_0 and so on. Complex poles of $S(\varepsilon)$ always arise in conjugate pairs (corresponding to k and $-\overline{k}$) while poles on the negative real axis correspond to either bound or antibound states.

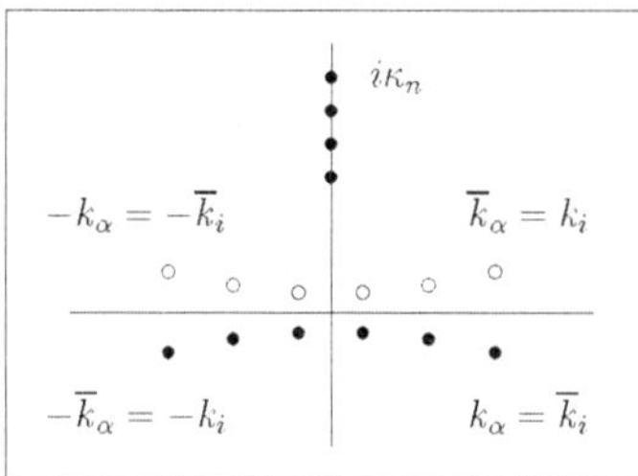

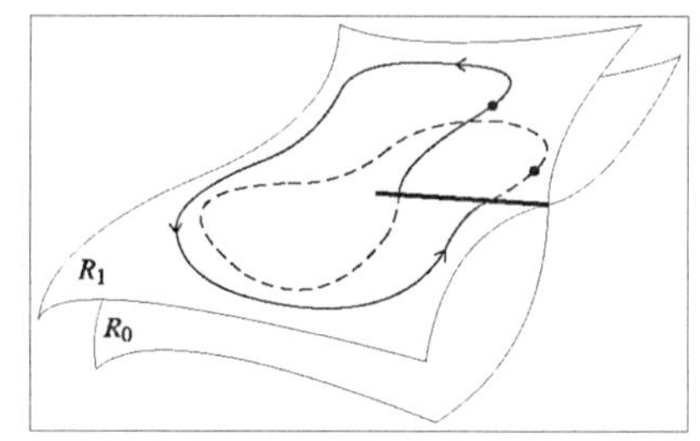

FIGURE 4. Left: Schematic representation of the poles (disks) and the zeros (circles) of the transmission amplitude $S(k)$ in the complex k-plane. Bounded energies correspond to poles located on the positive imaginary axis **Right:** The two-sheet Riemann surface $\sqrt{\varepsilon} = k$. The lower edge of the slit in R_0 is joined to the upper edge of the slit in R_1, and the lower edge of the slit in R_1 is joined to the upper edge of the slit in R_0. The picture is based on the description given by J.M. Brown and R.V. Churchill in ref. [13].

Observe that density (76) increases exponentially for either large $|x|$ or large negative values of t. The usual interpretation is that the compound (u_ε, V) represents a decaying system which emitted waves in the remote past $t - x/v$. As it is well known, the long lifetime limit ($\Gamma \to 0$) is useful to avoid some of the complications connected with the limit $t \to -\infty$ (see discussions on time asymmetry in [22]). In this context, one usually imposes the condition:

$$\frac{\Gamma/2}{\Delta E} << 1. \tag{77}$$

Thus, the level width Γ must be much smaller than the level spacing ΔE in such a way that closer resonances imply narrower widths (longer lifetimes). In general, the main difficulty is precisely to find the adequate E and Γ. However, for one-dimensional stationary short range potentials, in [21, 23] it has been shown that the superposition of a denumerable set of FBW distributions (each one centered at each resonance $E_n, n = 1, 2, \ldots$) entails an approximation of the coefficient T such that the larger the number N of close resonances involved, the higher the precision of the approximation (see Fig. 5):

$$T \approx \omega_N(\varepsilon_R) = \sum_{n=1}^{N} \omega(\varepsilon_R, E_n) \tag{78}$$

with

$$\omega(\varepsilon_R, E) = \frac{(\Gamma/2)^2}{(\varepsilon_R - E)^2 + (\Gamma/2)^2}. \tag{79}$$

Processes in which the incident wave falls upon a single scatterer are fundamental in the study of more involved interactions [24]. In general, for a single target the scattering amplitude is a function of two variables (e.g. energy and angular momentum). The above model corresponds to the situation in which one of the variables is held fixed (namely, the angular momentum). A more realistic three dimensional model is easily obtained from these results: even functions are dropped while an infinitely extended, impenetrable wall is added at the negative part of the straight line [23, 25]. Such a situation corresponds to s-waves interacting with a single, spherically symmetric, square scatterer (see e.g. [26]).

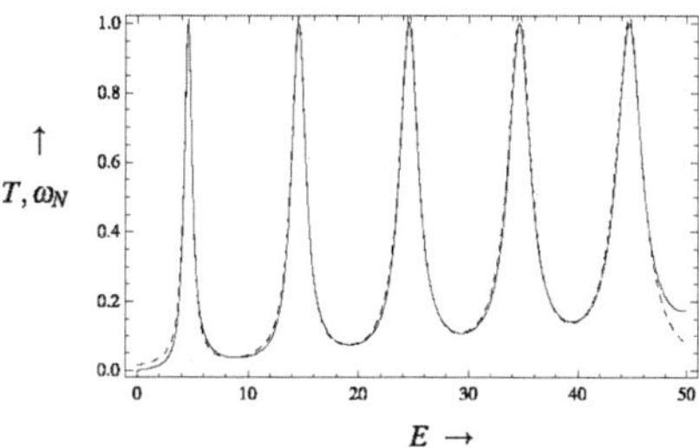

FIGURE 5. Functions T and ω_N (dotted curve) for a square well with strength $V_0 = 992.25$ and weigh $b = 20$ for which the FWB sum matches well the transmission coefficient for the first five resonances [23] (see also [21]).

Complex-Scaling Method

Some other approaches extend the framework of quantum theory so that quasi–stationary states can be defined in a precise form. For example, the complex–scaling method [7, 8, 27] (see also [28]) embraces the transformation $H \rightarrow UHU^{-1} = H_\theta$, where U is the complex–scaling operator $U = e^{-\theta XP/\hbar}$, with θ a dimensionless parameter and $[X,P] = i\hbar$. The transformation is achievable by using the Baker-Campbell-Hausdorff formulae [29]:

$$e^A B e^{-A} = \left\{e^A, B\right\} = \sum_{n=0}^{\infty} \frac{1}{n!} \{A^n, B\} \tag{80}$$

with A and B two arbitrary linear operators and

$$\{A^n, B\} = \underbrace{[A,[A,\ldots[A}_{n \text{ times}}, B]\ldots]].$$

The identification $A = -\theta XP/\hbar$ and $B = X$ leads to

$$UXU^{-1} = \sum_{n=0}^{\infty} \frac{1}{n!}(i\theta)^n X = e^{i\theta} X, \qquad UPU^{-1} = \sum_{n=0}^{\infty} \frac{1}{n!}(-i\theta)^n P = e^{-i\theta} P, \tag{81}$$

where we have used

$$\{(XP)^n, X\} = (-i\hbar)^n X, \qquad \{(XP)^n, P\} = (i\hbar)^n P. \tag{82}$$

The following calculations are now easy

$$\begin{aligned} UP^2U^{-1} &= \left(UPU^{-1}\right)\left(UPU^{-1}\right) = \left(UPU^{-1}\right)^2 = e^{-2i\theta} P^2, \\ UV(X)U^{-1} &= U\left(\sum_{k=0}^{\infty} \frac{1}{k!} V_k X^k\right) U^{-1} = V\left(e^{i\theta} X\right). \end{aligned} \tag{83}$$

So that we finally get

$$UHU^{-1} \equiv H_\theta = e^{-2i\theta} P^2 + V(e^{i\theta} X). \tag{84}$$

Remark that in the Schrödinger's representation we have

$$X = x, \quad P = -i\hbar\frac{d}{dx}, \quad U = e^{i\theta x d/dx} \quad \Rightarrow \quad Uf(x) = f(xe^{i\theta}). \tag{85}$$

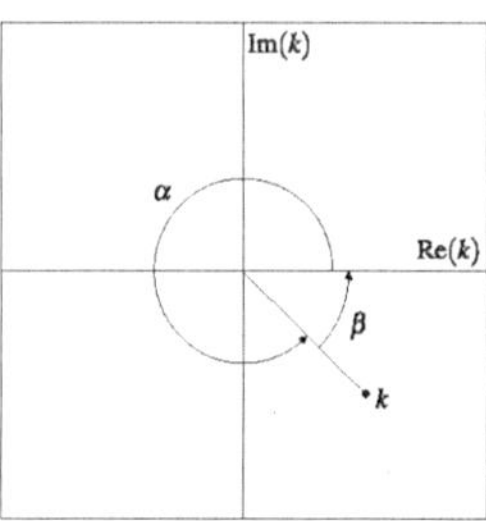

FIGURE 6. Polar form of an arbitrary point on the complex k-plane.

This transformation converts the description of resonances by non–integrable Gamow-Siegert functions into one by square integrable functions. Let $k = |k|e^{i\alpha} \equiv |k|e^{-i\beta}$ be a point on the complex k-plane (see Figure 6). If k lies on the fourth quadrant then $0 < \beta < \pi/2$ and the related Gamow-Siegert function u_ε behaves as follows

$$u_\varepsilon(x \to \pm\infty) \sim e^{\pm i|k|x\cos\beta} e^{\pm|k|x\sin\beta}. \tag{86}$$

That is, u_ε diverges for large values of $|x|$. The behavior of the complex-scaled function $\widetilde{u}_\varepsilon = U(u_\varepsilon)$, on the other hand, reads

$$\widetilde{u}_\varepsilon(x \to \pm\infty) \sim e^{\pm i|k|x\cos(\theta-\beta)} e^{\mp|k|x\sin(\theta-\beta)}. \tag{87}$$

Thereby, $\widetilde{u}_\varepsilon$ is a bounded function if $\theta - \beta > 0$, i.e., if $\tan\theta > \tan\beta$. The direct calculation shows that complex-scaling preserves the square-integrability of the bounded states φ_n, whenever $0 < \theta < \pi/2$. Then one obtains

$$0 < \theta - \beta < \pi/2. \tag{88}$$

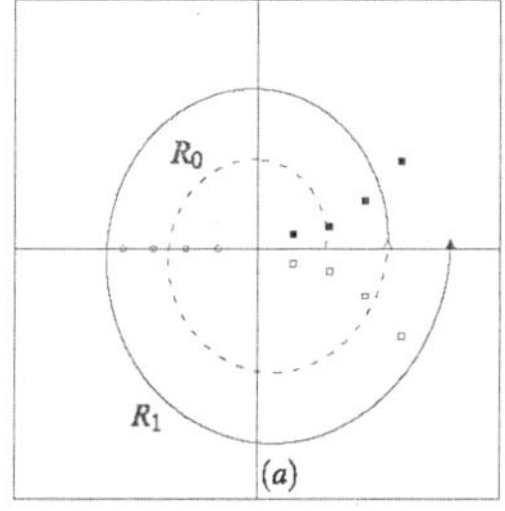

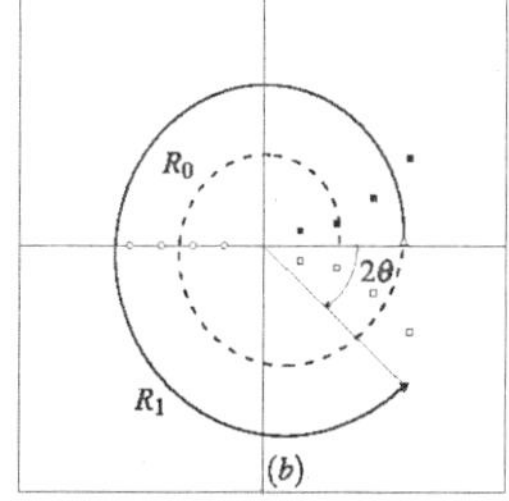

FIGURE 7. Left: Schematic representation of the two-sheet Riemann surface showed in Figure 4. Dashed curves and empty squares lie on the first sheet R_0, continuous curves and fulled squares lie on the second sheet R_1 **Right:** The complex rotated plane 'exposing' the resonant poles lying on the first sheet R_0.

As regards the complex-scaled scattering states we have

$$e^{\pm i|k|x} \rightarrow e^{\pm i|k|x\cos\theta} e^{\mp|k|x\sin\theta}. \tag{89}$$

So that plane waves are transformed into exponential decreasing or increasing functions for large values of $|x|$. To preserve the oscillating form of scattering wave-functions the kinetic parameter k has to be modified. That is, the transformation $k = |k| \rightarrow |k|e^{-i\theta}$ reduces (89) to the conventional plane-wave form of the scattering states. This transformation, however, induces a rotation of the positive real axis in the clockwise direction by the angle 2θ: $E \propto k^2 \rightarrow |k|^2 e^{-i2\theta} \propto Ee^{-i2\theta}$. That is, the rotated energy is complex $\varepsilon = E_R - i\Gamma/2$ with $E_R = E\cos(2\theta)$, $\Gamma/2 = E\sin(2\theta)$. In summary, complex rotation is such that: 1) Bound state poles remain unchanged under the transformation 2) Cuts are now rotated downward making an angle of 2θ with the real axis 3) Resonant poles are 'exposed' by the cuts (see Figure 7). Another relevant aspect of the method is that it is possible to construct a resolution to the identity [30]. Moreover, as the complex eigenvalues are θ–independent, the resonance phenomenon is just associated with the discrete part of the complex–scaled Hamiltonian [31] (but see [28]). As a final remark, let us emphasize that complex-scaling 'regularizes' the divergent Gamow-Siegert functions u_ε at the cost of introducing a non-Hermitian Hamiltonian H_θ. From equation (84) we get

$$H_\theta^\dagger = e^{i2\theta}P^2 + V(e^{-i\theta}R) \neq H_\theta. \tag{90}$$

In other words, the 'regularized' solutions $\widetilde{u}_\varepsilon$ are square-integrable eigenfunctions of a complex potential $V(e^{i\theta}x)$ belonging to the complex eigenvalue ε.

Darboux-Gamow Transformations

In a different survey, complex eigenvalues of Hermitian Hamiltonians have been used to implement Darboux (supersymmetric) transformations in quantum mechanics [32, 33, 34, 35, 36, 37, 38] (see also the discussion on 'atypical models' in [39]). The transformed Hamiltonians include non-Hermitian ones, for which the point spectrum sometimes has a single complex eigenvalue [33, 34, 36, 21, 23, 25, 38]. This last result, combined with appropriate squeezing operators [40], could be in connection with the complex-scaling technique. In general, supersymmetric transformations constitute a powerful tool in quantum mechanics [39]. However, as far as we know, until the recent results reported in [21, 25, 36, 37] the connection between supersymmetric transformations and resonant states has been missing. In this context and to throw further light on the complex function β we may note that (70) transforms the Schrödinger equation (52) into a Riccati one

$$-\beta' + \beta^2 + \varepsilon = V, \tag{91}$$

where we have omitted the units. Remark that (91) is not invariant under a change in the sign of the function β:

$$\beta' + \beta^2 + \varepsilon = V + 2\beta'. \tag{92}$$

These last equations define a Darboux transformation $\widetilde{V} \equiv \widetilde{V}(x,\varepsilon) = V(x) + 2\beta'(x)$ of the initial potential V. This transformation necessarily produces a complex function if u in equation (70) is a Gamow-Siegert function u_ε. That is, a *Darboux-Gamow deformation* is defined as follows [21]:

$$\widetilde{V} = V + 2\beta' \equiv V - 2\frac{d^2}{dx^2}\ln u_\varepsilon. \tag{93}$$

The main point here is that the purely outgoing condition (69) leads to $\beta' \to 0$ so that $\widetilde{V} \to V$, in the limit $|x| \to +\infty$. In general, according to the excitation level of the transformation function u_ε, the real $\widetilde{V}_R$ and imaginary $\widetilde{V}_I$ parts of $\widetilde{V}$ show a series of maxima and minima. Thus, the new potential behaves as an optical device emitting and absorbing probability flux at the same time, since the function $I_I(x)$ shows multiple changes of sign [21, 23, 25, 36, 37, 38]. On the other hand, the solutions $y \equiv y(x, \varepsilon, \mathscr{E})$ of the non-Hermitian Schrödinger equation

$$-y'' + \widetilde{V}y = \mathscr{E}y \tag{94}$$

are easily obtained

$$y \propto \frac{W(u_\varepsilon, \psi)}{u_\varepsilon}, \tag{95}$$

where $W(*,*)$ stands for the Wronskian of the involved functions and ψ is eigensolution of (94) with eigenvalue $\mathscr{E}$. It is easy to show that scattering waves and their Darboux-Gamow deformations share similar transmission probabilities [21]. Now, let us suppose that the Hamiltonian H includes a point spectrum $\sigma_d(H) \subset \text{Sp}(H)$. If ψ_n is a (square-integrable) eigenfunction with eigenvalue $\mathscr{E}_n$, then its Darboux-Gamow deformation (95) is bounded:

$$\lim_{x \to \pm\infty} y_n = \mp(\sqrt{\mathscr{E}_n} + ik)(\lim_{x \to \pm\infty} \psi_n). \tag{96}$$

Thereby, y_n is a normalizable eigenfunction of $\widetilde{H}$ with eigenvalue $\mathscr{E}_n$. However, as ε is complex, although the new functions $\{y_n\}$ may be normalizable, they will not form an orthogonal set [34] (see also [41] and the 'puzzles' with self orthogonal states [42]). There is still another bounded solution to be considered. Function $y_\varepsilon \propto \varphi_\varepsilon^{-1}$ fulfills equation (94) for the complex eigenvalue ε. Since $\lim_{x \to \pm\infty} |y_\varepsilon|^2 = e^{\pm 2k_I x}$ and $k_I < 0$, we have another normalizable function to be added to the set $\{y_n\}$.

In summary, one is able to construct non-Hermitian Hamiltonians $\widetilde{H}$ for which the point spectrum is also $\sigma_d(H)$, extended by a single complex eigenvalue $\text{Sp}(\widetilde{H}) = \text{Sp}(H) \cup \{\varepsilon\}$. As we can see, the results of the Darboux-Gamow deformations are quite similar to those obtained by means of the complex-scaling method. This relationship deserves a detailed discussion which will be given elsewhere.

CONCLUSIONS

We have studied the concept of *resonance* as it is understood in classical mechanics by analyzing the motion of a forced oscillator with damping. The resonance phenomenon occurs for steady state oscillations when the driving force oscillates at an angular frequency equal to the natural frequency w_0 of the oscillator. Then the amplitude of the oscillation is maximum and w_0 is called the resonance frequency. The spectral energy distribution corresponds to a Fock-Breit-Wigner (FBW) function, centered at w_0 and having a line breadth equal to the damping constant γ. The resonance phenomenon is present even in the absence of external forces (transient oscillations). In such case the energy decreases exponentially with the time so that the damping constant γ is a measure of the lifetime of the oscillation $\tau = 1/\gamma$. Similar phenomena occur for the electromagnetic radiation. In this context, the effects of radiative reaction can be approximated by considering the reaction force $\vec{F}_{\rm rad}$ as a friction force which damps the oscillations of the electric field. Thus, the concept of resonance studied in classical mechanics is easily extended to the Maxwell's electromagnetic theory. The model also applies in vibrating elastic bodies, provided that the displacement is now a measure of the degree of excitation of the appropriate vibrational mode of the sample. *Acoustic resonances* are then obtained when the elastic bodies vibrate in such a way that standing waves are set up (Some interesting papers dealing with diverse kinds of resonances in metals can be consulted in [43]). Since the profile of atomic phenomena involving a high number of quanta of excitation can be analyzed in the context of classical mechanics, cyclotron and electron spin resonances can be studied, in a first approach, in terms of the above model (see, e.g. the paper by A.S. Nowick in [43], pp 1-44, and references quoted therein). The quantum approach to the problem of spinning charged particles showing magnetic resonance is discussed in conventional books on quantum mechanics like the one of Cohen-Tannoudji et. al. [44]. Resonant light-atom interactions as well as resonances occurring in cavity quantum electrodynamics can be consulted in [45].

We have also shown that the resonance phenomenon occurs in quantum decaying systems. According to the Fock's approach, the corresponding law of decay depends only on the energy distribution function $\omega(E)$ which is meromorphic and acquires the form of a FBW, bell-shaped curve. The introduction of $\varepsilon = E - i\Gamma/2$, a complex eigenvalue of the energy, is then required in analyzing the resonances which, in turn, are identified with the decaying states of the system. The inverse of the lifetime is then in correspondence with $\Gamma/2$. In a simple model, the related exponential decreasing probability can be justified by introducing a complex potential $U = U_R + U_I$, the imaginary part of which is the constant $-\Gamma/2$. Thus, U_I works like a sink of probability waves. The situation resembles the absorption of electromagnetic waves by a medium with complex refractive index so that $U = U_R + U_I$ is called an optical potential. The treatment of complex energies in quantum mechanics includes non-square integrable Gamow-Siegert functions which are outside of the Hilbert spaces. In this sense, the complex-scaling method is useful to 'regularize' the problem by complex-rotating positions x and wavenumbers k. As a consequence, bounded and scattering states maintain without changes after the rotation while the Gamow-Siegert functions become square-integrable. Another important aspect of the method is that the positive real axis of the complex energy plane is clockwise rotated by an angle 2θ, so that resonant energies are exposed by the cuts of the corresponding Riemann surface. The method, however, produces complex potentials. That is, the Gamow-Siegert functions are square-integrable solutions of a non-Hermitian Hamiltonian belonging to complex eigenvalues. Similar results are obtained by deforming the initial potential in terms of a Darboux transformation defined by a Gamow-Siegert function. In this sense, both of the above approaches could be applied in the studying of quantum resonances. A detailed analysis of the connection between the complex-scaling method and the Darboux-Gamow transformation is in progress.

ACKNOWLEDGMENTS

ORO would like to thank to the organizers for the kind invitation to such interesting Summer School. Special thanks to Luis Manuel Montaño and Gabino Torres. SCyC thanks the members of Physics Department, Cinvestav, for kind hospitality The authors wish to thank Mauricio Carbajal for his interest in reading the manuscript. The support of CONACyT project 24233-50766-F and IPN grants COFAA and SIP (IPN, Mexico) is acknowledged.

REFERENCES

1. Gamow G, *Z Phys* **51** (1928) 204-212
2. Breit G and Wigner EP, *Phys Rev* **49** (1936) 519-531
3. Bohm A, Gadella M and Mainland GB, *Am J Phys* **57** (1989) 1103-1108;
   Bohm A and Gadella M, *Lecture Notes in Physics* Vol 348 (New York: Springer, 1981)
4. de la Madrid R and Gadella M, *Am J Phys* **70** (2002) 6262-638
5. de la Madrid R, *AIP Conf Proc* **885** (2007) 3-25
6. Civitarese O and Gadella M, *Phys Rep* **396** (2004) 41-113
7. Aguilar J and Combes JM, *Commun Math Phys* **22** (1971) 269-279;
   Balslev E and Combes JM, *Commun Math Phys* **22** (1971) 280-294
8. Simon B, *Commun Math Phys* **27** (1972) 1-9
9. Fock VA, *Fundamentals of Quantum Mechanics*, translated from the Russian by E Yankovsky (Moscow: URSS Publishers, 1976)
10. Hecht E and Zajac A, *Optics* (Massachusetts: Addison-Wesley 1974)
11. Jackson JD, *Classical Electrodynamics*, 3rd edition (USA: John Wiley and Sons, 1999)
12. Landau LD and Lifshitz EM, *The Classical Theory of Fields*, 4th revised English edition, translated from the Russian by M. Hamermesh (Oxford: Pergamon Press, 1987)
13. Brown JW and Churchill RV, *Complex Variables and Applications*, 7th edition (New York: McGraw Hill, 2003)
14. Fowles G.R., *Introduction to Modern Optics* (New York: Dover Pub, 1975)
15. Siegert AJF, *Phys. Rev.* **56** (1939) 750-752
16. Heitler W and Hu N, *Nature* **159** (1947) 776-777
17. de la Madrid R, *Quantum Mechanics in Rigged Hilbert Space Language* (Spain: PhD dissertation, Theoretical Physics Department, Universidad de Valladolid, 2003)
18. Dirac PAM, *The principles of quantum mechanics* (London: Oxford University Press, 1958)
19. Maurin K, *Generalized eigenfunction expansions and unitary representations of topological groups* (Warsav: Pan Stuvowe Wydawn, 1968)
20. Gelfand IM and Vilenkin NY *Generalized functions* Vol 4 (New York: Academic Press, 1968)
21. Fernández-García N and Rosas-Ortiz O, *Ann Phys* **323** (2008) 1397-1414
22. Bohm AR, Scurek R and Wikramasekara S, *Rev Mex Fís* **45 S2** (1999) 16-20
23. Fernández-García N, *Estudio de Resonancias y Transformaciones de Darboux-Gamow en Mecánica Cuántica* (Mexico: PhD dissertation, Departamento de Física, Cinvestav, 2008)
24. Nussenzveig HM, *Causality and Dispersion Relations* (New York: Academic Press, 1972)
25. Fernández-García N and Rosas-Ortiz O, "Optical potentials using resonance states in Supersymmetric Quantum Mechanics", *J Phys Conf Ser* (2008) in press.

26. Feshbach H, Porter CE and Weisskopf VF, *Phys Rev* **96** (1954) 448-464
27. Giraud BG and Kato K, *Ann Phys* **308** (2003) 115-142
28. Sudarshan ECG, Chiu CB and Gorini V, *Phys Rev D* **18** (1978) 2914-2929
29. Mielnik B and Plebański J, *Ann Inst Henry Poincaré* **XII** (1970) 215
30. Berggren T, *Phys Lett B* **44** (1973) 23-25
31. Moiseyev N, *Phys Rep* **302** (1988) 211-293
32. Cannata F, Junker G and Trost J, *Phys Lett A* **246** (1998) 219-226;
Andrianov AA, Ioffe MV, Cannata F and Dedonder JP, *Int J Mod Phys A* **14** (1999) 2675-2688;
Bagchi B, Mallik S and Quesne C, *Int J Mod Phys A* **16** (2001) 2859-2872
33. Fernández DJ, Muñoz R and Ramos A, *Phys Lett A* **308** (2003) 11-16
34. Rosas-Ortiz O and Muñoz R, *J Phys A: Math Gen* **36** (2003) 8497-8506
35. Muñoz R, *Phys Lett A*, **345** (2005) 287-292;
Samsonov BF and Pupasov AM, *Phys Lett A* **356** (2006) 210-214;
Samsonov BF, *Phys Lett A* **358** (2006) 105-114
36. Rosas-Ortiz O, *Rev Mex Fís* **53 S2** (2007) 103-109
37. Fernández-García, *Rev Mex Fís* **53 S4** (2007) 42-45
38. Cabrera-Munguia I and Rosas-Ortiz O, "Beyonf conventional factorization: Non-Hermitian Hamiltonians with radial oscillator spectrum", *J Pgys Conf Ser* (2008) in press
39. Mielnik B and Rosas-Ortiz O, *J Phys A: Math Gen* **37** (2004) 10007-10036
40. Fernández DJ and Rosu H, *Rev Mex Fís* **46 S2** (2000) 153-156;
Fernández DJ and Rosu H, *Phys Scr* **64** (2001) 177-183
41. Ramírez A and Mielnik B, *Rev Mex Fís* **49 S2** (2003) 130-133
42. Sokolov AV, Andrianov AA and Cannata F, *J Phys A: Math Gen* **9** (2006) 10207-10227
43. Vogel FL (Editor), *Resonance and Relaxation in Metals*, Second (Revised) Edition (New York: Plenum Press, 1964)
44. Cohen-Tannoudji C, Diu B and Laloë B, *Quantum Mechanics*, Vols 1 and 2 (New York: John Wiley and Sons, 1977)
45. Fox M, *Quantum Optics. An Introduction* (New York: Oxford University Press, 2006)

II. PARTICLES AND FIELDS

Status of the ALICE experiment at the LHC

G. Herrera Corral[1 a]

[a]*Departamento de Física, Cinvestav, P. O. 14 740, Mexico 07300, D.F. Mexico.*

Abstract. The Large Hadron Collider will provide soon, beams of protons and collisions at high energy to the experiments. ALICE stands for A Large Ion Collider Experiment. It is one of the experiments at the Large Hadron Collider. ALICE will be dedicated to the study of heavy ion collisions. The main goal of ALICE is the observation of the transition of ordinary matter into a plasma of quarks and gluons. ALICE consists of 16 systems of detection. Two of them were designed and constructed in Mexico: i) The V0A detector, located at 3.2 mts. from the interaction point and ii) The cosmic ray detector on the top of the magnet. After a quick review of the LHC and the ALICE experiment we will focus on the description of these systems.

THE LARGE HADRON COLLIDER

The Large Hadron Collider (LHC) will accelerate protons in a 27 Kms long tunnel located in the European Center for Nuclear Research (CERN). In a couple of months the first proton-proton collisions will be provided to the experiments that will study many different aspects of the high energy reaction. LHC will also accelerate lead ions.

The acceleration process starts in LINAC 2 for protons and LINAC 3 for lead ions. The produced protons are injected to a Proton Synchrotron Booster with an energy of 50 MeV. In the Synchrotron, protons reach an energy of 1.4 GeV.

The LINAC 3 produces lead ions. LINAC 3 was commissioned in 1994 by an international collaboration and upgraded in 2007 for the LHC. It delivers 4.2 MeV/u lead ions. The Low Energy Injector Ring (LEIR) will be used as a storage and cooler unit providing ions to the Proton Synchrotron with an energy of 72 MeV/nucleon. Ions will be further accelerated by the Proton Synchrotron and the Super Proton Synchrotron before they are injected into the LHC where they will reach an energy of 2.76 TeV/nucleon.

The Super Proton Synchrotron has been modified to deliver high brightness proton beams required by the LHC. It takes 26 GeV protons from the PS and bring them to 450 GeV before extraction.

[1] A complete list of the members of ALICE – Mexico group appears at the end of this article.

CP1077, *Advanced Summer School in Physics 2008, Frontiers in Contemporary Physics—EAV08*
edited by L. M. Montaño Zetina, G. Torres Vega, M. García Rocha, L. F. Rojas Ochoa, and R. López Fernández

Ions are obtained from purified lead that is heated to 550 ^0C . The lead vapor is then ionized with an electric current that produces various charge states. The

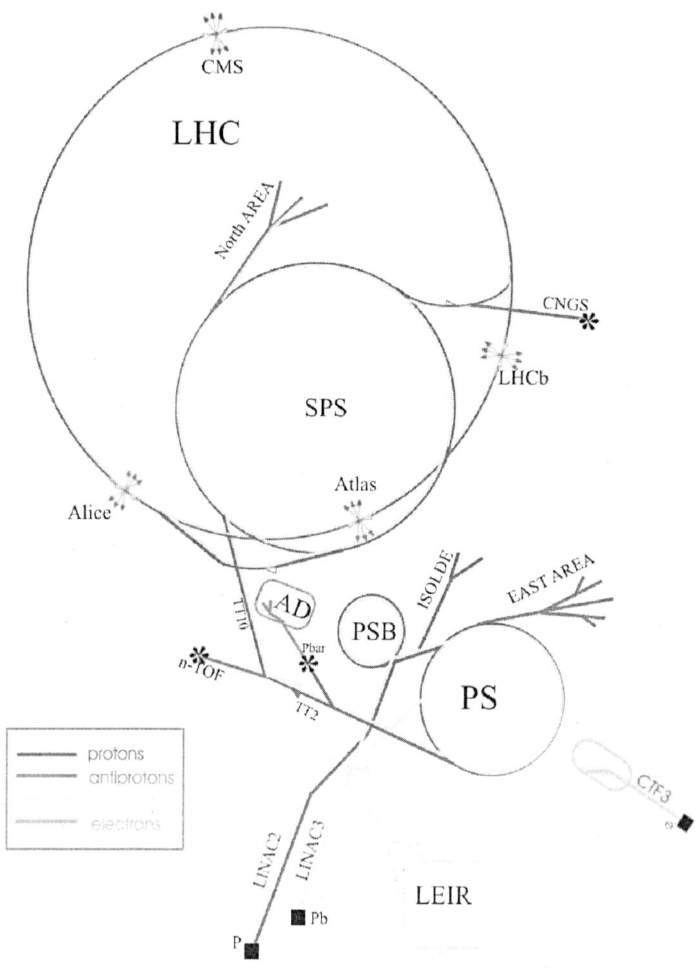

FIGURE 1. Accelerators at CERN. The relevant labels are: LINAC for Linear Accelerator, LEIR for Low Energy Ion Ring, PS for Proton Synchrotron, PSB for Proton Synchrotron Booster, SPS for Super Proton Synchrotron, LHC for Large Hadron Collider.

Pb^{27+} ions are then selected with magnetic fields. After acceleration, these ions go through a carbon foil that strips them to Pb^{54+}, which are accumulated in LEIR. At

the SPS, the ions go once more through a thin carbon foil which strips them to Pb^{82+}. Fig.1 shows the accelerators at CERN that are in use for the LHC.
The total cross section of proton proton interaction at 7 TeV is 110 mbarn. It corresponds to 60 mbarn of inelastic scattering cross section. The accelerator, at its design operation, will reach a luminosity of $10^{34}\sec^{-1} cm^{-1}$, it means that the interaction rate will be:

$$rate = 10^{34} \times 60 \times 10^{-3} \times 10^{-24} = 600 \times 10^{6} \sec$$

During the second week of september this year, bunchs of protons will be injected into the LHC ring. In the following weeks first collisions are expected to be observed in the detectors. LHC forsee to have 5 GeV protons in a first phase.

THE ALICE EXPERIMENT

The ALICE experiment has been designed to observe the transition of ordinary matter into a plasma of quarks and gluons. At the energies achieved by the LHC, the density, the size and the lifetime of the excited quark matter will be high enough as to allow a careful investigation of the properties of this new state of matter. The temperature will exceed by much the critical value predicted for the transition to take place.

The ALICE detector will have a tracking system over a wide range of transverse momentum which goes from 100 MeV/c to 100 GeV/c as well as particle identification able to separate pions, kaons, protons, electrons and photons. Topics like parton energy loss in a dense medium, heavy quark and jet production, prompt photons etc. will be addressed providing important information in that energy regime.

Here we will not give a description of the physics that will be studied with the ALICE detector. We refer the interested reader to ref. [1].

A longitudinal view of the ALICE detector is shown in Fig. 2. A detailed description of the ALICE detector can be found in ref. [2].

In the forward direction a set of tracking chambers inside a dipole magnet will measure muons. Electrons and photons are measured in the central region: photons will be measured in a high resolution calorimeter 5 m below from the interaction point. The PHOS is built from $PBWO_4$ crystals which have a high light output.

The track measurement is performed with a set of six barrels of silicon detectors and a large Time Projection Chamber (TPC). The TPC has an effective volume of 88 m^3. It is the largest TPC ever built. These detectors will make available information on the energy loss allowing particle identification too. In addition to this, a Transition Radiation Detector and a Time of Flight system will provide excellent particle separation at intermediate momentum respectively. The Time of Flight system uses Multi-gap Resistive Plate Chambers (MRPC) with a total of 160,000 readout channels. A Ring Imaging Cherenkov will extend the particle identification capability to higher momentum particles. It covers 15% of the acceptance in the central area and will separate pions from kaons with momenta up to 3 GeV/c and kaons from protons with momenta up to 5 GeV/c.

A Forward Multiplicity Detector consisting of silicon strip detectors and a Zero Degree Calorimeter will cover the very forward region providing information on the charge multiplicity and energy flow.

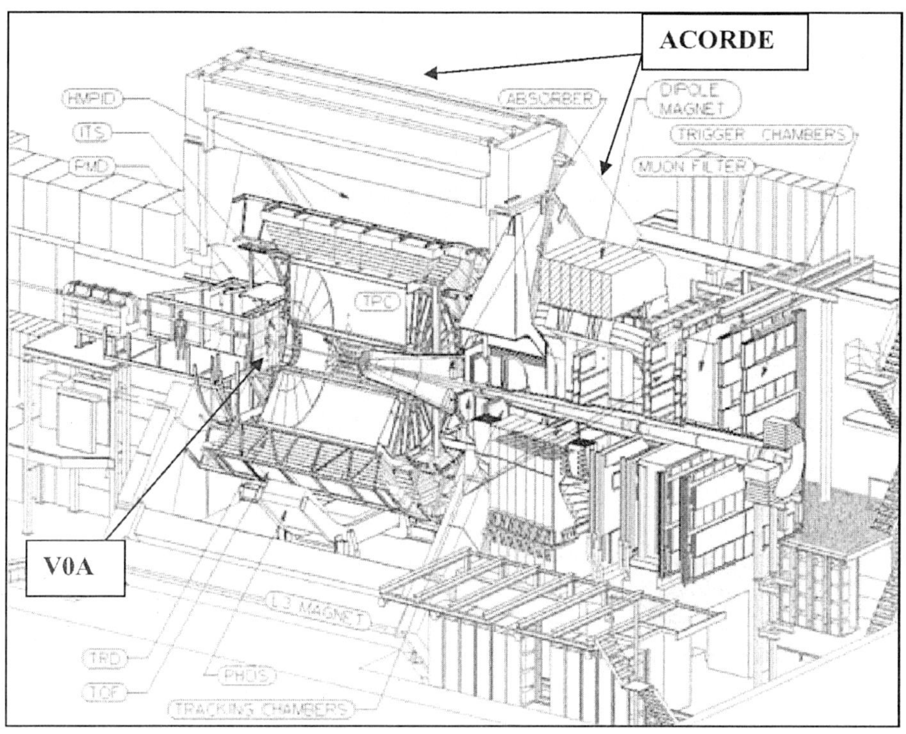

FIGURE 2. ALICE detector. The components mentioned in the text are shown. The Cosmic Ray Detector (ACORDE) on the top of the magnet and the V0 system on the left side of the interaction points are indicated.

The V0 system consisting of two scintillation counters on each side of the interaction point will be used as the main interaction trigger. In the top of the magnet a Cosmic Ray Detector will signal cosmic muons arrival.

THE V0A DETECTOR

The V0 system consists of two detectors: V0A and V0C, which will be located in the central part of ALICE. The V0A is installed at a distance of 328 cm from the interaction point as shown in Fig. 2, mounted in two rigid half boxes around the beam pipe. Each detector is an array of 32 cells of scintillator plastic, distributed in 4 rings forming a disc with 8 sectors. For the V0C, the cells of rings 3 and 4 are divided into two identical pieces that will be read with a single photo-multiplier. This is done to achieve uniformity of detection and a small time fluctuation.

In proton - proton mode the mean number of charged particles within 0.5 units of rapidity is about 3. Each ring covers approximately 0.5 units of rapidity. The particles coming from the main vertex will interact with other components of the detector generating secondary particles. In general, each cell of the V0 detector will, on the average register one hit. For this reason the detector should have a very high efficiency. In Pb-Pb collisions the number of particles in a similar pseudo-rapidity range could be up to 4000 once secondary particles are included. Comparing the number of hits in the detector for proton - proton versus Pb-Pb mode, we can see that the required dynamic range will be 1 – 500 minimum ionizing particles.

FIGURE 3. The V0A detector. The photo multipliers are grouped in shoe boxes at the edge of the V0 container. Front End Electronics are located also in aluminum boxes attached to the V0A structure in a similar array as the shoe boxes.

The Hamamatsu photomultiplier tubes (PMT) are installed inside the magnet not far from the detector. In order to tolerate the magnetic field, fine mesh tubes have been chosen.
The segments of the V0A detector were constructed with a megatile technique (see ref. [3]). This technique consists of machining the scintillator plastic and filling the grooves with TiO_2 loaded epoxy in order to separate one sector from the other.
A detailed description of the V0 system can be found in ref. [4]. Fig. 3 shows the V0A detector in its mechanical structure.

THE COSMIC RAY DETECTOR

The Cosmic Ray Detector consists of an array of 60 scintillator counters located in the upper part of the ALICE magnet.
The plastic used for the construction of the detector was part of the DELPHI detector [5]. The material was carefully studied and the design of the detector was done according to the capabilities of the plastic available. The material was transported to Mexico where the construction was done.
Each module has a sensitive area of $1.9 \times .195 m^2$ and is built with two superimposed plastics. The doublet has an efficiency around 90 % along the module.

The Cosmic Ray Detector:

- Generates a single muon trigger to calibrate the Time Projection Chamber and other components of ALICE.

- Generates a multi-muon trigger to study cosmic rays.

- Provides a wake-up signal for the Transition Radiation Detector.

The geometry is shown in Fig. 4. Modules on the far ends of the inner and outer faces of the magnet were moved to the center of the upper face in order to have a much better efficiency for single muons since, the central part of ACORDE were used to align the Inner Tracking System.
Fig.4 shows a real cosmic ray event reconstructed with the Time Projection Chamber and projected to ACORDE on the top of the magnet. The hit in a central module is visible.

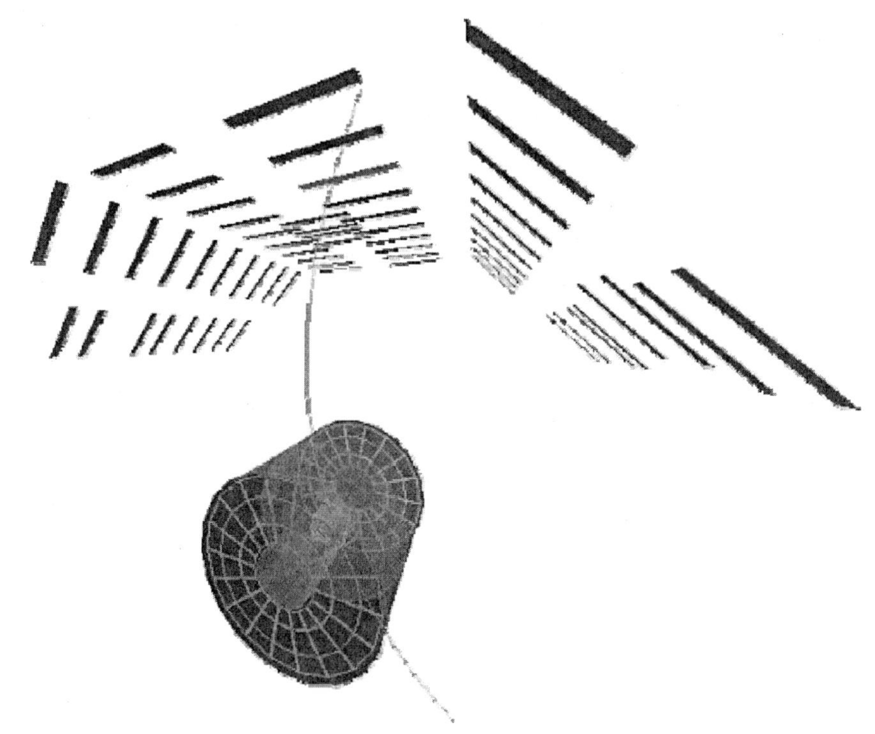

FIGURE 4. A real event. A single muon deflected in the magnetic field of ALICE. ACORDE triggered the Time Projection Chamber (TPC) which in turn, reconstructed the track back to the module of the cosmic ray detector that started the process.

ALICE - MEXICO GROUP

Benemérita Universidad Autónoma de Puebla: A. Fernández, H. González, M. I. Martínez, J. Muñoz, M. Rodríguez, S. Román, G. Tejeda, M. A. Vargas, S. Vergara.

CINVESTAV, Departamento de Física Aplicada: Guillermo Contreras.

CINVESTAV, Departamento de Física: G. Herrera Corral, L. Montaño, A. Zepeda.

Universidad Autónoma de Sinaloa: I. León Monzón, P. Podesta.

Universidad Nacional Autónoma de Mexico, Instituto de Ciencias Nucleares: E. Cuautle, L. Díaz, I. Domínguez, I. Maldonado, L. Nellen, A. Ortiz, G. Paic, L. Serkin, O. Sokolov.

Universidad Nacional Autónoma de Mexico, Instituto de Física: R. Alfaro, E. J. Almaraz, A. Anzo, E. Belmont, L. H. González, V. Grabski, A. Martínez, A. Menchaca, A. Sandoval, L. Valencia.

ACKNOWLEDGMENTS

This project was supported by the World Bank through the Iniciativa Científica del Milenio 2001 as well as by several research projects supported along the years by CONACYT. Mobility and presence at CERN has been a key issue for the success of the project, in this regard a big support was obtained from the High Energy Physics Latin American–European Network - HELEN - with the leadership of Luciano Maiani and coordinated by Veronica Riquer.

REFERENCES

1. ALICE Collaboration, J. of Phys. G: Nucl. Part. Phys. 30 (2004)1517-1763
   J. of Phys. G: Nucl. Part. Phys. 32 (2006)1295-2040
2. ALICE Technical Proposal, Ahmad *et al.* CERN/LHCC/95-71, 1995
3. S. Kim, *et al.* Nucl. Inst. Meth. Phys. Res. **A360**(1995)206
4. ALICE Technical Design Report, Forward Detectors FMD, T0, V0. CERN/LHCC-2004-025, 2004.
5. R.I. Dzhelyadin, et al., DELPHI Internal Note 86- 108, TRACK 42 , CERN 1986.

B Physics @ the DØ experiment

Eduard De La Cruz Burelo

Department of Physics,
Centro de Investigaciones y de Estudios Avanzados
del Instituto Politécnico Nacional,
Av. IPN 2508, México D.F., México.

Abstract.
At the beginning of RunII of the Tevatron and after more than 30 years of the discovery of the b quark at Fermilab, the lack of statistics had restricted our knowledge on b-baryons to the observation of the lightest b-baryon, the Λ_b, and to its lifetime measured in decays which did not allow a fully reconstruction of this particle. I present results of the search for b-baryons in the DØ experiment. As part of this program, a precise measurement of the Λ_b lifetime was performed, and the discovery of the Ξ_b^- resulted from an analysis of 1.3 fb^{-1} of data collected with the D0 detector during 2002–2006.

Keywords: Bottom baryons
PACS: 14.20.Mr, 14.40.Nd, 13.30.Eg, 13.25.Hw

INTRODUCTION

The observation of B hadrons[1] and the measurement of its properties, has been an important part of the rich physics program of Tevatron experiments, CDF [1] and DØ [2]. Properties like the lifetime of B hadrons provides information on the b quark decay, and what is the role that play the lighter quarks which along with the b-quark form the B hadron. Also, b-quark decays are related to the V_{ub} and V_{cb} elements of the CKM matrix, and B rare decays are always considered a playground to test extensions of the Standard Model of particles.

Since the discovery of the b quark at Fermilab in 1977, and before RunII of Tevatron, all b-mesons in the ground state ($B^+(\bar{b}u)$, $B^0(\bar{b}d)$, $B_s(\bar{b}s$, and $B_c^+(\bar{b}c)$) have been experimentally established, but only the lightest b-baryon, the $\Lambda_b(bud)$, had been observed. During RunII of Tevatron a significant progress has been made in the b-baryon sector. The Λ_b has been observed by both experiments, CDF and DØ, and its lifetime has been measured in different decays channels. Other b-baryons has been finally observed, and properties as their masses have been established with high precision [3, 4, 5].

In the first part of this note, a description of the Λ_b observation and measurement of its lifetime [6] in the DØ experiment, by reconstructing $\Lambda_b \rightarrow J/\psi\Lambda$ decays, is presented. An earlier difference between measurement and prediction created a great deal of interest on the Λ_b lifetime, and it was named as the Λ_b *lifetime puzzle*[7]. Recent theoretical calculations of $\tau(\Lambda_b)/\tau(B^0)$, which include next-to-leading order

[1] Particles which heavy constituent quark is a b quark

CP1077, *Advanced Summer School in Physics 2008, Frontiers in Contemporary Physics—EAV08*
edited by L. M. Montaño Zetina, G. Torres Vega, M. García Rocha, L. F. Rojas Ochoa, and R. López Fernández

effects in QCD [8], corrections at $\mathcal{O}(1/m_b^4)$ in HQET [9], and lattice QCD studies [10], have led to a prediction of $\tau(\Lambda_b)/\tau(B^0) = 0.88 \pm 0.05$ [11], that significantly reduced this difference ($\tau(\Lambda_b)/\tau(B^0) = 0.800 \pm 0.053$ was the world average in 2004 [12]). However, a recent precise measurement [13] by CDF experiment reports a value of the Λ_b lifetime consistent with b meson lifetimes, and the ratio $\tau(\Lambda_b)/\tau(B^0)$ consistent with unity. This measurement, which statistically dominates the current world average, has turned around the Λ_b lifetime puzzle, by making the prediction to be lower than the experimental measurement.

The second part of this note is dedicated to describe the discovery of the $\Xi_b^-(bsd)$, the first observed particle formed by a quark of each known family of matter. In the quark model the Ξ_b^-[2] is formed by the combination of a b, a d, and a s quarks, and it is expected to have $J^P = 1/2^+$ although I, J or P have yet to be measured. Evidence for the Ξ_b^- has been inferred from an excess of same sign $\Xi^\pm \ell^\pm$ events in jets which are interpreted as $\Xi_b^- \to \Xi^- \ell^- \bar{\nu}_\ell X$ [14]. From this decay mode, the average lifetime of the Ξ_b^- is measured to be $1.42^{+0.28}_{-0.24}$ ps [15]. These semileptonic decays of the Ξ_b^- did not allow for a mass measurement, but theoretical calculations of heavy quark effective theory [16] and nonrelativistic QCD [17] predict the Ξ_b^- mass in the range $5.7 - 5.8$ GeV [18]. DØ experiment observed for first time the Ξ_b^- baryon fully reconstructed in an exclusive decay, $\Xi_b^- \to J/\psi \Xi^-$, with $J/\psi \to \mu^+\mu^-$, $\Xi^- \to \Lambda\pi^-$, and $\Lambda \to p\pi^-$.

THE DØ DETECTOR

The D0 detector is described in detail elsewhere [2]. The components most relevant to these analyses are the central tracking system and the muon spectrometer. The central tracking system consists of a silicon microstrip tracker (SMT) and a central fiber tracker (CFT) that are surrounded by a 2 T superconducting solenoid. The SMT is optimized for tracking and vertexing for the pseudorapidity region $|\eta| < 3$ ($\eta = -\ln[\tan(\theta/2)]$ and θ is the polar angle) while the CFT has coverage for $|\eta| < 2$. Liquid-argon and uranium calorimeters in a central and two end-cap cryostats cover the pseudorapidity region $|\eta| < 4.2$. The muon spectrometer is located outside the calorimeter and covers the pseudorapidity region $|\eta| < 2$. It comprises a layer of drift tubes and scintillator trigger counters in front of 1.8 T iron toroids followed by two similar layers behind the toroids.

Λ_b LIFETIME MEASUREMENT

Event selection

The search for $\Lambda_b \to J/\psi\Lambda$ decays begins with the reconstruction of J/ψ candidates in the decay channel $J/\psi \to \mu^+\mu^-$, where the muon system of the DØ detector is used to

[2] Charge conjugation is always assumed for this state

confirm the muon candidates. Then these events containing a J/ψ meson are searched for $\Lambda \to p\pi^-$ decays. J/ψ and Λ candidates are then combined to search for those decays coming from Λ_b particles. Although specific triggers are not required for event selection, most of the selected events satisfy dimuon or muon triggers due to the present of $J/\psi \to \mu^+\mu^-$ candidates. To avoid bias in the lifetime measurement, we reject events that fired a trigger based on impact parameter measurements. In addition, selected events must have at least one reconstructed interaction vertex.

To form a J/ψ candidate, two oppositely charged tracks must originate from a common vertex and to have an invariant mass in the range of 2.8 – 3.35 GeV , that approximately corresponds to $M_{J/\psi} \pm 3\sigma_{J/\psi}$, where $\sigma_{J/\psi}$ is the width of the J/ψ signal observed in DØ and $M_{J/\psi}$ is the mass of the J/ψ measured in DØ. Each track must either match hits in the muon system, or have calorimeter energies consistent with a minimum-ionizing particle. In addition, a minimum p_T of 1.5 GeV and at least one CFT hit for each track is required, and at least one of the muons must have signatures in the three layers of the muon detector. Both tracks are required to have p_T >2.5 GeV if they are in the region $|\eta|$ <1. The J/ψ candidate is required to have a $p_T > 5$ GeV, and in addition, the distance between primary vertex and the dimuon vertex is required to be less than 10 cm.

The $\Lambda \to p\pi^-$ candidates are reconstructed from two oppositely charged tracks which must originate from a common vertex and an invariant mass between $1.105 - 1.125$ GeV, that similar to the J/ψ selection, approximately corresponds to three times the width of the observed Λ signal around the mass of the Λ. The two tracks are collectively required to have no more than two hits in the tracking system before the vertex, and the track with the higher p_T is assigned to be proton if positive and to be antiproton if negative. Monte Carlo studies show that this assignment gives nearly 100% correct combination. The Λ is a particle that travels long distance in the detector before it decays, then tracks of its decay products must have a considerable impact parameter with respect to the primary vertex. To exploit this feature, the axial ε_T and stereo ε_L impact parameter projections of each track with respect to the primary vertex and their uncertainties are calculated, and the combined significance $(\varepsilon_T/\sigma(\varepsilon_T))^2 + (\varepsilon_L/\sigma(\varepsilon_L))^2$ is required to be greater than 9 for both tracks and greater than 16 for at least one of the tracks. To reduce combinatoric backgrounds, the Λ decay distance from the primary vertex is required to be greater than 4 times its estimated uncertainty when this uncertainty is less than 0.5 cm.

The Λ_b candidates are reconstructed by performing a constrained fit to a common vertex for the Λ and the two tracks forming the J/ψ candidate, with the latter constrained to the J/ψ mass of 3.097 GeV/c^2 [19]. To suppress contamination from cascade decays of more massive baryons such as $\Sigma^0 \to \Lambda\gamma$ or $\Xi^0 \to \Lambda\pi^0$, the cosine of the angle between the $\boldsymbol{p}_T$ vector of the Λ and the vector in the perpendicular plane from the J/ψ vertex to the Λ decay vertex is required to be larger than 0.9999. For Λ's that decay from Λ_b the cosine of this angle is very close to one. If more than one $J/\psi + \Lambda$ candidate is found in the event, the candidate with the best χ^2 probability is selected as the Λ_b. The mass of $J/\psi + \Lambda$ combination is required to be within the range 5.1–6.1 GeV/c^2. For the choice of the final selection criteria, $S/\sqrt{S+B}$ is optimized, where S and B are the number of signal (Λ_b) and background candidates, respectively, by using Monte Carlo estimates for S and data for B. As a result of this optimization, the p_T of the Λ is required to be greater than 2.4 GeV/c, and the total momentum for the Λ_b is required to be greater than 5 GeV$/c$.

In order to test the selection and lifetime measurement procedure, simultaneously to the reconstruction of $\Lambda_b \to J\psi(\mu^+\mu^-)\Lambda(p\pi^-)$ decays, the well known B^0 meson is reconstructed in the topologically similar decay $B^0 \to J\psi(\mu^+\mu^-)K^0_S(\pi^+\pi^-)$, where a pion takes the place of the proton in the Λ reconstruction to form a K^0_S. Same selection is applied for the J/ψ reconstruction, and the $K^0_S \to \pi^+\pi^-$ selection follows the same criteria as for the Λ reconstruction, except that for the K^0_S, the mass window is 0.460–0.525 GeV/c^2, and pion mass assignments are used. In addition, the p_T of the K^0_S is required to be greater than 1.8 GeV/c and $J/\psi + K^0_S$ candidates within 4.9–5.7 GeV/c^2 are considered for the B^0 search. Finally, any event which has been selected in the Λ_b reconstruction is removed from the B^0 sample.

Technique for lifetime measure

The decay time of a Λ_b or B^0 is determined by measuring the distance traveled by the b hadron candidate in a plane transverse to the beam direction, and then applying a correction for the Lorentz boost. The transverse decay length is defined as $L_{xy} = \boldsymbol{L}_{xy} \cdot \boldsymbol{p}_T / p_T$ where $\boldsymbol{L}_{xy}$ is the vector that points from the primary vertex to the b hadron decay vertex and $\boldsymbol{p}_T$ is the transverse momentum vector of the b hadron. The event-by-event value of the proper transverse decay length, λ, for the b hadron candidate is given by:

$$\lambda = \frac{L_{xy}}{(\beta\gamma)^B_T} = L_{xy}\frac{cM_B}{p_T}, \tag{1}$$

where $(\beta\gamma)^B_T$ and M_B are the transverse boost and the mass of the b hadron. In this measurement, the value of M_B in Eq. 1 is set to the Particle Data Group (PDG) mass value of Λ_b or B^0 [19]. In order to reduce background from mismeasurement of vertices, the uncertainty on λ is required to be less than 500 μm.

A simultaneous unbinned maximum likelihood fit is performed to the mass and proper decay length distributions. The likelihood function $\mathscr{L}$ is defined by:

$$\begin{aligned} \mathscr{L} &= \frac{(n_s+n_b)^N}{N!}\exp(-n_s-n_b) \times \\ &\prod_{j=1}^{N}\left[\frac{n_s}{n_s+n_b}\mathscr{F}^j_{sig} + \frac{n_b}{n_s+n_b}\mathscr{F}^j_{bkg}\right] \end{aligned} \tag{2}$$

where n_s and n_b are the expected number of signal and background events in the sample, respectively. N is the total number of events. $\mathscr{F}^j_{sig}$ ($\mathscr{F}^j_{bkg}$) is the product of three probability density functions that model the mass, proper decay length, and uncertainty on proper decay length distributions for the signal (background). The background is divided into two categories, prompt and non-prompt. The prompt background is primarily due to direct production of J/ψ's which are then randomly combined with a Λ or K^0_S candidate in the event. The non-prompt background is mainly produced by the combination of J/ψ mesons from b hadron decays with Λ or K^0_S candidates present in the event.

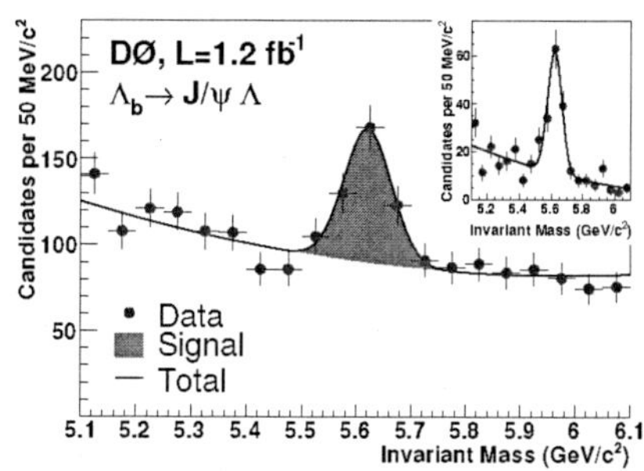

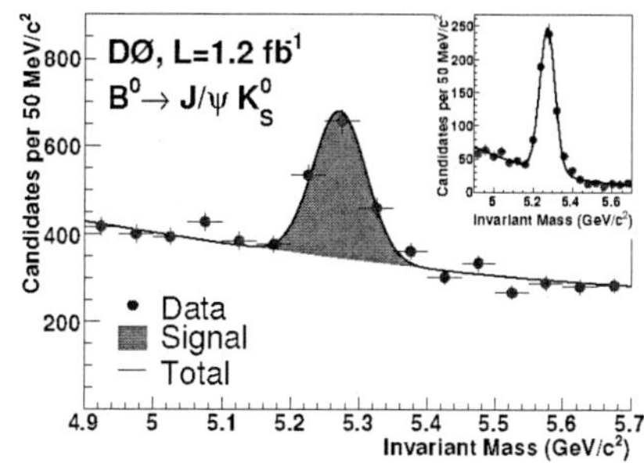

FIGURE 1. Invariant mass distribution for Λ_b (left) and B^0 (right) candidates with the fit results superimposed. The inserts show the mass distributions after requiring $\lambda/\sigma > 5$.

For the signal, the mass distribution is modeled by a Gaussian function, and the λ distribution is parametrized by an exponential decay convoluted with the resolution function:

$$G(\lambda_j, \sigma_j) = \frac{1}{\sqrt{2\pi} s\sigma_j} \exp\left[\frac{-\lambda_j^2}{2(s\sigma_j)^2}\right], \tag{3}$$

where λ_j and σ_j represent λ and its uncertainty, respectively, for a given decay j, and s is a common scale parameter introduced in the fit to account for a possible mis-estimate of σ_j. The convolution is defined by:

$$S_\lambda(\lambda_j, \sigma_j) = \frac{1}{\lambda_B} \int_0^\infty G(x - \lambda_j, \sigma_j) \exp\left(\frac{-x}{\lambda_B}\right) dx, \tag{4}$$

where $\lambda_B = c\tau_B$, and τ_B is the lifetime of the Λ_b (B^0). The distribution of the uncertainty of λ is modeled by an exponential function convoluted by a Gaussian.

For the background, the mass distribution of the prompt component is assumed to follow a flat distribution as observed in data when a cut of $\lambda > 100$ μm is applied. The non-prompt component is modeled with a second-order polynomial function. The λ distribution is parametrized by the resolution function for the prompt component, and by the sum of negative and positive exponential functions for the non-prompt component. A positive and a negative exponential functions model combinatorial background, and an exponential function accounts for long-lived heavy flavor decays. The distribution of the uncertainty of λ is modeled by two exponential functions convoluted by a Gaussian.

By minimizing $-2\ln\mathscr{L}$, it is found: $c\tau(\Lambda_b) = 365.1^{+39.1}_{-34.7}$ μm and $c\tau(B^0) = 450.0^{+23.5}_{-22.1}$ μm. From the fits, it is obtained $s = 1.41 \pm 0.05$ for the Λ_b and $s = 1.41 \pm 0.03$ for the B^0. The numbers of signal decays are 171 ± 20 Λ_b and 717 ± 38 B^0. Figures 1 and 2 show the mass and λ distributions for the Λ_b and B^0 candidates. Fit results are superimposed.

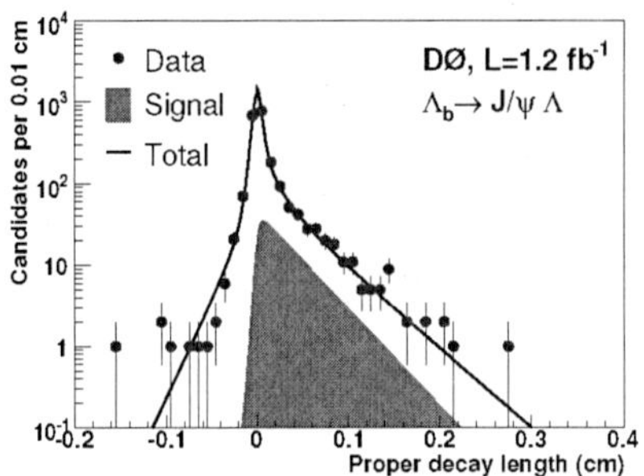

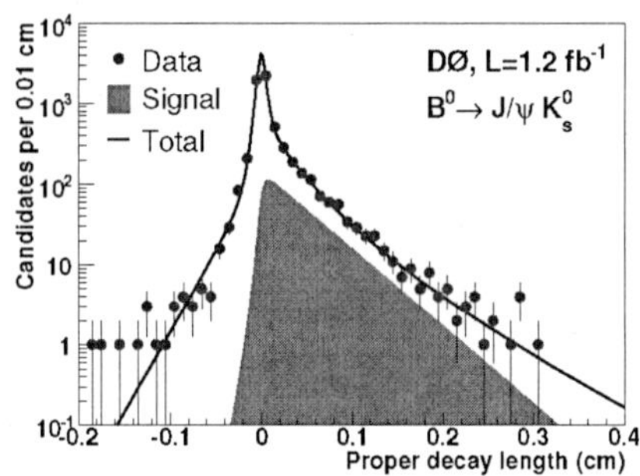

FIGURE 2. Proper decay length distribution for Λ_b(left) and B^0(right) candidates, with the fit results superimposed. The shaded region represents the signal.

Systematic uncertainties

Table 1 summarizes the systematic uncertainties considered in these lifetime measurements. The contribution from possible misalignment of the SMT detector is estimated to be 5.4 μm [20]. Systematic uncertainties due to the modeling of the λ and mass distributions are estimated by varying the parametrizations of the different components. To take into account correlations between the effects of the different models, a fit that combines all different model changes is performed, and the difference between the result of this fit and the nominal fit is quoted as the systematic uncertainty.

The lifetime of the background events under the $\Lambda_b(B^0)$ signal is mostly modeled by events in the low and high mass sideband regions with respect to the peak. To estimate the effect of any difference between the lifetime distributions of these two regions, separate fits are performed to the Λ_b (B^0) mass regions of 5.1–5.8 and 5.4–6.1 GeV/c^2 (4.9–5.45 and 5.1–5.7 GeV/c^2) where the contributions from high and low mass background events are reduced, respectively. The largest difference between these fits and the nominal fit is quoted as the systematic uncertainty due to this source.

Contamination of the Λ_b sample by B^0 events that pass the Λ_b selection is also considered. From Monte Carlo studies, it is estimated that 6.5% of B^0 events can pass the Λ_b selection criteria. In order to estimate any effect due to this possible contamination, any event which also passes the B^0 selection criteria is removed from the Λ_b sample, and the lifetime fit is repeated to the remaining events. The difference between this and the nominal fit is quoted as a systematic uncertainty. For the B^0 lifetime, this source of systematic uncertainty is not considered since any event with a Λ_b candidate is removed from the B^0 sample.

Results

The results of the measurement of the Λ_b and B^0 lifetimes are summarized as:

$$c\tau(\Lambda_b) = 365.1^{+39.1}_{-34.7}\ \text{(stat)} \pm 12.7\ \text{(syst)}\ \mu\text{m}, \tag{5}$$

TABLE 1. Summary of systematic uncertainties in the measurement of $c\tau$ for Λ_b and B^0 and their ratio. The total uncertainties are determined by combining individual uncertainties in quadrature.

Source	Λ_b (μm)	B^0 (μm)	Ratio
Alignment	5.4	5.4	0.002
Distribution models	6.6	2.8	0.020
Long-lived components	6.0	13.6	0.022
Contamination	7.2	–	0.016
Total	12.7	14.9	0.034

$$c\tau(B^0) = 450.0^{+23.5}_{-22.1}\,(\text{stat}) \pm 14.9\,(\text{syst})\ \mu\text{m},$$

from which:

$$\begin{aligned}\tau(\Lambda_b) &= 1.218^{+0.130}_{-0.115}\,(\text{stat}) \pm 0.042\,(\text{syst})\ \text{ps},\\ \tau(B^0) &= 1.501^{+0.078}_{-0.074}\,(\text{stat}) \pm 0.050\,(\text{syst})\ \text{ps}.\end{aligned} \tag{6}$$

These can be combined to determine the ratio of lifetimes:

$$\frac{\tau(\Lambda_b)}{\tau(B^0)} = 0.811^{+0.096}_{-0.087}\,(\text{stat}) \pm 0.034\,(\text{syst}), \tag{7}$$

The systematic uncertainty on the ratio is computed by calculating the ratio for each systematic source and quoting the deviation in the ratio as the systematic uncertainty due to that source. All systematics are combined in quadrature as shown in Table 1. The main contribution to the systematic uncertainty of the lifetime ratio is due to the long-lived component of the B^0 sample. This is expected since the B^0 is more likely than the Λ_b to be contaminated by mis-reconstructed b mesons due to its lower mass.

The measurement is consistent with the world average [19], and the ratio of Λ_b to B^0 lifetimes is consistent with the most recent theoretical predictions [11].

Ξ_b^- DISCOVERY

The decay $\Xi_b^- \to J/\psi\Xi^-$ has topological characteristics which help to reduce combinatorial background, but these characteristics also considerably reduce the reconstruction efficiency for this particle. This decay includes the reconstruction of three particles: the $J/\psi \to \mu^+\mu^-$, $\Lambda \to p\pi^-$, and $\Xi^- \to \Lambda\pi^-$. The Λ and Ξ^- have a lifetime of the order of centimeters, while the Ξ_b^- should have a lifetime of the order of microns. The Ξ_b^- has a decay vertex displaced from the primary vertex and its decay product includes a particle, the Ξ^-, which travels a long distance in the detector before it decays to a pion and another very long-lived particle, the Λ. Three charged particles in the final state, (p,π^-,π^-) should have a significant impact parameter with respect to the primary vertex. This final characteristics is what considerably reduces the reconstruction efficiency of the $\Xi_b^- \to J/\psi\Xi^-$ decay. This led to a reprocessing of dimuon events with an extended version of the tracking algorithm to allow the reconstruction of tracks with very high impact parameters and low p_T.

Event selection

The Ξ_b^- reconstruction starts by searching for events with J/ψ mesons. Then these events are searched for Λ candidates. These Λ candidates are combined with an extra charged track in the event to reconstruct $\Xi^- \to \Lambda\pi^-$ decays. Then J/ψ and Ξ^- candidates are combined to reconstruct the $\Xi_b^- \to J/\psi\Xi^-$ decays.

The reconstruction of $J/\psi \to \mu^+\mu^-$ and $\Lambda \to p\pi^-$ decays follows the same selection as for the Λ_b lifetime measurement. However, events containing a J/ψ candidate were reprocessed with a version of the track reconstruction algorithm that improves the efficiency for tracks with low p_T and high impact parameters. Consequently, the efficiencies for K_S^0, Λ, and Ξ^- reconstruction are significantly increased. Figure 3(a) shows for comparison, the Ξ^- reconstruction (as described below) before and after reprocessing events with at least a J/ψ candidate.

The reconstruction of $\Xi^- \to \Lambda\pi^-$ decays is inspired on the selection criteria applied to $\Lambda \to p\pi^-$ decays, which is as the Ξ^-, a particle that travels long distance in the detector before it decays. The Ξ^- candidates are reconstructed by combining Λ candidates with negatively charged tracks (excluding from the list of tracks the proton and pion tracks from Λ decays) with a pion mass assigned. The Λ and the π^- must originate from a common vertex and the Λ decay distance in the transverse plane (the plane perpendicular to the beam direction) measured with respect to the Ξ^- vertex is required to have a uncertainty less than 0.5 cm. The combined significance $(\varepsilon_T/\sigma(\varepsilon_T))^2 + (\varepsilon_L/\sigma(\varepsilon_L))^2$ of the π^- is required to be greater than 9. To reduce combinatoric backgrounds, the Ξ^- decay distance from the primary vertex is required to be greater than 4 times its estimated uncertainty when this uncertainty is less than 0.5 cm. In addition, the decay distance in the transverse plane measured with respect to the J/ψ vertex is required to have a uncertainty less than 0.5 cm.

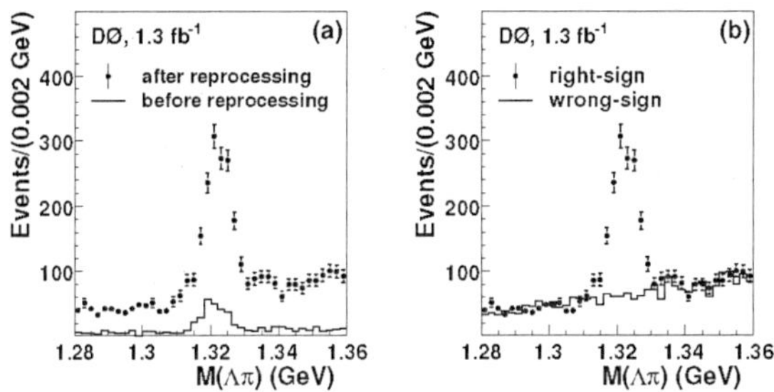

FIGURE 3. Invariant mass distributions of the $\Lambda\pi$ pair before the Ξ_b^- reconstruction for (a) the right-sign $\Lambda\pi^-$ combinations before and after reprocessing and (b) the right-sign $\Lambda\pi^-$ and the wrong-sign $\Lambda\pi^+$ combinations after reprocessing. The reprocessing significantly increases the Ξ^- yield. Fits to the post-reprocessing distributions of the right-sign combination with a Gaussian signal and a first-order polynomial background yield 603 ± 34 Ξ^-'s and 548 ± 31 $\overline{\Xi}^+$'s.

The two pions from $\Xi^- \to \Lambda\pi^- \to (p\pi^-)\pi^-$ decays (right-sign) have the same charge. Consequently, the combination $\Lambda\pi^+$ (wrong-sign) events form an ideal control sample for background studies. Figure 3(b) compares mass distributions of the right-sign $\Lambda\pi^-$ and the wrong-sign $\Lambda\pi^+$ combinations. The Ξ^- mass peak is evident in the distribution of the right-sign events. A $\Lambda\pi^-$ pair is considered to be a Ξ^- candidate if its mass is within the range $1.305 < M(\Lambda\pi^-) < 1.340$ GeV.

To reconstruct $\Xi_b^- \to J/\psi\Xi^-$ decays, J/ψ and Ξ^- candidates are required to be in the same semi-hemisphere in the transverse plane and must have a common vertex. Backgrounds from mismeasurements are reduced by requiring uncertainties of the proper decay length of the $J/\psi\,\Xi^-$ vertex to be less than 0.05 cm in the transverse plane.

Several distinctive features of the $\Xi_b^- \to J/\psi\,\Xi^- \to J/\psi\,\Lambda\pi^- \to (\mu^+\mu^-)\,(p\pi^-)\pi^-$ decay are utilized to further suppress backgrounds. The wrong-sign background events $(\Lambda\pi^+)$, Λ and J/ψ signals from $\Lambda_b \to J/\psi(\mu^+\mu^-)\Lambda(p\pi^-)$ decays from data, and Monte Carlo signal Ξ_b^- events are used for studying additional event selection criteria. Protons and pions from the Ξ^- decays of the Ξ_b^- events are expected to have higher momenta than those from most of the background processes. Therefore, protons are required to have $p_T > 0.7$ GeV. Similarly, minimum p_T requirements of 0.3 and 0.2 GeV are imposed on pions from Λ and Ξ^- decays, respectively. These requirements remove 91.6% of the wrong-sign background events while keeping 68.7% of the Monte Carlo Ξ_b^- signal events.

Backgrounds from combinatorics and other b hadrons are reduced by using topological decay information. Contamination from decays such as $B^- \to J/\psi K^{*-} \to J/\psi K_S^0\pi^-$ and $B^0 \to J/\psi K^{*-}\pi^+ \to J/\psi\,(K_S^0\pi^-)\pi^+$, where all tracks in the combination $\mu^+\mu^-\Lambda\pi^-$ originate from the same vertex, are suppressed by requiring the Ξ^- candidates to have decay lengths greater than 0.5 cm and $\cos(\theta) > 0.99$, as the Ξ^- baryons in Monte Carlo have an average decay length of 4.8 cm. Here θ is the angle between the Ξ^- direction and the direction from the Ξ^- production vertex to its decay vertex in the transverse plane. Figure 4 shows the comparison of $\cos(\theta)$ distributions from wrong-sign background and Ξ_b^- Monte Carlo events. From this comparison the cut $\cos(\theta) > 0.99$ is selected. These two requirements on the Ξ^- reduce the background by an additional 56.4%, while removing only 1.7% of the Monte Carlo signal events.

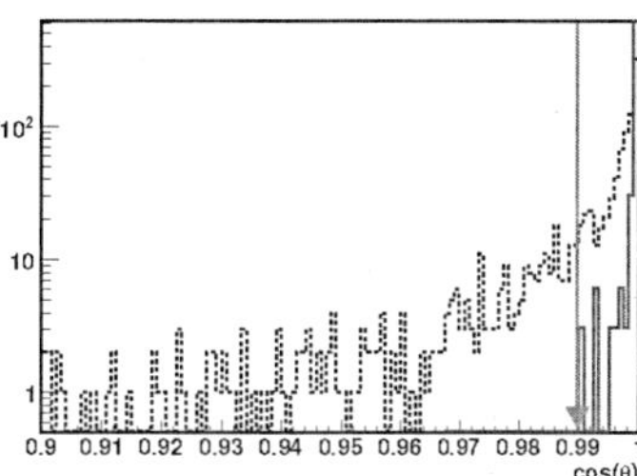

FIGURE 4. Comparison of the Ξ^- collinearity distribution in wrong-sign background events and Monte Carlo signal events, after pre-selection cuts. Solid line in blue represents the signal and the dashed black line the background. Signal has been scaled to the same number of background events. The red arrow shows the selected cut.

Finally, Ξ_b^- baryons are expected to have a sizable lifetime. To reduce prompt backgrounds, the transverse proper decay length significance of the Ξ_b^- candidates is required to be greater than two. This final criterion retains 83.1% of the Monte Carlo signal events but only 43.9% of the remaining background events.

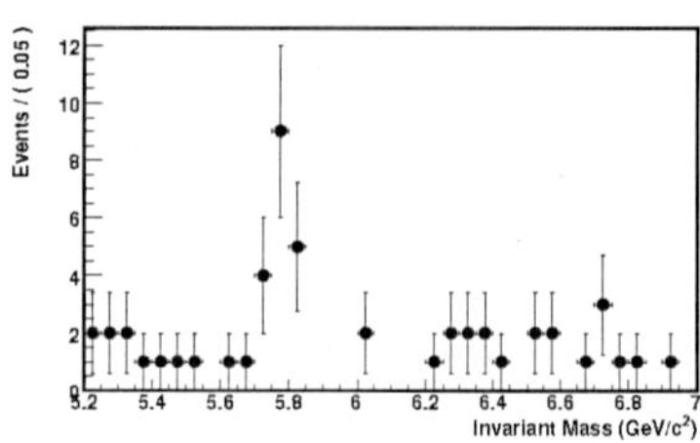

FIGURE 5. Ξ_b^- mass distribution from data after all selection cuts have been applied.

In the data, 51 events with the Ξ_b^- candidate mass between 5.2 and 7.0 GeV pass all selection criteria. Figure 5 shows the Ξ_b^- candidates in the remaining events. An excess of events is observed near to 5.8 GeV. The mass range 5.2–7.0 GeV is chosen to be wide enough to encompass masses of all known b hadrons as well as the predicted mass of the Ξ_b^- baryon. The candidate mass, $M(\Xi_b^-)$, is calculated as $M(\Xi_b^-) = M(J/\psi\,\Xi^-) - M(\mu^+\mu^-) - M(\Lambda\pi^-) + M_{\rm PDG}(J/\psi) + M_{\rm PDG}(\Xi^-)$ to improve the resolution. Here $M(J/\psi\,\Xi^-)$, $M(\mu^+\mu^-)$, and $M(\Lambda\pi^-)$ are the reconstructed masses while $M_{\rm PDG}(J/\psi)$ and $M_{\rm PDG}(\Xi^-)$ are taken from Ref. [19]. A number of cross checks are performed to ensure the observed peak is not due to artifacts of the analysis: (1) The $J/\psi\,\Lambda\pi^+$ mass distribution of the wrong-sign events, shown in Fig. 6(top), is consistent with a flat background. (2) The event selection is applied to the sideband events of the Ξ^- mass peak, requiring $1.28 < M(\Lambda\pi^-) < 1.36$ GeV but excluding the Ξ^- mass window. Similarly, the selection is applied to the J/ψ sideband events with $2.5 < M(\mu^+\mu^-) < 2.7$ GeV. The high-mass sideband is not considered due to potential contamination from ψ' events. As shown in Fig. 6(middle and bottom), no evidence of a mass peak is present for either $(\mu^+\mu^-)\,(p\pi^-)\pi^-$ distribution. (3) Though most backgrounds from other b hadron production are removed by the Ξ^- reconstruction and its selection, the possibility of a fake signal due to the residual b hadron background is investigated by applying the final Ξ_b^- selection to high statistics Monte Carlo samples of $B^- \to J/\psi K^{*-} \to J/\psi K_S^0\pi^-$, $B^0 \to J/\psi K_S^0$, and $\Lambda_b \to J/\psi\,\Lambda$. No indication of a mass peak is observed in the reconstructed $J/\psi\,\Xi^-$ mass distributions. (4) The mass distributions of J/ψ, Ξ^-, and Λ are investigated by relaxing the mass requirements on these particles one at a time for events both in the Ξ_b^- signal region and the sidebands. The numbers of these particles determined by fitting their respective mass distribution are fully consistent with the quoted numbers of signal events plus background contributions. (5) The robustness of the observed mass peak is tested by varying selection criteria within reasonable ranges. All studies confirm the existence of the peak at the same mass.

Ξ_b^- mass measurement and signal significance

Interpreting the peak as Ξ_b^- production, candidate masses are fitted with the hypothesis of a signal plus background model using an unbinned likelihood method. The signal

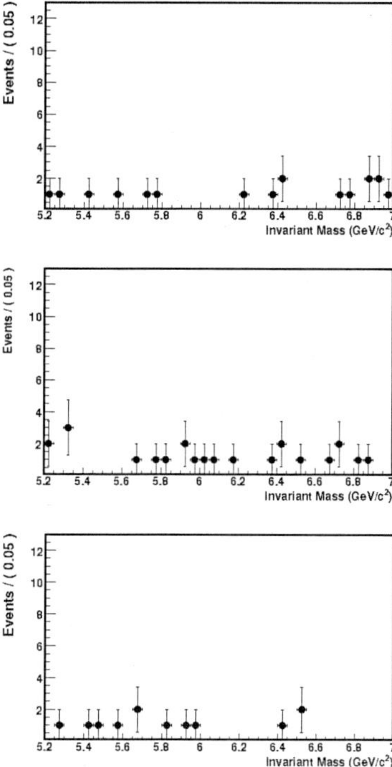

FIGURE 6. Invariant mass distribution of background events from wrong-sign combinations (top), Ξ^- sideband events (middle), and J/ψ sideband events (bottom).

and background shapes are assumed to be Gaussian and flat, respectively. The fit results in a Ξ_b^- mass of 5.774 ± 0.011 GeV with a width of 0.037 ± 0.008 GeV and a yield of 15.2 ± 4.4 events. Where all uncertainties are statistical. Following the same procedure, a fit to the Monte Carlo Ξ_b^- events yields a mass of 5.839 ± 0.003 GeV, in good agreement with the 5.840 GeV input mass. The fitted width of the Monte Carlo mass distribution is 0.035 ± 0.002 GeV, consistent with the 0.037 GeV obtained from the data. Since the intrinsic decay width of the Ξ_b^- baryon in the Monte Carlo is negligible, the width of the mass distribution is thus dominated by the detector resolution.

To assess the significance of the signal, the likelihood, $\mathscr{L}_{s+b}$, of the signal plus background fit above is first determined. The fit is then repeated using the background-only model, and a new likelihood $\mathscr{L}_b$ is found. The logarithmic likelihood ratio $\sqrt{2\ln(\mathscr{L}_{s+b}/\mathscr{L}_b)}$ indicates a statistical significance of 5.5σ, corresponding to a probability of 3.3×10^{-8} from background fluctuation for observing a signal that is equal to or more significant than what is seen in the data. Including systematic effects from the mass range, signal and background models, and the track momentum scale results in a minimum significance of 5.3σ and a Ξ_b^- yield of $15.2 \pm 4.4\,(\text{stat.})^{+1.9}_{-0.4}\,(\text{syst.})$. The significance can also be estimated from the numbers of candidate events and

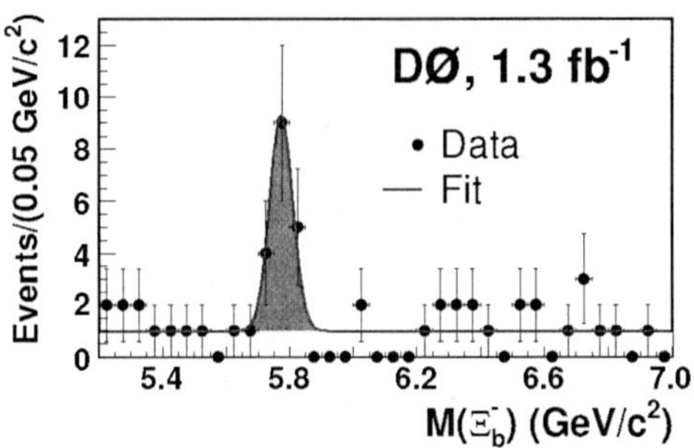

FIGURE 7. Decay $\Xi_b^- \to J/\psi\Xi^-$.

estimated background events. In the mass region of 2.5 times the fitted width centered on the fitted mass, 19 candidate events (8 $J/\psi\,\Xi^-$ and 11 $J/\psi\,\overline{\Xi}^+$) are observed while $14.8 \pm 4.3\,(\text{stat.})^{+1.9}_{-0.4}\,(\text{syst.})$ signal and $3.6 \pm 0.6\,(\text{stat.})^{+0.4}_{-1.9}\,(\text{syst.})$ background events are estimated from the fit. The probability of backgrounds fluctuating to 19 or more events is 2.2×10^{-7}, equivalent to a Gaussian significance of 5.2σ.

Potential systematic biases on the measured Ξ_b^- mass are studied for the event selection, signal and background models, and the track momentum scale. Varying cut values and using a multivariate technique of different variables for event selection leads to a maximum change of 0.020 GeV in the Ξ_b^- mass. Subtracting an estimated statistical contribution to the change, a conservative ± 0.015 GeV systematic uncertainty is assigned due to the event selection. Using double Gaussians for the signal model, a first-order polynomial for the background model, or fixing the mass resolution to that obtained from the Monte Carlo Ξ_b^- events all lead to negligible changes in the mass. The mass, calculated using the world average values [19] of intermediate particle masses above, is found to have a weak dependence on the track momentum scale. This has been verified using the $\Lambda_b \to J/\psi\,\Lambda$ and $B^0 \to J/\psi\,K_S^0$ events observed in the data. A systematic uncertainty of ± 0.002 GeV is assigned, corresponding to the mass difference between our measurement and the world average [19] for the Λ_b and B^0 hadrons. Adding in quadrature, a total systematic uncertainty of ± 0.015 GeV is obtained to yield the measured Ξ_b^- mass: $5.774 \pm 0.011\,(\text{stat.}) \pm 0.015\,(\text{syst.})$ GeV.

Relative production ratio

In addition to the observation and mass measurement of the Ξ_b^- baryon, its $\sigma \times \mathscr{B}$ relative to that of the Λ_b baryon is calculated using

$$\frac{\sigma(\Xi_b^-) \times \mathscr{B}(\Xi_b^- \to J/\psi\,\Xi^-)}{\sigma(\Lambda_b) \times \mathscr{B}(\Lambda_b \to J/\psi\,\Lambda)} = \frac{\varepsilon(\Lambda_b \to J/\psi\,\Lambda)}{\varepsilon(\Xi_b^- \to J/\psi\,\Xi^-)} \frac{N_{\Xi_b^-}}{N_{\Lambda_b}}$$

where $N_{\Xi_b^-}$ and N_{Λ_b} are the numbers of Ξ_b^- and Λ_b events reconstructed in data. Analyzing the same data (reprocessed data in difference with that used for the Λ_b lifetime measurement) and using the similar event selection criteria and fitting procedure as the Ξ_b^- analysis, a yield of $240 \pm 30\,(\text{stat.}) \pm 12\,(\text{syst.})$ Λ_b baryons is determined. The efficiencies to reconstruct the decays, $\varepsilon(\Xi_b^-)$ and $\varepsilon(\Lambda_b)$, are determined by Monte Carlo simulation, and the efficiency ratio, $\varepsilon(\Lambda_b)/\varepsilon(\Xi_b^-)$, is found to be 4.4 ± 1.3. The uncertainty on $\varepsilon(\Lambda_b)/\varepsilon(\Xi_b^-)$ arises from Monte Carlo modeling (27%), Monte Carlo statistics (10%), the reconstruction of the additional pion in the Ξ_b^- decay (7%), and the Ξ_b^- mass difference between data and Monte Carlo (5%). The largest component, Monte Carlo modeling uncertainty, is due to the difference in the efficiency ratio with and without Monte Carlo reweighting. This reweighting is needed to match Monte Carlo and data momentum distributions. The efficiency ratio is found to be insensitive to changes in Λ_b and Ξ_b^- production models. Many other systematic uncertainties on the efficiencies themselves tend to cancel in the ratio of the efficiencies. The relative production ratio is found to be $0.28 \pm 0.09\,(\text{stat.})^{+0.09}_{-0.08}\,(\text{syst.})$.

SUMMARY

The Λ_b lifetime measurement by the DØ experiment presented here is consistent with a Λ_b lifetime shorter than the lifetime of b mesons. In addition, DØ also measured the Λ_b lifetime in semileptonic decays [21]. This independent measurement is consistent with the lifetime measured in the exclusive $\Lambda_b \to J/\psi\Lambda$ decay channel. When both DØ measurements are combined, it is found $\tau(\Lambda_b) = 1.251^{+0.102}_{-0.096}$ ps, what is less consistent with the CDF measurement $\tau(\Lambda_b) = 1.593^{+0.089}_{-0.085}$ ps. However, within the current precision it is not possible to settle the question of the lifetime puzzle, at least not from the experimental side. More statistics is needed for this.

Before RunII of the Tevatron, the experimental situation of b baryons was not yet quite conclusive, with only the Λ_b experimentally established. However, the observation of new heavy baryon states at Tevatron experiments has been prolific in Run II. In addition to the Ξ_b^- baryon observed by DØ and also reported by CDF [5], other states, $\Sigma_b^{\pm}$ and $\Sigma_b^{*\pm}$, have been also observed [3]. These observations have allowed to compare experimental measurements with QCD potential models and lattice QCD predictions for mass of heavy baryons. Prediction and experiment agree very well, proving once more the success of the quark model. With more statistics, other properties besides masses could be measured, and search of physics beyond the Standard Model in b baryons [22, 23] would be possible.

ACKNOWLEDGMENTS

The author would like to thanks the organizers of this School for the invitation to participate and for an excellent organization.

REFERENCES

1. CDF Collaboration, F. Abe *et al.*, Nucl. Instrum. Methods Phys. Res. A **271**, 388 (1998);
2. V.M. Abazov *et al.* (D0 Collaboration), Nucl. Instrum. Methods A **565**, 463 (2006).
3. A. Abulencia *et al.*, Phys. Rev. Lett. **99**, 202001 (2007).
4. V.M. Abazov *et al.* (D0 Collaboration), Phys. Rev. Lett. **99**, 052001 (2007).
5. A. Abulencia *et al.*, Phys. Rev. Lett. **99**, 052002 (2007).
6. V.M. Abazov *et al.* (D0 Collaboration), Phys. Rev. Lett. **99**, 142001 (2007).
7. F. Gabbiani *et al.* Phys. Rev. Lett. **D** 68, (2003) 114006-1-4.
8. E. Franco *et al.*, Nucl. Phys. B **633**, 212 (2002).
9. F. Gabbiani, A. I. Onishchenko, A. A. Petrov, Phys. Rev. D **68**, 114006 (2003).
10. M. Di Pierro *et al.*, Phys. Lett. B **468**, 143 (1999).
11. C. Tarantino, Nucl. Phys. B **156**, (Proc. Suppl.), 33 (2006).
12. S. Eidelman et al. (PDG 2004), Phys. Lett. B **592**, 1 (2004).
13. A. Abulencia *et al.*, Phys. Rev. Lett. **98**, 122001 (2007).
14. J. Abdallah *et al.* (DELPHI Collaboration), Eur. Phys. J. **C44**, 299 (2005); D. Buskulic *et al.* (ALEPH Collaboration), Phys. Lett. B **384**, 449 (1996).
15. E. Barberio *et al.* (Heavy Flavor Averaging Group Collaboration), arXiv:0704.3575.
16. N. Isgur and M.B. Wise, Phys. Rev. Lett. **66**, 1130 (1991).
17. G.T. Bodwin, E. Braaten, G.P. Lepage, Phys. Rev. D **51**, 1125 (1995); erratum-*ibid*, Phys. Rev. D **55** 5853 (1997).
18. E. Jenkins, Phys. Rev. D **55**, R10 (1997); *ibid*, Phys. Rev. D **54**, 4515 (1996); N. Mathur, R. Lewis and R.M. Woloshyn, Phys. Rev. D **66**, 014502 (2002).
19. W. M. Yao *et al.*, J. Phys. G **33**, 1 (2006).
20. V. Abazov *et al.*, Phys. Rev. Lett. **94**, 102001 (2005).
21. V.M. Abazov *et al.* (D0 Collaboration), Phys. Rev. Lett. **99**, 182001 (2007).
22. I. Dunietz, z. Phys. C **56**, 129 (1992).
23. C.Q. Gen *et al.*, Phys. Rev. D **65**, 0191502 (2002).

The solar neutrino problem and Non Standard Interactions

A. Bolaños

Departamento de Física, Centro de Investigación y de Estudios Avanzados del IPN, Apartado Postal 14-740 07000 México D F, Mexico

Abstract. We analize the solar neutrino data to obtain constraints on physics beyond the Standard Model. We use the framework of non standard interactions in the neutrino sector. This formalism is an useful tool to study physics beyond the Standard Model. In particular we concentrate on the diagonal neutrino couplings to electrons. We perform a phenomenological analysis by considering data from Super-Kamiokande, SNO and radiochemical solar experiments and combine them with the KamLAND results. We conclude that this analysis can improve current constraints on non standard interactions parameters.

The oscillation neutrino phenomena is one of the most important current research topics and, consequently, many different and sophisticated neutrino oscillation experiments have been developed to confirm it and study it. Since we are entering in a new period in neutrino physics, where precision experiments are at order, it is very important to distinguish between a chance for discovering new physics or a potential confusion problem due to effects from new interactions with with the small mixing angle θ_{13} [1]. In this work we concentrate our attention in the new physics parameters. We study the solar neutrino data, with the main focus on the Super-Kamiokande detector, in order to obtain constraints on new physics. We will show that it is possible to use solar neutrino experiments to improve the constrains coming from neutrino-electron interactions.

In order to get model-independent results on new physics in the neutrino sector we will consider a Lagrangian that includes a general type of non standard interactions (NSI) [2]:

$$\mathscr{L}_{NS} = -\varepsilon^{fP}_{\alpha\beta}\sqrt{2}G_F(\bar{\nu_\alpha}\gamma_\mu L\nu_\beta)(\bar{f}\gamma^\mu Pf) \tag{1}$$

Here $P = L, R$ and f is a first generation fermion: e, u, d. The strength of this non standard interactions are given by $\varepsilon^{fP}_{\alpha\beta}$, where the neutrino flavor is denoted by α and β. In this work we study the non-universal non standard interactions (NU-NSI), with $\alpha = \beta$. In this case, the usual coupling constants for the neutrino electron scattering, g_R and g_L, will be changed to be

$$g_R \to \tilde{g}_R = g_R + \varepsilon^{eR}_{\alpha\alpha}\,, \tag{2}$$

$$g_L \to \tilde{g}_L = g_L + \varepsilon^{eL}_{\alpha\alpha}\,. \tag{3}$$

CP1077, *Advanced Summer School in Physics 2008, Frontiers in Contemporary Physics—EAV08*
edited by L. M. Montaño Zetina, G. Torres Vega, M. García Rocha, L. F. Rojas Ochoa, and R. López Fernández

In the Standard Model the $\nu_e e$ differential cross section scattering involves both neutral and charged currents and is given by

$$\frac{d\sigma}{dT} = \frac{2G_F^2 m_e}{\pi}[g_L^2 + g_R^2(1-\frac{T}{E_\nu})^2 - g_L g_R \frac{m_e T}{E_\nu^2}] \tag{4}$$

with G_F the Fermi constant, m_e the electron mass, T the kinetic recoil electron energy, and E_ν the neutrino energy. The Standard Model coupling constants g_L and g_R, at three level, are expressed as:

$$g_L = \frac{1}{2} + \sin^2\theta_W \tag{5}$$

$$g_R = \sin^2\theta_W. \tag{6}$$

The differential cross section that includes NSI is

$$\begin{aligned}\frac{d\sigma(E_\nu,T)}{dT} &= \frac{2G_F^2 m_e}{\pi}[\left(\tilde{g_L}^2 + \sum_{\alpha\neq\beta}|\varepsilon^L_{\alpha\beta}|^2\right) + \left(\tilde{g_R}^2 + \sum_{\alpha\neq\beta}|\varepsilon^R_{\alpha\beta}|^2\right)\left(1-\frac{T}{E_\nu}\right)^2 \\ &- \left(\tilde{g_L}\tilde{g_R} + \sum_{\alpha\neq\beta}|\varepsilon^L_{\alpha\beta}||\varepsilon^R_{\alpha\beta}|\right) m_e\frac{T}{E_\nu^2}],\end{aligned} \tag{7}$$

Since Super-Kamiokande detect solar neutrinos via the neutrino elastic scattering off electrons, it is possible to test NSI by performing a fit to the solar neutrino data considering the neutrino cross section introduced above. The neutrino spectrum and composition is governed both by the Solar Standard Model and the oscillation mechanism, therefore, although the NSI affects only the neutrino electron scattering detection in Super-Kamiokande, we perform a global analysis considering the neutrino mass difference and mixing angle as a free parameter. We consider the BP05 solar model [3] and include the data from Chlorine, Gallex-GNO, SAGE, SNO [4], and KamLAND [5] in order to perform a global fit together with the Super-Kamiokande [6] data and obtain a more reliable constraint on the NSI parameters.

The Super-Kamiokande measurements are based on recoiled electrons in the energy range from 5.0 to 20.0 MeV, detected via Cherenkov light production. A two neutrino mixing picture is a good approximation for describing solar neutrino oscillations. Since the dominant $\nu_e e$ scattering cross section is decreased by about 2% due to radiative corrections [7], we take them into account. In this case the neutrino-electron scattering is expressed as

$$\begin{aligned}\frac{d\sigma}{dT} &= \frac{2G_f^2 m_e}{\pi} g_L^2(T)[1+\frac{\alpha}{\pi}f_-(z)] + g_R^2(T)(1-z)^2[1+\frac{\alpha}{\pi}f_+(z)] \\ &- g_R(T)g_L(T)\frac{m}{E_\nu}z[1+\frac{\alpha}{\pi}f_{+-}(z)]\end{aligned} \tag{8}$$

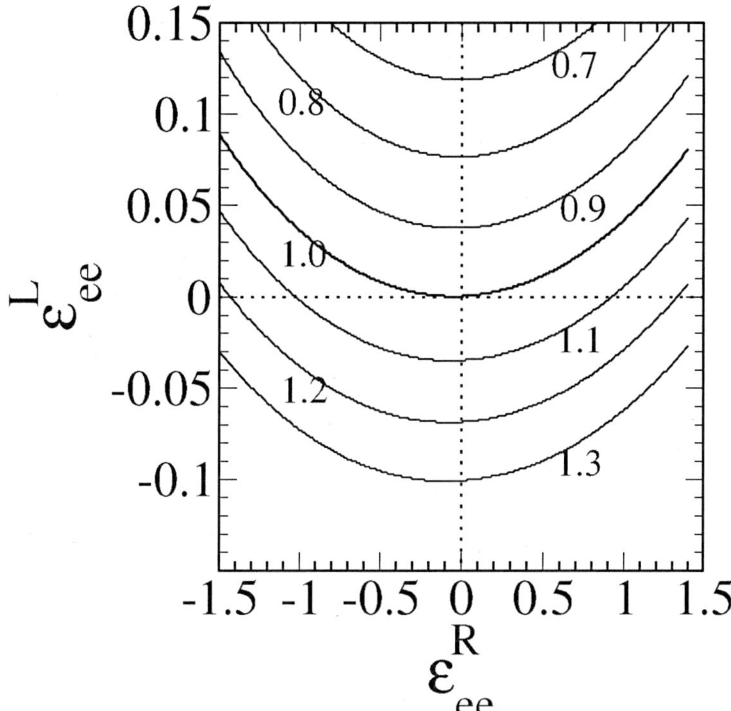

FIGURE 1. Ratio of the averaged cross section for $\nu_e - e$ scattering in the case of non standard interactions to the Standard Model one. The cross sections have been integrated in both the ^{8}B spectrum and the Super-Kamiokande energy resolution have been taken into account. It is possible to notice that smaller values of ε_{ee}^{eL} predicts an increase of the cross section.

where $f_{\pm}$, f_{+} and f_{-} describe QED effects [7]. The number of events expected in Super-Kamiokande is given by:

$$N_{\rm ev} = \text{Time}\phi_B N_e \int_0^{T_{max}} \int_{T_i}^{T_i+1} c(T)\lambda(E_\nu)R(T,T') \times \tag{9}$$

$$\left[P \quad (\tan^2\theta, \frac{\triangle m^2}{4E_\nu})\frac{d\sigma_{\nu_e - e}}{dT} + (1 - P(\tan^2\theta, \frac{\triangle m^2}{4E_\nu}))\frac{d\sigma_{\nu_l - e}}{dT}\right] \mathrm{d}T' \mathrm{d}T \mathrm{d}E_\nu \tag{10}$$

where Time is the experiment running time, ϕ_B is the total ^{8}B neutrino flux, estimated to be 5.69×10^6 $\nu/(cm^2 s)$[3]; N_e is the number of target electrons in Super-Kamiokande; $\lambda(E_\nu)$ is the ^{8}B spectrum energy; $P(\tan^2\theta, \frac{\triangle m^2}{4E_\nu})$ is the electron neutrino survival probability; $c(T)$ is the detector efficiency (73 % for $T \geq 6.5$ MeV and 52 % for $T \leq 6.5$ MeV [8]). The reconstructed kinetic energy take values of $T = 5$ MeV $-m_e$ to $T_{max} = 20$ MeV $-m_e$. The resolution function that describes the possible difference between the measured T and the real energy T' is: $R(T,T') = \frac{1}{\sqrt{2\pi}\sigma} \exp[-\frac{(T-T')^2}{2\sigma^2}]$,

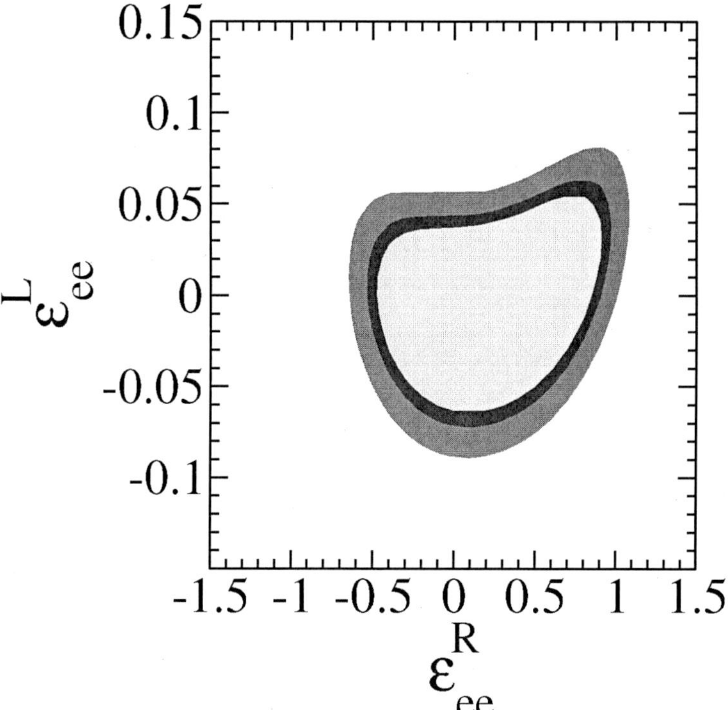

FIGURE 2. Allowed regions at 90 %, 95 % and 99 % C. L. for the non-universal non standard parameters, ε^{eL}_{ee} and ε^{eR}_{ee}, from the global analysis of solar and KamLAND experiments. account for the χ^2 analysis and marginalized it.

where $\sigma = \sigma_0(\frac{T'}{MeV})^{\frac{1}{2}}$ with $\sigma_0 = 0.\ 47MeV$. Finally, $\frac{d\sigma_{\nu_e - e}}{dT}$ and $\frac{d\sigma_{\nu_l - e}}{dT}$, are the cross sections corresponding to the $\nu_e - e$ and $\nu_l - e$ interactions, where l stands for either μ or τ neutrino. In principle, the NU-NSI could be present in any of these cross sections. However, here we will restrict ourselves only to the electron neutrino case and, therefore, we will consider that $\frac{d\sigma_{\nu_l - e}}{dT}$ has the usual SM expression.

In order to have an idea about the NU-NSI behaviour in Super-Kamiokande, we integrate the $\nu_e e$ differential cross section over the whole Super-Kamiokande energy range (from 5 to 20 MeV) convoluted with the energy resolution function and efficiency, both for the NSI case as well as for the Standard Model one. The ratio of NSI *vs* SM cross section is shown in Figure 1. We can see that the general tendency is that for negative values of ε^L_{ee} the predicted cross section increases, while there is also a kind of 'parabolic' dependence on ε^R_{ee}.

In order to do a complete χ^2 analysis, we have also calculated the expected number of events in Super-Kamiokande per bin. We obtained the ratio of the expected events for each couple of values ε^{eL}_{ee}, ε^{eR}_{ee} to the Solar Standard Model without oscillations. We have also calculated the systematic errors associated to the energy uncertainty scale evaluated changing the resolution function by 0.64 %, that is: $T \rightarrow T'(1.0+.0064)$ [9]. The uncertainty in the resolution is calculated by changing the resolution function by 2.5 %, that is $\sigma_T \rightarrow \sigma_T(1+0.025)$. Finally the systematic error in the Boron spectral shape is also taken into account. Besides, we performed a global analysis of the solar data, taking into account all the correlations in systematic errors, and we combined it with the KamLAND results in order to take into account the oscillation parameters. The procedure that we followed is very similar to that developed in [10] where a detailed explanation of the χ^2 analysis can be found.

The results are shown in Figure 2, where we introduce the allowed region for the NSI parameters (the oscillation parameters have been marginalized). It is possible to see that this result could improve previous reported constraints [11] coming from reactor and accelerator data. In particular, the current constraint on ε_{ee}^{eL}, at 90 % C. L. lies in the range $-0.14 < \varepsilon_{ee}^{eL} < 0.09$. Clearly, from the plot shown here, the constraints coming from the solar analysis are more restrictive. On the other hand, for the right parameter, the solar analysis is not competitive since the current constraint is given by $-0.03 < \varepsilon_{ee}^{eR} < 0.018$.

Acknowledgements A more detailed analysis is under progress. This work has been done in collaboration with Omar Miranda, A.Palazzo, Mariam Tórtola, and José Valle. This work has been supported by CONACYT.

REFERENCES

1. P. Huber, T. Schwetz and J. W. F. Valle, Phys. Rev. D **66**, 013006 (2002) [arXiv:hep-ph/0202048].
2. Z.Berezhiani et al./ Nuclear Physics B 638 (2002) 62-80; N. Fornengo, M. Maltoni, R. T. Bayo and J. W. F. Valle, Phys. Rev. D **65**, 013010 (2002) [arXiv:hep-ph/0108043]; S. Davidson, C. Pena-Garay, N. Rius and A. Santamaria, JHEP **0303** (2003) 011 [hep-ph/0302093]; M. M. Guzzo, P. C. de Holanda and O. L. G. Peres, Phys. Lett. **B591**, 1 (2004), [hep-ph/0403134].
3. J. N. Bahcall, A. M. Serenelli and S. Basu, Astrophys. J. **621**, L85 (2005) [arXiv:astro-ph/0412440].
4. R. Davis, Prog. Part. Nucl. Phys. **32**, 13 (1994); B. T. Cleveland *et al.*, Astrophys. J. **496**, 505 (1998); SAGE Collaboration, J. N. Abdurashitov *et al.*, Phys. Rev. **C60**, 055801 (1999), [astro-ph/9907113]; GNO Collaboration, Nucl. Phys. Proc. Suppl. **110**, 311 (2002); GALLEX Collaboration, W. Hampel *et al.*, Phys. Lett. **B447**, 127 (1999); GNO Collaboration, M. Altmann *et al.*, Phys. Lett. **B490**, 16 (2000), [hep-ex/0006034]; SNO Collaboration, Q. R. Ahmad *et al.*, Phys. Rev. Lett. **89**, 011301 (2002), [nucl-ex/0204008]; SNO Collaboration, Q. R. Ahmad *et al.*, Phys. Rev. Lett. **89**, 011302 (2002), [nucl-ex/0204009]; SNO Collaboration, S. N. Ahmed *et al.*, Phys. Rev. Lett. , 041801 (2003), [nucl-ex/0309004].
5. KamLAND Collaboration, K. Eguchi *et al.*, Phys. Rev. Lett. **90**, 021802 (2003), [hep-ex/0212021].
6. Super-Kamiokande Collaboration, S. Fukuda *et al.*, Phys. Lett. **B539**, 179 (2002), [hep-ex/0205075];
7. J. N. Bahcall, M. Kamionkowski and A. Sirlin, Phys. Rev. D **51**, 6146 (1995) [arXiv:astro-ph/9502003].
8. The Super Kamiokande Collaboration, Phys. Rev. Lett. 86 (2001) 5651-5655.
9. G. L. Fogli, E. Lisi, A. Marrone, D. Montanino and A. Palazzo, Phys. Rev. D **66**, 053010 (2002) [arXiv:hep-ph/0206162]; M. Maltoni, T. Schwetz, M. A. Tortola and J. W. F. Valle, New J. Phys. **6**, 122 (2004) [arXiv:hep-ph/0405172].
10. O. G. Miranda, M. A. Tortola and J. W. F. Valle, JHEP **0610**, 008 (2006) [arXiv:hep-ph/0406280].
11. J. Barranco, O. G. Miranda, C. A. Moura and J. W. F. Valle, arXiv:0711.0698 [hep-ph].

Probing Unparticle Physics in Reactor Neutrinos

A. Bolaños

Departamento de Física, Centro de Investigación y de Estudios Avanzados del IPN, Apartado Postal 14-740 07000 México D F, Mexico

Abstract. Unparticle physics is studied by using reactor neutrino data. We obtain limits to the scalar unparticle couplings depending on different values for the parameter d. We found that, as has been already noticed, reactor neutrino data is a good tool to put constraints on unparticle physics. Thanks to a detailed analysis of the experimental characteristics of reactor data we find better constraints than the previously reported.

INTRODUCTION

With his new proposal made the last year [1, 2], Georgi opened a new window to study physics beyond the Standard Model (SM). In his proposal, he reinterpreted the physics of an exact scale invariant sector possessing a nontrivial infrared fixed point at a higher energy in the framework of unparticles. In this scheme the hidden sector with non trivial infrared fixed point is described by the Banks Zaks operators (O_{BZ}) [3] and interact throught the exchange of particles with a large mass scale M_U. The non renormalizable couplings, involving both fields, and suppressed by powers of M_u, have the form:

$$\frac{1}{M_U^k} O_{SM} O_{BZ} \tag{1}$$

In an effective theory below the scale Λ_U, the Banks Zaks operators match onto unparticle operator, and the interactions are of the form:

$$\frac{C_U \Lambda_U^{d_{BZ}-d_u}}{M_U^k} O_{SM} O_U \tag{2}$$

where d_{SM} and d_{BZ} are the dimensions for the SM and Banks Zaks operators, respectively, and C_U is a coefficient function.

On the other hand, there are several experiments that could test this type of new physics and a lot of work have already been done in this direction [4]. In this work, we focus in experimental results coming from reactor neutrinos, especially the MUNU detector, to place bounds to unparticle interactions of scalar type. We will show that this experiment could give very competitive constraints. Moreover, by analyzing the data from more reactor experiments a better constraint could be given.

CP1077, *Advanced Summer School in Physics 2008, Frontiers in Contemporary Physics—EAV08*
edited by L. M. Montaño Zetina, G. Torres Vega, M. García Rocha, L. F. Rojas Ochoa, and R. López Fernández

THE NEUTRINO-ELECTRON SCATTERING

We introduce the differential neutrino electron cross section for a neutrino with initial flavor a. The standard differential cross section, as a function of the incident neutrino energy E_ν and the recoil electron kinetic energy T_e, is given by:

$$\frac{d\sigma_a^{\rm std}}{dT_e}(E_\nu,T_e)=\frac{\sigma_0}{m_e}\left[(g_V^a+g_A^a)^2+(g_V^a-g_A^a)^2\left(1-\frac{T_e}{E_\nu}\right)^2-((g_V^a)^2-(g_A^a)^2)\frac{m_eT_e}{E_\nu^2}\right], \tag{3}$$

with $\sigma_0=G_F^2m_e^2/(2\pi)$. For μ and τ neutrinos, where only neutral current interactions are possible, the Standard Model of electroweak interactions predicts, at tree level, $g_V^{\mu,\tau}=2\sin^2\theta_W-\frac{1}{2}$ and $g_A^{\mu,\tau}=-1/2$, with $\sin^2\theta_W=0.23122$ [5]. For electron neutrinos, where also charge current interactions are present, we have $g_{V,A}^e\to g_{V,A}^{\mu,\tau}+1$.

On the other hand, the coupling for leptons to scalar unparticle is given through the Lagrangian:

$$\lambda_{0e}\frac{1}{\Lambda^{d-1}}\bar{e}e\mathscr{O}_\mathscr{U}+\lambda_{0\nu}^{\alpha\beta}\frac{1}{\Lambda^{d-1}}\bar{\nu^\alpha}\nu^\beta\mathscr{O}_\mathscr{U} \tag{4}$$

In this work we are concentrate in scalar interactions. Moreover, we will consider only unparticle contributions coming from the flavor changing procces $\nu_\alpha e\to\nu_\beta e$, with $\alpha\neq\beta$. In this case, the contribution to the amplitude will have no interference terms and it is given by

$$\mathscr{M}_\mathscr{U}=\frac{f(d)}{\Lambda_U^{2d-2}}[\bar{\nu_\beta}(k')\nu_\alpha(k)[\bar{e}(p')e(p)][-q^2-i\varepsilon]^{d-2}, \tag{5}$$

and the corresponding differential scattering cross section is

$$\frac{d\sigma}{dT}=\frac{f(d)^2(2^{(2d-6)})}{\pi E_\nu^2\Lambda_U^{4d-4}}(m_eT)^{(2d-3)}(T+2m_e) \tag{6}$$

where

$$f(d)=\frac{\lambda_{0\nu}^{\alpha\beta}\lambda_{0e}A_d}{2\sin(d\pi)}. \tag{7}$$

RESULTS

Now we perform an analysis of the impact of this contribution to the observed $\bar{\nu_e}e\to\bar{\nu}e$ considering the MUNU data. In order to estimate a constraint on the parameters

$$d,\qquad \lambda_0=\sqrt{\lambda_{0\nu}\lambda_{0e}} \tag{8}$$

we compute the integral

$$\sigma=\int dT'\int dT\int dE_\nu\frac{d\sigma}{dT}\lambda(E_\nu)R(T,T') \tag{9}$$

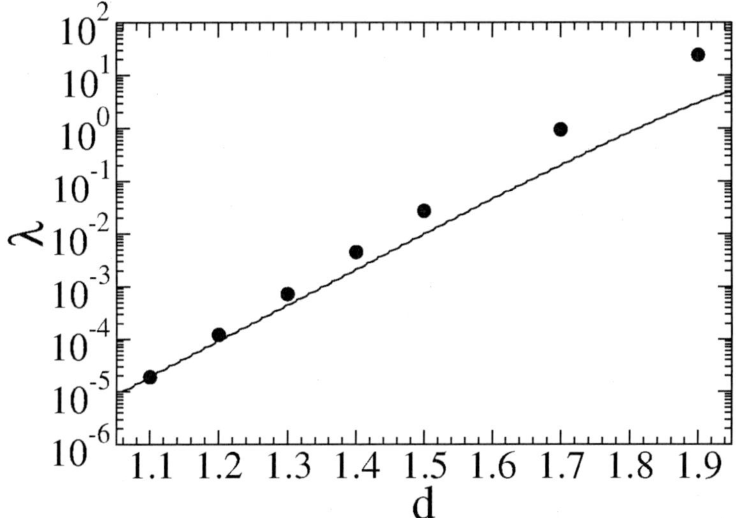

FIGURE 1. Limits on the parameters d and λ_0 (solid line) from the MUNU experiment at 90 % C. L. Previous bounds obtained by Balantekin and Ozansoy (dots) are shown for comparison. It is possible to see that the analysis coming from MUNU gives stronger constraints.

with $R(T,T')$ the energy resolution function for the MUNU detector,. The relative energy resolution in this detector was found to be 8 % and it scales with the power 0.7 of the energy [6]:

$$R(T,T') = \frac{MeV}{\sqrt{2\pi T'}\Delta_{1MeV}} exp\left(\frac{-(T-T')^2}{2\Delta_{1MeV}^2 T'}\right) \tag{10}$$

with

$$\Delta_{1MeV} = (0.08 \times T)^{0.7}. \tag{11}$$

We use an anti-neutrino energy spectrum $\lambda(E_\nu)$ given by

$$\lambda(E_\nu) = \sum_{k=1}^{4} a_k \lambda_k(E_\nu), \tag{12}$$

where a_k is the abundance of ^{235}U ($k=1$), ^{239}Pu ($k=2$), ^{241}Pu ($k=3$) and ^{238}U ($k=4$) in the reactor, $\lambda_k(E_\nu)$ is the corresponding neutrino energy spectrum which we take from the parametrization given in [7], with the appropriate fuel composition. For energies below 2 MeV there are only theoretical calculations for the antineutrino spectrum which we take from Ref. [8].

With this formula we can compute the number of events expected in MUNU in the case of a SM cross section, as well as in the case of an extra contribution due to unparticle physics, for the parameters d and λ_0. We have fixed, for simplicity$\Lambda_U = 1$ TeV.

We define the χ^2 function as

$$\chi^2 = \sum_i \frac{(N_i^{\text{theo}} - N_i^{\text{exp}})^2}{\Delta_i^2} \tag{13}$$

where the N_i^{exp} are given by the experimental measured events and Δ_i are the corresponding errors: $N_i^{\text{exp}} = (1.07 \pm 0.34)$ events/day [6], while $N_i^{\text{theo}} = N_i^{\text{SM}} + N_i^{\text{unparticle}}$ are the theoretical expectations considering the effects of unparticle physics. We show the results of our analysis in Fig. 1, where the maximum allowed values of the unparticles parameters are shown at 90 % C. L. We also show in the same plot the results obtained in previous analysis [9].

We can see that the detailed analysis of the MUNU experiment can already improve the constraints on the unparticle parameters. A more detailed analysis including more reactor data is underway in order to improve our constraints. Unparticle physics can be a very interesting alternative in order to find physics beyond the standard model, in fact, a lot of work has been done since this new proposal was born a few years ago.

Acknowledgements This work has been done in collaboration with Juan Barranco, Omar Miranda, Celio Moura and Timur Rashba. A more complete analysis will be published elsewhere. This work has been supported by CONACYT and PAPIIT project IN104208.

REFERENCES

1. H. Georgi, Phys. Rev. Lett. **98**, 221601 (2007) [arXiv:hep-ph/0703260].
2. H. Georgi, Phys. Lett. B **650**, 275 (2007) [arXiv:0704.2457 [hep-ph]].
3. T. Banks and A. Zaks, Nucl. Phys. **B206** 23 (1982).
4. See for example: K. Cheung, W. Y. Keung and T. C. Yuan, arXiv:0809.0995 [hep-ph].
5. C. Amsler et al., Physics Letters **B667** 1 (2008)
6. Z. Daraktchieva *et al.* [MUNU Collaboration], Phys. Lett. B **615**, 153 (2005) [arXiv:hep-ex/0502037].
7. P. Huber and T. Schwetz, Phys. Rev. **D70**, 053011 (2004), [hep-ph/0407026].
8. V. I. Kopeikin, L. A. Mikaelyan and V. V. Sinev, Phys. Atom. Nucl. **60** (1997) 172 [Yad. Fiz. **60** (1997) 230].
9. A. B. Balantekin and K. O. Ozansoy, Phys. Rev. D **76**, 095014 (2007) [arXiv:0710.0028 [hep-ph]].

III. STATISTICAL PHYSICS

Molecular Dynamics: from basic techniques to applications (A Molecular Dynamics Primer)[1]

E. R. Hernández

Institut de Ciència de Materials de Barcelona, ICMAB - CSIC, Campus de Bellaterra, 08193 Bellaterra, Barcelona, Spain

Abstract. It is now 50 years since the first papers describing the use of Molecular Dynamics (MD) were published by Alder and Wainright, and since then, together with Monte Carlo (MC) techniques, MD has become an essential tool in the theoretical study of materials properties at finite temperatures. In its early days, MD was used in combination with simple yet general models, such as hard spheres or Lennard-Jones models of liquids, systems which, though simple, were nevertheless not amenable to an analytical statistical mechanical treatment. Nowadays, however, MD is most frequently used in combination with rather sophisticated models, ranging all the way between empirical force fields to first-principles methods, with the aim of describing as accurately as possible any given material. From a computational aid in statistical mechanics and many-body physics, MD has evolved to become a widely used tool in physical chemistry, condensed matter physics, biology, geology and materials science. The aim of this course is to describe the basic algorithms of MD, and to provide attendees with the necessary theoretical background in order to enable them to use MD simulations in their research work. Also, examples of the use of MD in different scientific disciplines will be provided, with the aim of illustrating the the many possibilities and the wide spread use of MD simulation techniques in scientific research today.

Keywords: Atomistic Simulation, Condensed Matter, Thermodynamics.
PACS: 02.70.-c, 02.70.Ns

1. INTRODUCTION

As a scientific tool for the study of condensed matter, Computer Simulation really started in the 1950s with the development of the first sufficiently powerful computers. The two main families of simulation techniques, Monte Carlo (MC) methods [1] and Molecular Dynamics (MD) methods [2], were described for the first time in that decade. During the second half of the XX century, Computer Simulation methods have established themselves as a mature and powerful research tool in condensed matter and molecular physics and chemistry, and are at present slowly but steadily extending their usefulness and applicability to other, more challenging areas such as biological systems and soft condensed matter.

The aim of this series of lectures is to provide an introduction to MD techniques, illustrating the power of these simulation tools. The outline of the material to be covered

[1] Dedicated to my teachers and mentors C. R. A. Catlow and M. J. Gillan, in recognition of an unpayable debt.

CP1077, *Advanced Summer School in Physics 2008, Frontiers in Contemporary Physics—EAV08*
edited by L. M. Montaño Zetina, G. Torres Vega, M. García Rocha, L. F. Rojas Ochoa, and R. López Fernández

will be the following. First I will discuss, in very general terms, the aims and usefulness of Computer Simulation, and in particular MD methods, at least as I see it. A self-contained and to-the-point description of MD will be provided. Next, some historical background of MD will be reviewed, although inevitably we will not do justice to the extensive literature on this topic that has been published over the last five decades! In actual fact, I will just mention a few landmark papers, which in my own (admittedly personal) view are of key importance in the history of MD. Then we will move into the practical aspects of performing MD simulations, how to integrate the equations of motion, how to simulate bulk systems, etc. I will also make some general comments on the different approaches used to model the interactions between atoms and molecules, discussing some examples of potentials, and we will also see alternative approaches involving electronic structure calculations. After all that introductory material, we will get slightly more technical, and discuss how standard MD can be extended to simulate systems in contact with a thermal bath, emulating the conditions of constant temperature, or to simulate systems in which the volume and/or cell-shape fluctuate in such a way as to reproduce conditions of constant pressure. Illustrative examples of these different techniques will be provided along the way. We will wrap up with some discussions on the possible shortcomings and limitations of MD, and attempt to guesstimate some of the developments we are likely to see in the future, which inevitably will attempt to ameliorate some of those limitations.

Much of the material presented here (though not all) has been taken from standard references about computer simulation, and can be found there in more detail, together with lots of useful references to the literature. Standard text books on atomistic computer simulation are those of Allen and Tildesley [3], Frenkel and Smit [4] and Thijssen [5], which I recommend for more details on the topics discussed here.

2. MOLECULAR DYNAMICS IN A NUTSHELL

In a nutshell, MD consists of numerically solving the classical equations of motion for a collection of atoms[2]. For doing this, three basic ingredients are necessary; firstly, we must have some law describing the mutual interactions between the atoms in the system, from which we can calculate, given the atomic positions, the associated potential energy, the forces on the atoms, and if necessary the stress on the container walls. This law is in general unknown, but it can be approximated with different degrees of accuracy (and realism) by a force field, or it can be modelled by means of electronic structure calculations, which can also be done at different levels of theory. Secondly, we need an algorithm to integrate numerically the equations of motion for the atoms in the system. Over the years many different schemes have been put forward for doing this. Thirdly and finally, in order to solve the equations of motion, the integration scheme needs to be fed with some valid initial conditions, i.e. initial positions and velocities for all atoms in the system. With these three basic ingredients, one is set for performing MD simulations.

Before going into describing these different ingredients in somewhat more detail

[2] Here the word *atom* is used in a lose sense to refer indistinctly to atoms, ions or entire molecules

below, it is worthwhile to pause for a moment and consider why it may be useful to perform an MD simulation, and what can be extracted from it. With such a simulation, we are emulating, i.e. simulating in an approximate way, the real dynamics of the system under study, and in so doing we can keep track of the doings of individual atoms in an incredibly detailed way; so much so that one can easily feel like Big Brother among the atoms. In this way MD simulations can help us to gain new insight into important processes taking place at the atomic and molecular level, an insight which is often impossible to obtain purely from experiments, as these rarely have sufficient resolution. Furthermore, when performing simulations, one can easily prepare the conditions (temperature, pressure, atomic configuration, etc.) at will, and has a level of control over them that is much greater than is usually possible in experiments.

Aside from the numerical approximations involved in the integration of the equations of motion, there are two basic approximations inherent in MD simulations. The first one is that we assume that atoms behave like classical entities, i.e. they obey Newton's equations of motion. How much an approximation this is depends on the particular system under study, and on the actual conditions in which it is simulated. One can expect this approximation to be crude for light atoms at low temperatures, but in general it is not a bad approximation. In this respect it is fortunate that normally quantum effects on the atomic dynamics are relatively small, except in a few notable examples such as liquid He, and other light atoms. For those cases where quantum effects cannot be neglected, one should use the Path Integral approach [6] or some similar method.

The other key approximation is the model used to describe the interactions between the atoms in the system. It is clear that only through a sufficiently realistic description of those interactions one has any chance of getting useful and reliable information on the atomic processes taking place in the system. On the other hand, if one wants to address generic questions about a particular class of systems, such as low density gases, or liquid metals, say, one probably does not need to describe a particular example of such systems with a very accurate potential; it will be sufficient to use a generic model that captures the essential features, the defining physics, of that particular class of systems. To be too specific in this case can actually be counterproductive and obscure the general picture. It is therefore important to find the right level of description for the particular problem at hand.

This nutshell description may give you the idea that MD is simply solving Newton's equations for atoms and molecules. But in reality MD is much more than this: one can design rather artificial-looking forms of MD, which nevertheless serve a useful purpose, such as simulating a system under conditions of constant temperature and/or constant pressure (see section 7), something that is not possible to do by a straightforward solution of the standard equations of motion, or one can combine the physical dynamics of ions with a fictitious dynamics of electronic wave functions, which makes possible the effective realisation of atomic dynamics from first principles (the so-called Car-Parrinello method, see section 6). In essence, MD is extremely powerful and flexible, and far from being a simple numerical recipe for integrating the equations of motion for atoms and molecules.

3. SOME HISTORY

It is no surprise that the two most fundamental methodologies for simulating condensed matter systems, namely Monte Carlo (MC) and Molecular Dynamics (MD) made their first appearance in the 1950s. At this time the first computers, originally available only for classified military research, were made available to scientists in the US, and the possibility of performing fast automated calculations was immediately seen to have great potential for problems in statistical mechanics, for example. Ever since the first publication describing the MD technique, by Alder and Wainright [2] in 1957, applications of the technique have been growing in number, and nowadays MD is an extensively used research tool in disciplines which include physics, chemistry, materials science, biology and geology.

In the early days of MD, covering mostly the 1960s and 70s, the technique was mostly used as an aid in statistical mechanics. For the largest part, there was no attempt to model realistic systems, but rather the focus was on simple, generic model systems such as hard spheres or the Lennard-Jones fluid. The aim was not so much to address questions concerning specific systems, but rather to learn about entire families of systems, e.g. simple liquids. In time, models grew in complexity and in their degree of specificity. Empirical (i.e. derived from experimental information) models began to be developed for specific classes of systems, such as the CHARMM [7] or AMBER [8] force fields for organic and biological molecules, the ionic potentials for oxide materials [9], the embedded atom potentials for metals [10], or the bond-order inspired potentials for covalent materials [11].

At the same time, new methodological developments were being carried out. Since MD consists basically of integrating the classical equations of motion for the atoms or molecules of a system, it was implicitly accepted that MD could only be used to simulate systems in microcanonical conditions, i.e. conditions of constant number of particles, N, constant volume, V, and constant energy, E. This was somewhat limiting, as experiments are most often conducted on samples which are not isolated, but in thermal and/or mechanical contact with their surroundings. However, in an influential paper, Andersen [12] demonstrated that new, more general forms of MD could be devised. Andersen introduced two new tools, known as the Andersen thermostat and the Andersen barostat, which, as their name indicates, serve the purpose of controlling the temperature and the pressure during the simulation, respectively. In section 7 we will discuss the details of Andersen's thermostat and barostat; for now let us just remark that particularly the idea of the barostat has proved to be very influential in the subsequent history of MD. In essence, Andersen introduced a new variable into the dynamics of the system, namely the system's volume, with an associated velocity, a fictitious mass, and a potential energy term depending on the external pressure. The coupled dynamics of atoms and volume proposed by Andersen ensured that the system samples the isoenthalpic (constant enthalpy) ensemble, which is useful for analysing how the system may react to an externally imposed pressure. Andersen showed that, by introducing a small number of additional fictitious degrees of freedom (the volume) it was possible construct a new dynamics which effectively achieved the same effect as coupling the system to the infinitely many degrees of freedom of a reservoir. As pointed out above, this idea was to prove extremely influential.

Shortly after Andersen's paper was published, Nosé [13] showed that the introduction of an additional fictitious variable coupled to the atomic dynamics could be done in such a way as to obtain sampling in the canonical (constant temperature) ensemble. Contrary to the thermostat already introduced by Andersen [12], which affects the atomic dynamics in a stochastic way, Nosé's thermostat is fully deterministic. Nosé's approach, as later modified by Hoover [14], has now become perhaps the most commonly used scheme for performing MD simulations in the canonical ensemble.

The constant-pressure scheme of Andersen, originally conceived for the simulation of bulk fluids, was not generally applicable to crystalline solids, because only volume, and not shape fluctuations were considered. Parrinello and Rahman [15] generalised the method of Andersen by incorporating the components of the lattice vectors of the simulation cell as new fictitious dynamical variables, thus making possible the observation of solid-solid phase transitions in MD simulations. This scheme also made possible the study of systems under non-hydrostatic stress conditions.

Andersen's barostat and Nosé's thermostat proved that MD was potentially much more than simply a scheme for solving the equations of motion for a collection of atoms isolated from the rest of the universe. By adequately incorporating appropriately designed fictitious variables, these developments showed that more general and experimentally relevant statistical ensembles could be sampled. But the introduction of fictitious variables was soon to be found to have even wider possibilities: in 1985, i.e. only 5 years after Andersen's barostat had been introduced, Car and Parrinello [16] demonstrated a new use of fictitious dynamical variables. In their seminal paper, Car and Parrinello showed for the first time that it was possible to perform ab initio MD, i.e. MD in which the forces on the atoms are not extracted from an empirical force field, but rather from a full blown first principles electronic structure calculation. This combination of methods has been given the name first principles molecular dynamics (FPMD), also known as *ab initio* molecular dynamics (AIMD).

Before Car and Parrinello's paper, FPMD had been regarded as essentially impossible mostly due to the computational cost involved in performing a time-consuming electronic structure calculation for each time step of an MD simulation, i.e. thousands or even tens of thousands of times. Computers were simply not fast enough for the task in 1985. However, Car and Parrinello showed that with a clever introduction of new fictitious variables, the cost of FPMD could be brought down significantly, so much so as to make it a realistic undertaking, even with the computers of the day. Briefly, Car and Parrinello's idea consisted of bringing electrons and ions simultaneously into the picture, but in a very unusual and imaginative way. Just as in conventional force-field MD, ions moved subject to the forces acting on them, but these forces came not from an empirical potential, but from their mutual (coulombic) interaction, and from their interaction with the valence electron density around them. Car and Parrinello formulated their FPMD in the context of density functional theory (DFT) [17, 18] formalism. Within this formalism, the electron density is obtained from a series of so-called Kohn-Sham orbitals, which are the solutions of a Schrödinger-like equation, the Kohn-Sham equation. These orbitals must be obtained for the given ionic configuration before the total energy of the system and the forces of the ions can be calculated, and this process is considerably more costly than any calculation based on force fields. Typically, Kohn-Sham orbitals are represented by means of some basis set of appropriately chosen functions, such as atomic-like orbitals, or plane-waves, the latter being particularly convenient in the case of periodic systems. Then, solving the electronic structure problem consists of finding the appropriate expansion coefficients for the relevant Kohn-Sham orbitals in terms of the basis set functions. The break-through of Car and Parrinello was to incorporate the expansion coefficients of the Kohn-Sham orbitals in terms of the basis functions as fictitious dynamical variables, with associated fictitious masses. By choosing these masses appropriately (set to values much smaller than those of the ions), the Kohn-Sham orbitals evolve much more rapidly than the ions, and as a result a regime is established in which the orbitals adapt quasi-instantaneously to the comparatively slow change in the ionic positions. This is the Born-Oppenheimer approximation again, but in a new, imaginative setting.

The achievement of Car and Parrinello served the purpose of waking up the scientific community to the fact that FPMD was indeed viable, and soon many groups worldwide began to perform FPMD simulations, either directly employing the Car-Parrinello scheme, or alternative ones. Throughout the 1990s and this century, FPMD has now become a relatively standard tool, with an impressive showcase of applications. This is not to say that FPMD has completely supplanted the simpler, more approximate force-field based MD; far from it. There are many problems that remain too challenging to tackle via FPMD, either because the system is too large, too complex, or because it cannot be modelled accurately enough with DFT. In such cases empirical force fields continue to be the only viable option, and this is likely to remain the case for some time to come.

Up to here what is now the history of MD, according to an admittedly personal view. As for future developments, well, as the saying goes, making predictions is extremely difficult (especially about the future!), but at the end of this chapter I will try to summarise what we can already begin to see, or guess, for the relatively short term future of MD.

4. MD: BASIC TECHNIQUES

In this section we are going to review some practical aspects of MD simulations, such as how to integrate numerically the classical equations of motion, how to deal with infinite systems, how to start and run a simulation, and how to analyse the results.

4.1. Integrating the equations of motion

Much has been written about how to integrate the equations of motion of a dynamical system most effectively and accurately. This is more an issue of applied mathematics [19] than of physics (although some methods have a very physical inspiration), therefore we are not going to go in depth here. I just want to provide a simple recipe, which will be useful in most cases that we are likely to encounter.

The classical equations of motion have the general form

$$\dot{q} = G(p,q), \quad \dot{p} = F(p,q), \tag{1}$$

where $G(p,q) = \partial H/\partial p$ and $F(p,q) = -\partial H/\partial q$, and H is the Hamiltonian, which in the standard case is given by

$$\mathscr{H} = \sum_i \frac{\mathbf{p}_i^2}{2m_i} + U(\mathbf{q}), \tag{2}$$

where $\mathbf{q}_i$ represents the coordinates of atom i, and $\mathbf{p}_i$ is its conjugate momentum. The recipe which we will use is known as the *generalised leapfrog*, and is summarised as follows. First, we advance the momenta in time half a time step, then, with the momenta at half time step we move the coordinates forward in time by a full time step, recalculate the forces at the new positions, and with these new forces, advance the momenta to full time step. The algorithm is symbolically written down as:

$$\begin{aligned} p(t+\Delta t/2) &= p(t) + \Delta t F[p(t+\Delta t/2), q(t)]/2, \\ q(t+\Delta t) &= q(t) + \Delta t\{G[p(t+\Delta t/2), q(t)] + \\ &\quad G[p(t+\Delta t/2), q(t+\Delta t)]\}/2, \\ p(t+\Delta t) &= p(t+\Delta t/2) + \Delta t F[p(t+\Delta t/2), q(t+\Delta t)], \end{aligned} \tag{3}$$

where Δt is the time step. For the simple case of a separable Hamiltonian such as that of Eq. (2), the generalised leapfrog algorithm reduces to:

$$\begin{aligned} p(t+\Delta t/2) &= p(t) - \frac{\Delta t}{2}\frac{\partial U}{\partial q}(t), \\ q(t+\Delta t) &= q(t) + \Delta t\, p(t+\Delta t/2), \\ p(t+\Delta t) &= p(t+\Delta t/2) - \frac{\Delta t}{2}\frac{\partial U}{\partial q}(t+\Delta t), \end{aligned} \tag{4}$$

which is known simply as the leapfrog algorithm. Using Eqs. (4) repeatedly, one can map out a trajectory from specified initial conditions (coordinates and momenta of all atoms in the system). Provided the time step Δt is sufficiently small, this scheme conserves the energy reasonably accurately, and is time reversible, as the equations of motion are. The generalised leapfrog (and therefore the leapfrog) is accurate to second order in Δt. So, how large (or small) should Δt be? This depends on the system and how it is modelled. There are two opposing requirements; on the one hand Δt should be as large as possible, as in this way we can span, with the same cost, a longer simulation time. On the other hand Δt should be small enough that we can integrate the equations of motion with sufficient accuracy (remember that we are only doing this with an accuracy of second order in Δt). The rule of thumb is that Δt should be at least 10-20 times smaller than the shortest oscillation period in the system of interest. This will typically be a bond vibration, so in practice, Δt is frequently in the femto-second (10^{-15} seconds) scale. If this rule is not fulfilled, things will go wrong, e.g. the total energy will not be conserved, and the simulation results will not be correct.

The classical equations of motion have many properties, of which the most obvious one is time reversibility, but there are other ones. A particularly important symmetry is that known as *simplecticity* or *simplecticness*. Consider the following sum of infinitesimal areas:

$$\sum_i \delta \mathbf{r}_i \times \delta \mathbf{p}_i, \tag{5}$$

where the sum extends over all degrees of freedom of the system, and the deltas imply infinitesimally short vectors centred at each position and momentum. It can be easily shown that this infinitesimal area is a constant of motion of classical mechanics. It is important that any numerical scheme for integrating the equations of motion respects as many as possible of the intrinsic properties of the equations of motion; the more of such properties that are respected, the greater the guarantee that we will have that the numerical solution found will resemble a physically correct trajectory. The generalised leapfrog scheme described above is time reversible and simplectic, and due to this it is particularly stable. A more in-depth discussion of these issues can be found in the book by Sanz-Serna and Calvo [19].

There are many other schemes for integrating numerically the equations of motion, and excellent discussions can be found in the literature [3, 4, 5], but the generalised leapfrog will suffice as an example for us, and turns out to be one of the best schemes in the market anyway.

4.2. Periodic boundary conditions

Frequently one is confronted with the need to study a system, periodic or not, which contains large numbers of atoms or molecules, where large means of the order of N_A, Avogadro's number. Naturally, we cannot deal with such large numbers, so we must resort to some computational tricks in order to emulate a system in these conditions. The trick used in this case is referred to as *periodic boundary conditions* (PBC), and consists of assuming that the simulation box (i.e. the box containing the atoms in the simulation) is surrounded by identical copies of itself in all directions. In the simulation of periodic systems the simulation box is typically a (super)cell with the periodicity of the system. If the simulation does not involve studying the dynamics of the system (as when we do a structural relaxation) then there is no approximation involved in the use of PBC, unless we are doing an electronic structure calculation, in which case care has to be taken in order to sample the electronic states in regions of the Brillouin zone beyond the Γ point to ensure the convergence of the calculation[3]. If, however, we are interested in the dynamics of the system, the use of PBC involves an approximation, even if the system is periodic. Adopting PBC implies assuming that all periodic images of atoms in the central simulation box move in exactly the same way.

In a liquid, or in an amorphous solid, the use of PBC imposes an artificial symmetry, the consequences of which can be subtle. Further limitations arise in the study of defects and impurities in solids: the use of PBC generally implies that one is considering a large concentration of the defect or impurity under consideration, since we cannot

---

[3] The discussion of this very important point is however specific to electronic structure calculations, and we will not discuss it here; see e.q. [20].

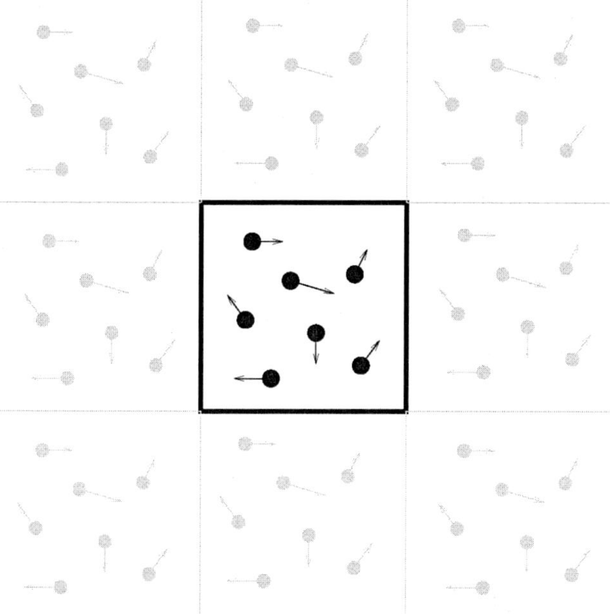

FIGURE 1. Periodic Boundary Conditions; illustration in two dimension. The simulation box is highlighted at the centre, and is surrounded by periodic images of itself.

always make the simulation box as large as we would like due to the computational cost involved in doing so. Furthermore, there are certain kinds of defects that cannot be easily accommodated in a periodic cell. This happens with dislocations. In such cases one has to include two dislocations of opposite sign so that they can both be included within a periodic cell, or renounce to the use of PBC altogether.

Another consideration when using PBC is that long-range interactions have to be dealt with appropriately. Electrostatic forces have such a long-range that it is necessary to include the contribution of far-away periodic images of the simulation box on the atoms contained in the central simulation box in order to get meaningful results. Simply truncating electrostatic interactions beyond a certain cutoff is crude, and generally frowned upon. Several ways have been described in the literature to deal with long-range interactions. Perhaps the most popular procedure is that due to Ewald, known as Ewald Summation. I will not go into details here, but I would like to describe the basic idea. The problem with electrostatic interactions is that terms of order $1/r$ decay very slowly with r, the interatomic distance. The Ewald summation method consists of splitting this term as follows:

$$\frac{1}{r} = \frac{\mathrm{erf}(\alpha r)}{r} + \frac{\mathrm{erfc}(\alpha r)}{r}, \tag{6}$$

where erf is the error function, erfc is the complementary error function, and α is a parameter. The idea is that the first term on the rhs of Eq. (6) can be shown to be short-ranged in reciprocal space, where it can be easily evaluated, and likewise, the second term is short-ranged in real space. The parameter α is chosen so that an optimal split between the real-space sum and the reciprocal-space sum is obtained.

A final consideration concerning PBC is the following. PBC are a useful device for calculating the energy and its derivatives (forces and stresses, see below) as if the atoms of the system were indeed in an infinite system, or at least sufficiently far away from any surface to notice its presence. However, in general it is not required, nor is it desirable, to modify the positions of the atoms as they move so that they all lie in the central simulation box, particularly if one is interested in monitoring the diffusion or other dynamical properties of the system. So, when using PBC, one calculates the total energy and the forces on the atoms as if all atoms were relocated in the central simulation box, regardless of their actual position in space, but they are not moved back to the simulation box if during the simulation they drift out of it.

4.3. Derivatives of the total energy

In order to integrate the equations of motion for the atoms constituting the system, we must be able to obtain the forces, i.e. the derivatives of the total energy with respect to the atomic positions:

$$\mathbf{f}_i = -\nabla_{\mathbf{r}_i} E_{tot}, \tag{7}$$

where E_{tot} is the total energy of the system. In the standard case, the kinetic energy does not depend on the atomic positions, and so only the derivative of the potential energy has to be considered.

Another useful derivative of the total energy, needed if one wishes to calculate the pressure of the system, or conduct a constant pressure simulation (constant pressure simulations will be discussed in section 7, is the stress. The stress is also useful because it can be related with the elastic constants of a crystal. The stress is defined as the derivative of the total energy with respect to the components of the strain tensor. The strain, ε, defines infinitesimal distortions of the simulation box. For example, consider that the simulation box (not necessarily cubic or orthorhombic) is defined by the three vectors $\mathbf{a}_\alpha$, with $\alpha = 1, 2, 3$. Then, a distortion of the simulation box defined by the strain tensor ε will lead to new cell vectors $\mathbf{a}'_\alpha$ given by

$$a'_\alpha = a_\alpha + \sum_\beta \varepsilon_{\alpha\beta} a_\beta. \tag{8}$$

Because a distortion of the cell will cause the distances between atoms and the angles between bonds in the cell to change, such a distortion changes the total energy, and that change is given to first order by the stress, defined as

$$\sigma_{\alpha\beta} = \frac{\partial E_{tot}}{\partial \varepsilon_{\alpha\beta}}. \tag{9}$$

Note that in this definition we have dropped a minus sign, so that the stress is not minus the derivative of the total energy with respect to the strain. With this definition, a negative applied stress is tensile, while a positive applied stress is compressive, resulting in an intuitive convention.

From Eq. (8) we see that the strain is a tensor with dimensionless components. Furthermore, we are only interested in symmetric strain tensors, because any asymmetric strain tensor involves rotation as well as distortion of the cell. Rigid rotations of the system are not interesting, however (at least in the absence of external fields), and complicate the dynamics, so, in practise, we will always be concerned with symmetric strain tensors, and, for the same reasons, the stress tensor will also be symmetric.

4.4. Start-up of an MD simulation

Imagine that you have written an MD code, which integrates the equations of motion for a given model, and everything works correctly. You are now in a position to perform a simulation. How does one start? Typically, one needs a starting configuration of the system. For a crystal, the perfect lattice will serve the purpose, although one can use also a slightly distorted version of the lattice (in fact this may have some advantages in achieving the thermalisation of the system; see below). For a liquid the initial configuration may be less obvious. In this case one can start with a lattice configuration known to be unstable at the temperature of the simulation, and hope that this will evolve rapidly toward configurations typical of the liquid phase.

As well as coordinates, one needs to generate initial velocities for the atoms. The almost universally adopted choice is to generate random velocities sampled from the Maxwell-Boltzmann distribution defined for the desired temperature of the simulation. These may need to be corrected, so that the centre of mass of the system has zero velocity (this avoids the drift of the system as the simulation proceeds). In the case of finite systems it is also useful to avoid the rotation of the system, by ensuring that its total angular momentum is zero.

Generally, one wishes to conduct a simulation at a given temperature. However, even if one generates initial velocities sampling from the Maxwell-Boltzmann distribution corresponding to the desired temperature, the evolution of the system will drive the temperature to other values, such that its average over the simulation run will not in general coincide with (or even be near) the desired temperature. This happens essentially because the starting atomic positions are not necessarily consistent with the desired temperature. It is therefore necessary to drive the system from its starting conditions to other conditions, compatible with the desired temperature. This is generally done by scaling the velocities during an initial period of the simulation, usually referred to as *thermalisation* or *equilibration*. Each atomic velocity is scaled by a factor $\sqrt{T_{ext}/T_{inst}}$, where T_{inst} is the *instantaneous* temperature (see below), and T_{ext} is the desired equilibrium temperature. This scaling will slowly drive the system toward the desired conditions; it can be done every time step of the equilibration period, or every few time steps. Obviously, this tampering with the velocities results in a lack of energy conservation. The dynamics is thus artificial, and only serves the purpose of preparing the system in conditions from which the real simulation can start. Therefore, no information obtained during this period is useful, and should not be included in the subsequent analysis of results.

How long should the equilibration period be really depends on the nature of the simulated system, but also on practical considerations, such as the cost involved. Ideally, one should run the equilibration period for long enough so that the system has lost any "memory" of its initial conditions, and is fully at equilibrium at the desired temperature. Once this is achieved, the average temperature should be close to (usually not more than a few degrees away from) the desired temperature. If this does not happen, then obviously the equilibration period was not sufficiently long.

5. ANALYSING THE RESULTS

MD simulations can produce a wealth of information, ranging from the time evolution of the coordinates and velocities of individual atoms to other so called "collective" properties such as the temperature, pressure, and so on. In this section we review the standard magnitudes that are monitored during an MD simulation.

5.1. Temperature

The temperature in a simulation can be calculated directly from the standard expression from statistical mechanics relating it to the kinetic energy of the atoms. This expression is

$$T_{inst} = \frac{2}{g k_B} E_{kin}, \tag{10}$$

where E_{kin} is the kinetic energy at the present time, g is the number of degrees of freedom of the system and k_B is Boltzmann's constant. This expression gives the *instantaneous* temperature of the simulation. This value will be different at different time steps; really only its average value over the length of the simulation gives a meaningful value to the temperature:

$$< T >= \frac{1}{N} \sum_n T_{inst}(n), \tag{11}$$

where the sum extends over all time steps (or a subset) of the simulation, N. Only if an appropriate equilibration period has been undertaken before the actual simulation (see 4.4 above) will one have that the temperature of the simulation will be close to the desired target temperature, i.e. $< T > \approx T_{target}$.

5.2. Pressure

The pressure is another useful magnitude to monitor through a simulation. Its average value will provide information on the mechanical state of the system, i.e. if the system is compressed or expanded with respect to its equilibrium volume at the temperature of the simulation. The instantaneous pressure of the system is given by

$$P_{inst} = \rho k_B T + \frac{1}{3V}\left\langle \sum_{i<j} \mathbf{f}_{ij} \cdot \mathbf{r}_{ij} \right\rangle, \tag{12}$$

where ρ is the number density, V is the volume, $\mathbf{f}_{ij}$ is the pair force between atoms i and j, and $\mathbf{r}_{ij}$ is the distance vector. This expression is actually only valid for the case of pair interactions, and must be generalised in more complicated models. Like in the case of the temperature, only the average value, $< P >$, makes sense, as the value of P_{inst} will fluctuate strongly in time.

If it is desired to perform the simulation at a pre-specified pressure, one has to adjust the volume of the simulation cell in such a way that the average pressure has the desired value. This usually requires performing several short simulations. The average pressure is a smooth function of the simulation volume, so it is usually sufficient to find two volumes which give average pressures which bracket the desired pressure, and then use linear interpolation to obtain the volume which would correspond to the desired average pressure. This procedure is simple if the shape of the simulation box is fixed, as is the case in a liquid, or in a solid where no phase transition is expected. However, if the crystal shape is complicated or unknown a priori, it is probably more desirable to conduct a variable-shape MD simulation at constant pressure, which allows the system to dynamically adopt volume and shape to the most favourable values at the conditions of the simulation. These are called constant-pressure MD simulations, and will be discussed in section 7.

5.3. Structure

Even though at finite temperatures the atoms of the system will never be at rest, the system will have a definite structure. Let us consider the atomic density of the system at each point in space, $\rho(\mathbf{r})$. We can write such a density as the following thermal average:

$$\rho(\mathbf{r}) = < \sum_{i=1}^{N} \delta(\mathbf{r} - \mathbf{r}_i) > . \tag{13}$$

If the system is crystalline or amorphous at the conditions of the simulation, $\rho(\mathbf{r})$ will peak at the average positions of the atoms, and will fall to low values close to zero at interstitial regions, which will be visited infrequently by the atoms. If, on the other hand, the system is fluid, then $\rho(\mathbf{r})$ should be constant everywhere, and equal to the bulk density. In the crystalline case $\rho(\mathbf{r})$ will have the periodicity of the lattice and the system will possess so-called long-range order. But even if the system is amorphous or fluid, it will possess short-range order, and it will be interesting to characterise it. Two quantities are frequently used to achieve this, known as the *radial distribution function* and the *bond-angle distribution function*.

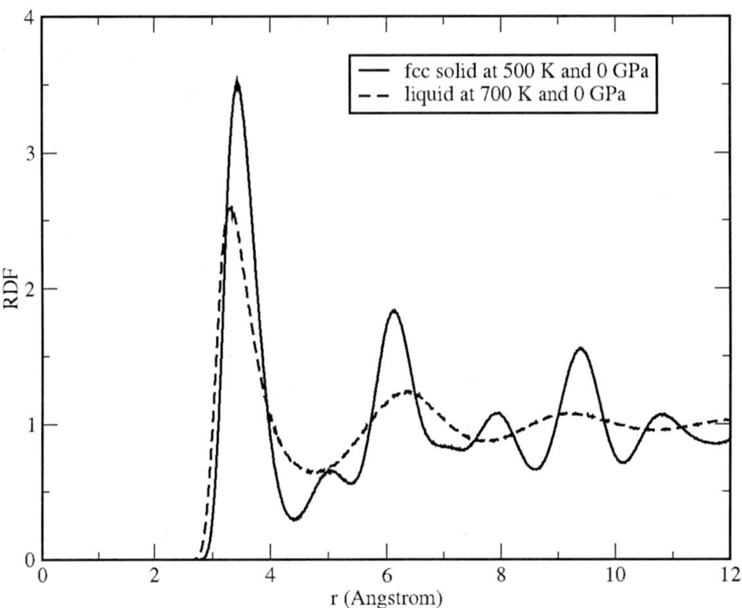

FIGURE 2. Radial distribution function calculated for Pb using the empirical potential due to Cleri and Rosato [21]. Results are shown for the solid fcc phase at 500 K and for the liquid at 700 K, both at zero pressure.

5.3.1. Radial distribution function

The radial distribution function (RDF) is constructed as a histogram of the distances between an atom and its neighbours during the simulation. Assuming that we are dealing with a one-component system, all atoms are equivalent, and the RDF is then averaged over all atoms. Suppose that we want to calculate the RDF, typically called $g(r)$, in the range r_0 to r_{max}. To do so, we divide this range into a number of equally spaced segments of length δr, and add a 2 in the appropriate segment of the $g(r)$ histogram for each pair of atoms separated by a distance r corresponding to that segment. Such a histogram will diverge for large distances, since the probability of finding two atoms separated by r when r is large grows very rapidly, so it is customary calculate $g(r)$ as the probability of finding two atoms separated by distance r relative to the probability of finding two atoms at the same distance in the ideal gas of the same density. According to this definition, $g(r)$ should tend to 1 as r becomes large.

In Fig. (2) a typical RDF function obtained from a simulation is shown for the particular case of solid and liquid Pb. As can be seen, the RDF is zero at short distances, reflecting the exclusion volume around a given atom; it then grows rapidly to reach a maximum at the nearest neighbour distance, falling down again to a first minimum, which can be followed by other (smaller) maxima and minima, or smoothly evolve to its large-distance limit of 1. The volume integral of the RDF from zero to the first minimum

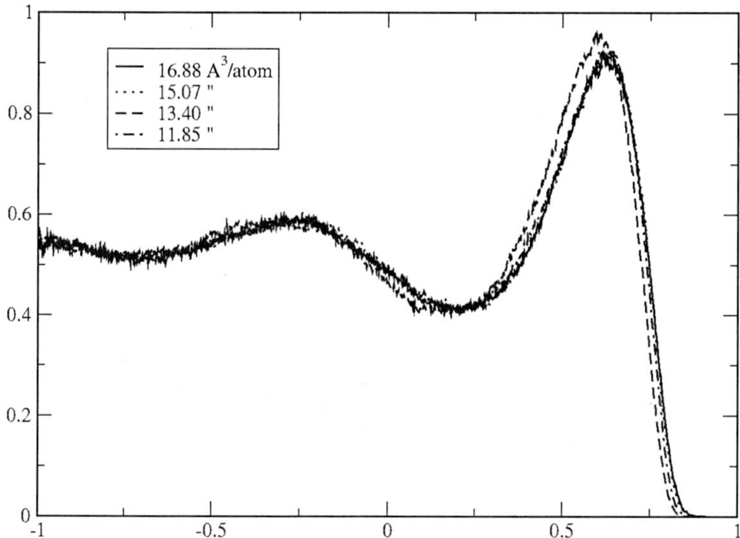

FIGURE 3. Bond-angle distribution function for liquid Na, calculated using ab initio simulations [22].

after then nearest neighbour peak gives the number of first neighbours of each atom, i.e. the average coordination of atoms through the simulation:

$$n_c = 4\pi\rho_0 \int_0^{r_{min}} r^2 g(r) dr, \tag{14}$$

where ρ_0 is the bulk density.

5.3.2. Bond-angle distribution function

Since the RDF gives a distribution of distances between atoms, it does not have any angular resolution. Therefore, in order to complete the picture of the short-range environment of atoms in the simulation, it is frequent to calculate what is called the Bond-angle distribution, or BAD. The BAD is exactly what its name implies, i.e. a distribution of the bond-angles, or actually, the cosine of the bond angles, found between an atom and its first shell of neighbours, taken two by two, with the atom in question forming the apex of the bond angle. A typical example of the bond-angle distribution found from a simulation is shown in Fig. (3), calculated for liquid Na.

5.4. Dynamics

The power of MD simulation as compared to MC is that it also provides information on the dynamics of the system, not just the structure. The dynamics of a system can

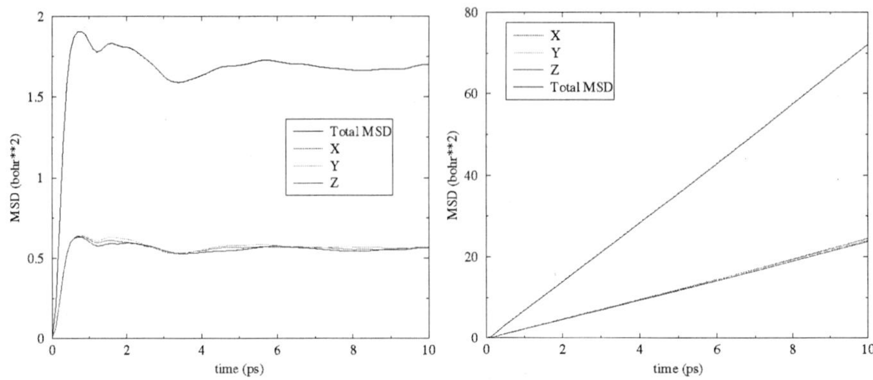

FIGURE 4. Mean-squared displacements of Pb at different temperatures, just before (500 K, left panel) and after melting (700 K, right panel) has taken place. The non-diffusive behaviour of the solid phase, and the diffusive behaviour of the liquid phase can be clearly appreciated. Taken from the simulations reported in [24].

be regarded from many angles: firstly, there is atomic motion, which can be vibrational around the equilibrium sites of atoms in a solid, or hopping from one site to another, or even diffusive. But we also have the so-called collective dynamics, such as density fluctuations, sound, or viscous flow. Much could be written (and has indeed been written, see for example [23]) about all these forms of dynamics and how to study them. But here we are going to limit ourselves to the most frequent kind of dynamical analysis, that related to diffusive motion.

5.4.1. Mean-squared displacements

In order to answer the question *how much does an atom move from its initial position in a given time t?* one can calculate the averaged mean-squared displacements (MSD), given by

$$< |\mathbf{r}(t) - \mathbf{r}(0)|^2 >= \sum_{t_0}\sum_{i=1}^{N} \frac{|\mathbf{r}_i(t+t_0) - \mathbf{r}_i(t_0)|^2}{N_{t_0} N}, \tag{15}$$

where we have calculated the average taking advantage of the fact that different time origins t_0 can be taken, N_{t_0} being the number of such time origins, and since all atoms of the same species are equivalent, we can also average over them. An example of MSDs as obtained from a typical simulation is illustrated in Fig. (4).

MSDs are important because they provide information on how fluid-like a system is. In Fig. (4) we show examples of how the MSD look in a system (in this particular case Pb) before and after the system has melted. It can be seen that while the system is in the solid phase, the MSD are flat, having zero slope at all times but very short ones. This is because in the solid atoms don't travel large distances, but rather oscillate around their equilibrium positions. The amplitude of those oscillations is related to the saturation value of the MSDs at long times. On the other hand, if the system is in the liquid phase, atoms can move and diffuse through the system. In this case the MSDs grow with time having a well-defined slope, which is related with the diffusion coefficient:

$$6Dt + b = < |\mathbf{r}(t) - \mathbf{r}(0)|^2 >, \tag{16}$$

where b is a constant, and D is the diffusion constant of the system. If there is more than one species of atoms in the system, one can define a diffusion constant for each species by generalising the above expression. For the particular example illustrated in Fig. (4), one obtains a value of D equal to 2.0×10^{-5}cm^2/s, very close to the experimental value of 2.0×10^{-5}cm^2/s found just after melting has taken place.

5.4.2. Velocity auto-correlation function

The last magnitude we will discuss is that of the velocity autocorrelation function (VAF). The VAF is defined as follows:

$$\mathrm{VAF}(t) = < \mathbf{v}(t) \cdot \mathbf{v}(0) >, \tag{17}$$

and is calculated from a simulation following the same scheme as for the MSD given in Eq. (15) above. It is also frequent to work with the normalised VAF, defined by

$$\hat{\mathrm{VAF}}(t) = \frac{< \mathbf{v}(t) \cdot \mathbf{v}(0) >}{< \mathbf{v}(0) \cdot \mathbf{v}(0) >}. \tag{18}$$

In actual fact, the VAF also provides information on the diffusivity of the different species in the system, as it is linked with the diffusion constant through the following relation:

$$D = \frac{1}{3} \int_0^\infty < \mathbf{v}(t) \cdot \mathbf{v}(0) > dt. \tag{19}$$

The VAF is really telling us how much time it takes for an atom in the system to "forget" its original velocity at time zero, through collisions with other atoms in the system. It starts at a positive large value, equal to 1 in the case of the $\hat{\mathrm{VAF}}$, and has an oscillating behaviour, falling more or less exponentially to zero as time increases. The Fourier transform of the VAF is given by

$$\mathrm{VAF}(\omega) = \frac{1}{\pi} \int_0^\infty \cos(\omega t) \mathrm{VAF}(t) dt, \tag{20}$$

and is related to the phonon density of states through

$$\mathrm{VAF}(\omega) \propto \mathrm{VDOS}(\omega) e^{-\beta \hbar \omega}, \tag{21}$$

where $\mathrm{VDOS}(\omega)$ is the vibrational density of states at frequency ω, and $e^{-\beta \hbar \omega}$ is the corresponding Boltzmann factor for that frequency ($\beta = (k_B T)^{-1}$).

5.5. Summary

In this section we have discussed some of the practical aspects of performing an MD simulation, ranging from the numerical integration of the equations of motion to how to deal with infinite systems, and how to start up the simulation. We have also discussed some of the typical properties that one analyses during or after an MD simulation has been performed, and how this analysis provides information on the properties of the system. Really, we have only skimmed the surface. There is much more to say about how to perform simulations and how to analyse the results than we can discuss in these notes, but at least I hope that you have got the basics, and can build from here if needed.

6. MODELLING INTERACTIONS IN ATOMIC SYSTEMS

A key issue in any for of modelling atomistic systems, be it by means of MD, MC, or any other technique, is the representation of the interactions between the atoms or molecules that make up the system. Ultimately, these interactions are the result of the subtle interplay of electrons and nuclei. This interplay can give rise to a wide variety of behaviours; some systems display a covalent type of bonding, while others would be better described as ionic, although more generally the situation is intermediate between these two extremes. Yet in other systems neither of these patterns takes place; rather, one has a metallic behaviour, where a significant portion of the electrons are free to move through the entire system, without being associated to a particular atom or bond. The opposite extreme to this is that of noble gases, where electrons are tightly attached to individual atoms, adopting a closed-shell electronic structure. To complicate matters further, several of these widely different behaviours can be displayed by one and the same system at different temperature and/or pressure conditions. For example, at ambient conditions, silicon is known to adopt the diamond structure, which is a covalent semi-conductor. But upon raising the temperature beyond $\approx$1670 K silicon melts, and in so doing it becomes metallic. Likewise, upon applying pressure at fixed temperature, the diamond structure eventually undergoes a phase transition to the so-called β-Sn structure, which is also metallic.

In general, we can find two different approaches to describing the interactions between atoms and molecules in a system. The first one is to employ some form of potential, i.e. a (in general complicated) function which depends on the relative interatomic positions (distances and angles) and on a series of parameters which must be fitted in order to reproduce as closely as possible some relevant properties of the system, such as a crystal structure, elastic or vibrational properties, etc. In this approach the electronic structure is obviated; rather one attempts to account for its effects with the potential function. The second approach involves retaining the picture of the system as composed by electrons and nuclei, and to obtain the energetics of the system as well as the forces on the atoms from a quantum mechanical treatment of the electronic structure, either at the semi-empirical level or through a fully first-principles treatment. This approach is theoretically more sound, but obviously more expensive.

The first approach is usually termed the *empirical potential* approach. The name makes reference to the fact that in general the form of the potential is *ad hoc*, i.e. there is no underlying guiding principle as to what the mathematical expression of the potential should be, beyond the fact that it should be repulsive at short distances, attractive at intermediate ones and decay to zero at infinite separation. It also refers to the fact that the potential has a series of disposable parameters that must be fitted, traditionally to empirical information on the system, though lately it is very common to parametrise potentials to results obtained with more accurate theoretical calculations, usually based on electronic structure calculations. To give an exhaustive review of the different types of potentials used in the literature would be a daunting task, far beyond the scope of what only aims to be an introduction to MD. Rather, I will just name a few common examples, and refer the interested reader to the appropriate literature.

It is customary to assume that the total potential energy of an atomic system in the absence of external fields can be written as a series of the form

$$U = \sum_{i,j} V_2(\mathbf{r}_i, \mathbf{r}_j) + \sum_{i,j,k} V_3(\mathbf{r}_i, \mathbf{r}_j, \mathbf{r}_k) + \sum_{i,j,k,l} V_4(\mathbf{r}_i, \mathbf{r}_j, \mathbf{r}_k, \mathbf{r}_l) + \ldots, \tag{22}$$

where V_2 represents the energy of interactions of pairs of atoms, V_3 that of triads, and so on. In practice this series is rarely taken beyond the sum containing the V_4 terms, and frequently it is truncated after the first or second sums.

Although much work has been done with discontinuous potentials such as hard spheres, here we will focus on continuous potentials. Of these, perhaps the simplest, though still extensively used, is the Lennard-Jones potential, which takes the form

$$V_{LJ} = 4\varepsilon \left[\left(\frac{\sigma}{r} \right)^{12} - \left(\frac{\sigma}{r} \right)^{6} \right]. \tag{23}$$

The potential is characterised by two parameters, namely ε, which has dimensions of energy, and which determines the minimum value of the potential, and σ, which has dimensions of length, and is related to the position of the minimum. The first term in the squared brackets of Eq. (23) causes the potential to be strongly repulsive at short distances, while the second term has the typical form expected for dispersion-type interactions, which decay as r^{-6}. This potential has been frequently used to model e.g. noble gases, and in such a case it is an example of a model in which Eq. (22) is truncated after the first sum. Eq. (23) is frequently used also to describe non-bonded type interactions (i.e. interactions between atoms that are not linked by a chemical bond) in more complex molecular systems.

Covalent systems generally have a more complex bonding, where interactions are not only distance dependent but also directional. Several potentials have been put forward for such systems, out of which perhaps the best well known are those of Stillinger and Weber [25] for silicon, and that of Tersoff [11], also for silicon, but which has been parametrised also for carbon. Both potentials have been extensively used for modelling covalent materials, and have inspired the formulation of more sophisticated models. For example, the Tersoff [11] potential has the following expression:

$$U = \frac{1}{2} \sum_i \sum_{i \neq j} f_C(r_{ij}) \left[f_R(r_{ij}) + b_{ij} f_A(r_{ij}) \right], \tag{24}$$

where $f_R(r_{ij}) = A_{ij}e^{-\lambda_{ij}r_{ij}}$ and $f_A(r_{ij}) = -B_{ij}e^{-\mu_{ij}r_{ij}}$ are repulsive and attractive pair potentials, respectively, and the parameters A_{ij}, B_{ij}, λ_{ij} and μ_{ij} depend on the chemical species of atoms i and j. It would appear from Eq. (24) that this model is a pair-wise potential, but this is not so. The third-body dependence of the potential is contained in the b_{ij} term, which is a function of θ_{ijk}, the angle defined by the vectors connecting one atom with every possible pair of its neighbours. The parametrisation for silicon is such that a tetragonal arrangement of each atom's neighbours, at appropriate first-neighbour distances, minimises the energy. In the case of carbon, a second minimum at 120° allows the obtain also the single-layer graphite structure (graphene).

Potentials have also been developed for metallic systems, such as the Embedded Atom Model (EAM) and its derivations, or the Cleri and Rosato [21] potential. In these models there is also an environment dependence of the potential, but it is not so strongly directional as in the case of covalent materials. This model, like many of its kind, is based on the observation that the energetics of d-band metals are largely dictated by the width and centre of gravity of the d-band density of states (d-DOS), but are fairly insensitive to its detailed shape. Since the width of the d-DOS is proportional to $\sqrt{\mu_2}$, where μ_2 is the second moment of the d-DOS, and μ_2 can be related to the hopping integrals of the Hamiltonian, the idea is then to write down the binding energy of the system as an expression reminiscent of this. Cleri and Rosato [21] proposed to use

$$E_b^i = -\left[\sum_j \eta^2 e^{-2q(r_{ij}/r_0-1)}\right]^{1/2}, \tag{25}$$

where η plays the role of a hopping integral, and r_0 is the nearest-neighbour distance. The binding energy is complemented by a pairwise repulsive energy of the form

$$E_r^i = \sum_j A e^{-p(r_{ij}/r_0-1)}. \tag{26}$$

The total potential energy of the system, $\mathscr{U}$, is then given as the sum over all atoms of Eqs. (25) and (26) above. This model has been parametrised for a series of metals, including Ni, Cu, Rh, Pd, Ag, Ir, Pt, Au, Al and Pb [21].

Organic and biological molecules are frequently simulated with potentials of the form

$$\begin{aligned} U &= \sum_{\text{bonds}} k_{\text{bond}}(d-d_0)^2 + \sum_{\text{angles}} k_{\text{angle}}(\theta-\theta_0)^2 + \\ &+ \sum_{\text{torsions}} k_{\text{torsion}}[1+\cos(n\chi-\delta)] + \\ &+ \sum_{ij}\left\{4\varepsilon_{ij}\left[\left(\frac{\sigma_{ij}}{r_{ij}}\right)^{12} - \left(\frac{\sigma_{ij}}{r_{ij}}\right)^{6}\right] + \frac{q_i q_j}{r_{ij}}\right\}, \end{aligned} \tag{27}$$

where the first term is a sum over the bonds, the energy of which is modelled by a harmonic spring or some similar potential (e.g. a Morse potential). The second term accounts for bond-angle vibrations, also modelled by a harmonic spring on the deviation from the equilibrium angle. The third term describes dihedral angles, and involves sequences of four atoms linked by three adjacent bonds. The last term encompasses the energy of interaction between pairs of atoms that are not directly bonded, and it includes a Lennard-Jones type potential (see above) and a Coulomb term to account for the electrostatic interaction between charged ions. Potentials similar to that of Eq. (27) form the core of such programs as CHARMM [7] and AMBER [8].

Tight-Binding (TB) models occupy the middle ground in the spectrum of models for materials simulation, between the extremes of empirical potentials and first-principles methods. TB models, unlike empirical potentials, do incorporate a description of the electronic structure, although they do so at a much more simplistic and approximate level than first-principles methods. In TB models the matrix elements of the electronic Hamiltonian are not evaluated rigorously from the Hamiltonian operator and a chosen basis set, but rather are assumed to have a certain parametrised dependence on the interatomic distances. This makes the cost of constructing the matrix representation of the TB Hamiltonian rather small, while it is a significant part of the calculation in first-principles methods. However, this is at the cost of assuming a given form of the matrix elements, which may be physically sound, but is ultimately *ad hoc*, just as the form of an empirical potential is. In spite of this, TB models have become extremely popular in materials modelling [26], due to their combination of methodological simplicity and comparative accuracy. We will not go into details on the different TB models; interested readers may find details of these techniques in several review papers [26, 27] and books [20, 28].

As discussed in Section 3, one of the landmarks in computational condensed matter physics was the development of first-principles MD by Car and Parrinello [16] (CP). These authors obviated the need to employ a potential for modelling the atomic interactions by means of an empirical potential. Rather, the potential energy and its derivatives were directly obtained from first-principles electronic structure calculation. Specifically, CP formulated FPMD within the context of Density Functional Theory (DFT). DFT was formulated in the 1960s by Kohn and collaborators. Hohenberg and Kohn [17] demonstrated that the energy of an ensemble of electrons moving in an external field, and in particular the field generated by the nuclei or ions, is a unique functional of the electron density, and furthermore, that this functional adopts a minimum value when the electron density is that corresponding to the ground state. Subsequently, Kohn and Sham [18] showed that the electronic structure problem could be cast into an independent particle problem in which the wave functions of each particle obey a Schrödinger-like equation of the form

$$\left[-\frac{1}{2}\nabla^2 + V_{KS}(\mathbf{r})\right]\psi_i = \varepsilon_i \psi_i, \tag{28}$$

where ψ_i and ε_i are the particle wave functions and energies, respectively, and V_{KS} is the Kohn-Sham potential, given by

$$V_{KS}(\mathbf{r}) = V_{ext}(\mathbf{r}) + \int d\,\mathbf{r}' \frac{n(\mathbf{r}')}{|\,\mathbf{r}-\mathbf{r}'\,|} + \frac{\delta E_{xc}[n(\mathbf{r})]}{\delta n[\mathbf{r}]}. \tag{29}$$

Here the first term on the rhs is the potential of interaction with the ions or nuclei, the second term is the potential due to the electrostatic interaction with the electron density, and the last term is the exchange-correlation potential. To cut a long story short, the Kohn-Sham orbitals must be obtained by self-consistently solving Eqs. (28) (note that V_{KS} depends on the ψ_i orbitals through the density $n(\mathbf{r}) = 2\sum_i |\psi_i(\mathbf{r})|^2$). Once the Kohn-Sham equations have been solved, the total energy, forces and stress can be obtained, and used in a conventional MD simulation. In order to solve Eqs. (28), it is customary to expand the orbitals ψ_i as a linear combination of basis functions, like so:

$$\psi_i = \sum_k c_{i,k}\phi_k, \tag{30}$$

where different choices of basis functions ϕ exist. The problem then is reformulated into finding out the coefficients of the expansion, $c_{i,k}$. This can be done by any of a number of techniques [20]. Let's imagine starting an MD simulation from an atomic configuration for which the $c_{i,k}$ coefficients in Eq. (30) had been previously obtained. For such a configuration the total energy and forces on the ions are available once the electronic structure problem is solved, so one can use these forces to perform a conventional MD step on the ions. Once the ions move, however, in principle one would have to go back and solve the electronic problem all over again. But CP proposed to do something different: they showed that it is possible to incorporate the $c_{i,k}$ expansion coefficients as fictitious classical variables in the dynamics, with adequately chosen fictitious masses. Thus one ends up with a combined dynamics of ions and wave function coefficients. This looks very strange indeed, but it is in fact a very clever trick: with a suitable choice of fictitious masses for the $c_{i,k}$ coefficients and a a small enough time step, it is possible to arrange things in such a way that the dynamics of the $c_{i,k}$ follows closely the Born-Oppenheimer surface, or in other words, the $c_{i,k}$ automatically adapt to the slowly varying ionic configuration, giving wave functions that are very close to the Kohn-Sham ground state for the current ionic configuration. The fictitious masses of the $c_{i,k}$ need to be small enough so that their dynamics is faster than that of the ions; this in turn imposes a smaller time step than would be required for a stable dynamics of the ions with a conventional force field, but the gain is that one has done away with the force field altogether!

Another consideration to take into account is that the dynamics of the $c_{i,k}$ must be subject to the constraints of orthonormality of the Kohn-Sham orbitals, i.e.

$$\sum_k c^{\star}_{i,k} c_{j,k} = \delta_{ij}, \tag{31}$$

where δ_{ij} is the Kroneker delta. There are standard techniques for performing MD subject to constraints[4] which can be applied to impose Eq. (31) at each time step. Imposing such constraints (N^2 of them, where N is the number of Kohn-Sham orbitals) is a significant bottle-neck of FPMD, as this carries a computational cost that grows as $\mathcal{O}(N^3)$.

[4] Two well-known algorithms for imposing constraints are the so-called *rattle* and *shake* methods. We will not discuss them here, but interested readers will find full accounts in refs. [3, 4].

There are many intricacies in CPMD and DFT calculations, which we cannot cover here in any detail, but fortunately all this is extensively documented in the literature (see e.g. [20, 29, 30, 31]). Suffice it to say that FPMD in the CP flavour and in others is now a fairly standard and frequently used simulation technique. It is still computationally very demanding compared to MD based on force fields, but the cost is affordable in many cases, thanks in part to the continuing improvement of algorithms and numerical techniques, and to the ever increasing tendency of computer power.

7. BEYOND THE MICROCANONICAL (NVE) ENSEMBLE

Let us now briefly discuss how MD has been extended beyond microcanonical conditions, so as to simulate systems in mechanical and thermal contact with their surroundings. As mentioned in Sec. 3, the first work to consider the possibility of performing MD simulations in conditions of constant pressure was that of Andersen [12]. Andersen proposed to couple the dynamics of the atoms with that of the volume, Ω, of the system, in such a way that they would be both described by the following Lagrangian:

$$\mathscr{L}_{Andersen} = \frac{1}{2}\sum_i m_i \Omega^{2/3}\dot{\mathbf{q}}_i^2 - \mathscr{U}(\Omega^{1/3}, \{\mathbf{q}\}) + \frac{1}{2}m_\Omega \dot{\Omega}^2 - P_{\text{ext}}\Omega. \tag{32}$$

The first two terms here are the usual kinetic and potential energy of the atoms, but now rewritten in such a way as to make their dependence on the volume of the system explicit. Note that instead of the usual Cartesian positions for the atoms, $\mathbf{r}_i$, we have now used the scaled, or lattice, coordinates, $\mathbf{q}_i = \Omega^{-1/3}\mathbf{r}_i$, which are more convenient in this case. The third and fourth terms in Eq. (32) above correspond to the kinetic and potential energies for the volume, which is now itself a dynamical variable. m_Ω is the thermostat fictitious mass, and P_{ext} is the external pressure which is exerted on the system. If the volume was to stay fixed, its kinetic energy would be zero (no volume momentum), and the volume potential energy would be constant. In this case $\mathscr{L}_{Andersen}$ reduces to the conventional microcanonical Lagrangian for the atoms. However, when the volume is free to move, it will react to the external pressure, increasing or decreasing as dictated by the pressure and the combined dynamics of atoms and volume. The volume will eventually settle and oscillate around an average value, which will be the mean volume for the imposed external pressure. It is easy to make the transition from the Lagrangian formulation of Eq. (32) to the Hamiltonian form, by simple application of the usual rules of classical mechanics [32], with momenta defined as

$$\mathbf{p}_i = \frac{\partial \mathscr{L}}{\partial \dot{\mathbf{q}}_i} \tag{33}$$

and defining the Hamiltonian function

$$\mathscr{H} = \sum_i \dot{\mathbf{q}}_i \cdot \mathbf{p}_i - \mathscr{L}. \tag{34}$$

It is a useful exercise for the reader to transform Eq. (32) to Hamiltonian form and then use the generalised leapfrog scheme discussed in 4.1 to obtain a constant pressure integration algorithm.

Andersen's approach only considers volume fluctuations, i.e. the size of the simulation box is allowed to change, but its shape is constrained to remain cubic. This is ok for liquids, but for crystalline solids it is actually restrictive. If a undergoes a phase transition to another solid phase, it will in general change not only its cell volume, but also its shape. In order to account for such situations, Parrinello and Rahman [15, 33] generalised the method of Andersen with the following Lagrangian:

$$\mathscr{L}_{PR} = \frac{1}{2}\sum_i m_i \dot{\mathbf{q}}_i \mathbf{H}^t \cdot \mathbf{H}\dot{\mathbf{q}}_i - \mathscr{U}(\mathbf{q},\mathbf{H}) + \frac{1}{2}M_H \mathrm{Tr}(\dot{\mathbf{H}}^t\dot{\mathbf{H}}) - P_{ext}V. \tag{35}$$

This can be seen to be somewhat similar to Andersen's Lagrangian, Eq. (32), though there are some differences. The key difference is that now, instead of the volume, the cell degrees of freedom are the components of the vectors defining the shape of the simulation box. These vectors can be arranged into a matrix which is labelled as $\mathbf{H}$; the Cartesian coordinates of the atoms are $\mathbf{r}_i = \mathbf{H}\mathbf{q}_i$. In Eq. (35) a fictitious kinetic energy term which now includes the kinetic energy of each of the cell vector components. There are several subtleties here, not least the fact that the fictitious dynamics of the cell in the formulation of Parrinello and Rahman allows not only for the deformation of the cell, but also for changes in its orientation in space. This is certainly rather inconvenient in simulations, as one would have to somehow distinguish between the motion of the atoms which is intrinsically due to the atomic dynamics, and that which results from cell rotation. There are cures for this, but we do not need to concern ourselves with such technicalities here [34, 35]. Again, it is possible to transform Eq. (35) to Hamiltonian form, to obtain the equations of motion from the Hamiltonian, and with them derive an algorithm for their numerical integration using the generalised leapfrog scheme. This would be a recommended exercise for anyone who wants to become familiar with MD techniques.

In his famous paper, Andersen [12] proposed not only a way to perform simulations in conditions of constant pressure (i.e. the isobaric-isoenthalpic, or NPH ensemble), but also in conditions of constant temperature (canonical or NVT ensemble). His strategy to attain canonical sampling consisted of selecting an atom at random, and changing its velocity to a new value, selected from the Maxwell-Boltzmann distribution corresponding to the desired temperature of the simulation. This process, repeated at regular intervals, was shown to sample the canonical ensemble, and is now known as Andersen's thermostat. However, there is one key difference between Andersen's procedure for sampling at constant pressure and that for constant temperature, and it is that while the first one is deterministic (i.e. the barostat obeys a certain equation of motion), Andersen's thermostat is stochastic. In this sense it is a bit like introducing ingredients of MC into MD. This is not a bad thing in itself, provided one is not interested in the dynamical properties of the system, such as transport properties. If this is, however, the case, one must be aware of the fact that the stochastic nature of Andersen's thermostat affects the dynamics of the atoms, and will cause an artificially rapid loss of coherence in their dynamics. In other words, the VAF (see 5.4.2) will decay faster than it would otherwise do, and this clearly affects the value of e.g. the diffusion coefficient [see Eq. (19)]. In general, if you are interested in obtaining dynamical information of the studied system, my recommendation would be to avoid the use of thermostats and other artifacts, which may affect the dynamics of the atoms in the system, and to stick to NVE sampling. This is not to say that thermostats and barostats are not useful; far from it! there are many situations in which one is not really interested in the atomic dynamics, and in which one needs to simulate systems in conditions of constant pressure and/or temperature. In such cases the use of these artful devices is highly recommended.

Andersen's paper was the starting point of a plethora of different methods to allow MD to sample other ensembles. Among these, a key development was that of Nosé [13], who proposed a new thermostat to achieve canonical sampling. Contrary to that of Andersen, though, Nosé's thermostat has the peculiarity of being deterministic, instead of stochastic. Nosé proposed the following Lagrangian,

$$\mathscr{L}_{\mathrm{N}} = \frac{1}{2}\sum_i m_i s^2 \dot{\mathbf{r}}_i^2 + \frac{1}{2} m_s \dot{s}^2 - U(\{\mathbf{r}\}) - g k_B T_{\mathrm{ext}} \ln s, \tag{36}$$

where g is the number of degrees of freedom of the system, k_B is Boltzmann's constant, s is the thermostat *position* variable, m_s is its associated fictitious mass, and T_{ext} is the temperature at which we desire the system to be. For once, we will miss the opportunity of proposing an exercise for the reader, and give directly the Hamiltonian function for the Nosé thermostat; this is

$$\mathscr{H}_{\mathrm{N}} = \sum_i \frac{\mathbf{p}_i^2}{2 m_i s^2} + U(\{\mathbf{r}\}) + \frac{p_s^2}{2 m_s} + g k_B T_{\mathrm{ext}} \ln s, \tag{37}$$

where the atomic momenta are $\mathbf{p}_i = m_i s^2 \dot{\mathbf{r}}_i$, and similarly, $p_s = m_s \dot{s}$. I will, however, suggest to the reader to derive equations of motion from Nosé's Hamiltonian Eq. (37). Then, it will be seen that the force acting on the thermostat is proportional to the difference between the kinetic energy of the atoms and the $g k_B T_{\mathrm{ext}}/2$; in words, what this means is that the thermostat variable s is going to increase when the kinetic energy grows to values above that corresponding to T_{ext}, and the opposite will happen when the kinetic energy is below the target value. In this way, one ensures that the average kinetic energy of the system, during a sufficiently long run, will correspond to the correct value at T_{ext}. Not only this, but it can actually be shown that because of the chosen form of the potential energy for the thermostat in Eqs. (36) and (37), the dynamics of the atoms samples the canonical ensemble, under the usual assumption of ergodicity.

The thermostat variable s in Eq. (36) is actually a time-scaling factor. The *real* time of the simulation is actually given by

$$t_{\mathrm{real}} = \int_0^t \frac{dt}{s(t)}, \tag{38}$$

which means that the actual length of the time step varies during the simulation. This is somewhat inconvenient, particularly if one wishes to calculate time dependent properties of the system. Motivated by this, Hoover [14] modified Nosé's original formulation through a change of variables which resulted in a scheme in which the length of the time step is constant. The resulting scheme is frequently referred to in the literature as the Nosé-Hoover thermostat. There is one small caveat with this, though, which is perhaps of little practical significance, but it is worth commenting. Hoover's modifications amounted to a non-canonical (non-Hamiltonian) transformation of Eq. (37), and as a result of this the method looses its Hamiltonian structure. This means that the Nosé-Hoover equations of motion cannot be derived from a Hamiltonian, and as a consequence, one cannot use the generalised leapfrog scheme to obtain an integration algorithm. This is not such a serious problem, because alternative integration schemes can be used which do not rely on having a Hamiltonian structure (see e.g. [4]).

In more recent times Bond *et al.* [36] have shown that it is indeed possible to reformulate the Nosé thermostat by means of a canonical transformation (thus respecting the underlying Hamiltonian structure), a transformation which is designed to counteract the troublesome time scaling implicit in Nosé's original formulation. They did this by using a so-called Poincaré transformation, resulting in a new Hamiltonian given by:

$$\mathscr{H}_{\mathrm{NP}} = s(\mathscr{H}_{\mathrm{N}} - H_0), \tag{39}$$

where H_0 is a constant, and H_{N} is given by Eq. (37). Because this is a Hamiltonian, one can use the generalised leapfrog scheme, and this would be my recommended option for canonical sampling MD.

The use of new variables such as the barostats and thermostats discussed above has been called the *extended system approach*. These extended variables are designed in such a way that they emulate the effect of having the system placed in contact with its surroundings, i.e. with an essentially infinite number of degrees of freedom. It is quite remarkable that one can do this with just a few additional degrees of freedom. But in introducing these artificial variables one must assign values to their corresponding fictitious masses [m_Ω and m_s in Eqs. (32) and (36), respectively]; the dynamics of the extended variables and to some extent that of the atoms of the system do depend on the values chosen for these fictitious masses, and some tuning of their values may be necessary in order to achieve sensible results. The values of average properties are in general not very sensitive to the values of m_Ω and m_s; their choice should be guided by an efficiency of sampling, while at the same time trying to affect minimally the dynamics of the atoms.

Finally, before leaving this section, let us remark that both Andersen [12] and Nosé [13] considered sampling the isothermal-isobaric or NPT ensemble by simultaneously coupling the system to a thermostat and a barostat. This combination is also considered by Sturgeon and Laird [37] and Hernández [35].

8. PROBLEMS, CHALLENGES, ..., AND ALL THAT!

It would not be fair to conclude these introductory notes on MD without making some comments about the limitations of MD, which indeed exist and are not few. The most obvious one is the issue of time scales. Depending on the level at which you model your system (first-principles or empirical force field) MD may be limited to time scales ranging from a few pico-seconds to up to a few nano-seconds at most. Yet many processes of chemical and physical interest happen over time scales which can be many orders of magnitude larger than this (slow diffusion problems in solids, dynamics of glassy or polymer systems, or protein folding, to name but a few), and straightforward

MD simply can get you nowhere in such cases. In recent years, Voter and others have developed several techniques to try to address this problem, such as Hyperdynamics [38, 39], Temperature-accelerated dynamics [40] or the Parallel-replica method [41].

One of the reasons why the time scale that can be covered is limited is related to the computational cost involved in calculating the energy and forces necessary to perform MD simulations. It has been recently suggested [42] that specially designed neural networks may be trained to predict with sufficient accuracy the energetics and forces of a given system after being fed with a sufficient training data set obtained from simulations. This could potentially reduce significantly the cost of performing accurate simulations, and thus also extend the length of time scales accessible with such simulations. However, this methodology is still rather new, and its full potential is yet to be demonstrated.

MD is ultimately a sampling technique, like MC, with the added bonus of providing dynamical information, at the extra cost of calculating the forces. Systems with complicated energy landscapes are inherently more problematic to sample adequately, so special care has to be taken in such cases. In such systems one may have to wait for a long time for the dynamics to explore the configuration space. To ameliorate this problem, Parrinello and co-workers have proposed the technique known as metadynamics [43, 44]. In this technique, a dynamical trajectory is followed which is discouraged from visiting regions of configuration space that have already been visited by adding a Gaussian potential of a pre-specified height and width to each visited point. In this way potential energy minima are gradually filled up, facilitating the escape of the system from such trapping funnels, and improving the configuration space sampling.

Yet another challenge for MD is the issue of varying length-scales. In many systems, the phenomena under observation cover many different length scales. A typical example of this is crack propagation, where a material is loaded (stressed) until a crack tip forms and starts to propagate in the material. Close to the crack tip, chemical bonds are being broken, and atoms strongly rearrange. A bit further away from the tip the material may be severely deformed, but without bonds being actually broken, and yet further away the atomic positions may hardly deviate from those in the perfect crystal. To model such systems directly at the atomistic level requires extremely large simulation cells, soon growing into six orders of magnitude figures and beyond. For tackling such problems effectively it is necessary to treat different length scales at different levels of theory, effectively embedding a quantum mechanical description of the tip crack into a force field description for the atoms a certain distance away from the crack. This in turn must be matched at some point with a continuum mechanics description, valid for large length scales. A similar situation is encountered e.g. in enzymatic reactions, where the active site of the protein and the reactants must be described at a quantum mechanical level, while the rest of the protein and perhaps the solvent (typically water) may be accounted for at a lower level of theory.

To summarise, both challenges and exciting times lie ahead; MD in particular, and simulation in general, are very open fields, in constant evolution, and responding to the new issues which are continuously being raised by experimental progress in the physics and chemistry of materials and by nanotechnology. I have no doubt that very exciting times lie ahead in this field, a field full of opportunities for unfolding a productive and fulfilling career in science.

ACKNOWLEDGEMENTS

Many of the results of MD simulations used here as illustrations of the different aspects of the technique have been conducted in collaboration with or directly by my former students O. Natalia Bedoya-Martínez, Marcin Kaczmarski and Riccardo Rurali. I also wish to thank the organisers of this summer-school. My work is funded by the Spanish Ministry of Science and Innovation through grants FIS2006-12117-C04 and by the AGAUR through grant 2005SGR683.

REFERENCES

1. N. Metropolis, A. W. Rosenbluth, M. N. Rosenbluth, A. H. Teller, and E. Teller, *J. Chem. Phys.* **21**, 1087 (2005).
2. B. J. Alder, and T. E. Wainright, *J. Chem. Phys.* **27**, 1207–1208 (1957).
3. M. P. Allen, and D. J. Tildesley, *Computer Simulation of Liquids*, Clarendon Press, Oxford, 1987.
4. D. Frenkel, and B. Smit, *Understanding Molecular Simulation*, Academic Press, San Diego, 1996.
5. J. M. Thijssen, *Computational Physics*, Cambridge University Press, Cambridge, 1999.
6. D. M. Ceperley, *Rev. Mod. Phys.* **67**, 279–355 (1995).
7. B. R. Brooks, R. E. Bruccoleri, B. D. Olafson, D. J. States, S. Swaminathan, and M. Karplus, *J. Comput. Chem.* **4**, 187–217 (1983).
8. D. A. Case, T. E. Cheatham, T. Darden, H. Gohlke, R. Luo, K. M. Merz, A. Onufriev, C. Simmerling, B. Wang, and R. Woods, *J. Computat. Chem.* **26**, 1668–1688 (2005).
9. G. V. Lewis, and C. R. A. Catlow, *J. Phys. C: Solid State Physics* **18**, 1149–1161 (1985).
10. B. J. Lee, J. H. Shim, and M. I. Baskes, *Phys. Rev. B* **68**, 144112–1,11 (2003).
11. J. Tersoff, *Phys. Rev. B* **37**, 6991–7000 (1988).
12. H. C. Andersen, *J. Chem. Phys.* **72**, 2384–2393 (1980).
13. S. Nosé, *J. Comput. Phys.* **81**, 511–519 (1984).
14. W. G. Hoover, *Phys. Rev. A* **31**, 1695–1697 (1985).
15. M. Parrinello, and A. Rahman, *Phys. Rev. Lett.* **45**, 1196–1199 (1980).
16. R. Car, and M. Parrinello, *Phys. Rev. Lett.* **55**, 2471–2474 (1985).
17. P. Hohenberg, and W. Kohn, *Phys. Rev.* **136**, 864 (1964).
18. W. Kohn, and J. J. Sham, *Phys. Rev.* **140**, 1133 (1965).
19. J. M. Sanz-Serna, and M. P. Calvo, *Numerical Hamiltonian Problems*, Chapman and Hall, New York, 1995.
20. R. M. Martin, *Electronic structure: basic theory and practical methods*, Cambridge University Press, Cambridge, 2004.
21. F. Cleri, and V. Rosato, *Phys. Rev. B* **48**, 22–33 (1993).
22. E. R. Hernández, and J. Íñiguez, *Phys. Rev. Lett.* **98**, 055501–(1,4) (2007).
23. J. P. Hansen, and I. R. McDonald, *Theory of simple liquids*, Elsevier, London, 1986, 2nd edn.
24. O. N. Bedoya-Martínez, M. Kaczmarski, and E. R. Hernández, *J. Phys.: Condens. Matter* **18**, 8049–8062 (2006).
25. F. H. Stillinger, and T. A. Weber, *Phys. Rev. B* **22**, 3790–3794 (1980).
26. C. M. Goringe, D. R. Bowler, and E. R. Hernández, *Rep. Prog. Phys.* **60**, 1447–1512 (1997).
27. L. Colombo, *Nuovo Cimento* **28**, 1–59 (2005).
28. A. P. Sutton, *Electronic Structure of Materials*, Clarendon Press, Oxford, 1993.
29. G. Kresse, and J. Furthmüller, *Phys. Rev. B* **54**, 11169–11186 (1996).
30. M. C. Payne, M. P. Teter, D. C. Allan, T. A. Arias, and J. D. Joannopoulos, *Rev. Mod. Phys.* **64**, 1045–1097 (1992).
31. D. K. Remler, and P. A. Madden, *Mol. Phys.* **70**, 921–966 (1990).
32. H. Goldstein, *Classical Mechanics*, Addison-Wesley, Reading, Massachusetts, 1980.
33. M. Parrinello, and A. Rahman, *J. Appl. Phys.* **52**, 7182–7190 (1981).
34. I. Souza, and J. L. Martins, *Phys. Rev. B* **55**, 8733–8742 (1997).
35. E. Hernández, *J. Chem. Phys.* **115**, 10282–10290 (2003).

36. S. D. Bond, B. J. Leimkuhler, and B. B. Laird, *J. Comput. Phys.* **151**, 114–134 (1999).
37. J. B. Sturgeon, and B. B. Laird, *J. Chem. Phys.* **112**, 3474–3482 (2000).
38. A. F. Voter, *Phys. Rev. Lett.* **78**, 3908–3911 (1997).
39. A. F. Voter, *J. Chem. Phys.* **106**, 4665–4677 (1997).
40. M. R. Sorensen, and A. F. Voter, *J. Chem. Phys.* **112**, 9599–9606 (2000).
41. A. F. Voter, *Phys. Rev. B* **57**, 13985–13988 (1998).
42. J. Behler, and M. Parrinello, *Phys. Rev. Lett.* **98**, 146401–(1,4) (2007).
43. A. Laio, S. Bernard, G. L. Chiarotti, S. Scandolo, and E. Tosatti, *Science* **287**, 1027–1030 (2000).
44. R. Martoňák, A. Laio, and M. Parrinello, *Phys. Rev. Lett.* **90**, 075503–1,4 (2004).

An Alternative Approach to the Problem of Biomolecular Folding

Mauricio D. Carbajal-Tinoco[1]

Departamento de Física, Centro de Investigación y de Estudios Avanzados del IPN, A.P. 14-740, 07000 México D.F., Mexico

Abstract. After a brief overview that is focused on the importance of biological molecules like RNA and proteins, we present a model that can be used to predict the three-dimensional structure of RNA sequences. An appropriate version of our model was first used in the description of polypeptide folding. These coarse-graining models are based on a set of effective pair potentials that were extracted from experimental data. Such interaction potentials are then used as the main input of Monte Carlo simulations which are characterized by requiring a reasonable computer time, in comparison with other approaches. The resulting structures obtained from our method are clearly similar to the experimental ones.

INTRODUCTION

The prediction of the three-dimensional (3D) structure of biomolecules is of great importance in molecular biology and in biophysics. The solution of this problem will be of outstanding value for future applications like the design of new materials or new drugs with very specific purposes. However, any modification of an existing biomolecule or eventually the creation of a new biopolymer will require a deep understanding of them.

The first part of this work consists of a brief overview of certain situations showing that the 3D conformation of biomolecules is crucial to the proper function of living organisms. As a consequence, for more than 30 years, scientists have tested a variety of models intended to predict the spatial structure of biomolecules. The purpose of this work is to present an extension of a model, based on effective pair potentials, to infer the spatial structure of RNA, provided that our model was first utilized in the case of small polypeptides [1, 2, 3].

Let us first start by describing the basic features of biomolecules. In addition to the well studied DNA, proteins and RNA have been recognized as some of the most important biomolecules of life. The connection between these three classes of biopolymers was first stated by Francis Crick in the so-called *central dogma of molecular biology* [4], which describes the transfer of sequence information between them (see Fig. 1). Although the dogma includes all different transfer possibilities, the normal flow of biological information is the following: DNA can be copied to DNA (DNA replication), DNA may be copied into messenger RNA (transcription), and proteins can be synthesized us-

[1] mdct@fis.cinvestav.mx

CP1077, *Advanced Summer School in Physics 2008, Frontiers in Contemporary Physics—EAV08*
edited by L. M. Montaño Zetina, G. Torres Vega, M. García Rocha, L. F. Rojas Ochoa, and R. López Fernández

ing the information in messenger RNA as a template (translation). The translation is the first stage of protein biosynthesis and it occurs in the cytoplasm where the ribosomes are located. Ribosomes are made of a small and large subunit which surrounds the messenger RNA (mRNA). In translation, mRNA is decoded to produce a specific polypeptide according to the rules specified by the genetic code [5]. As the amino acids are linked into the growing peptide chain, they begin folding into the correct conformation. Mature proteins are then released from the ribosome to perform their divers set of functions. For a long time, the general thought was that the genes were like repositories of information about how to build proteins.

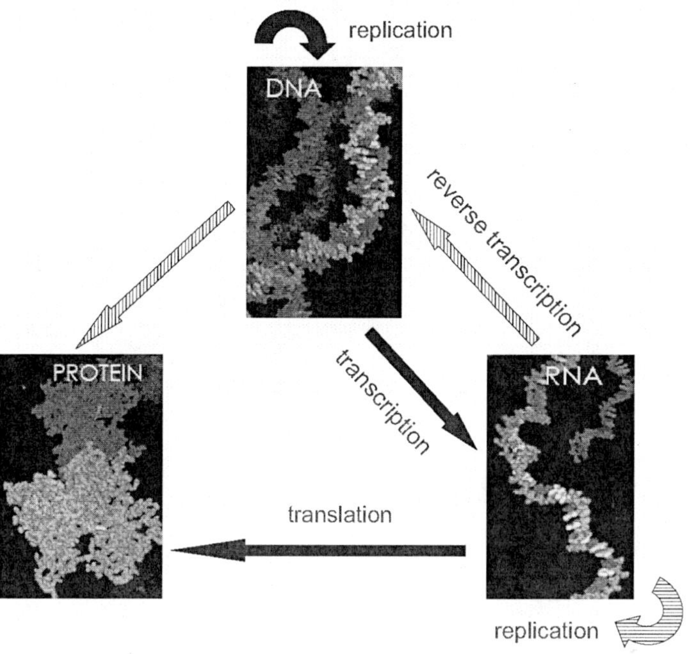

FIGURE 1. Central dogma of molecular biology as stated by F. Crick [4]. Solid arrows are probable events, while dashed arrows indicate possible events.

Although the over mentioned processes have been verified in numerous experiments, this knowledge is far from being the full story of life. For instance, geneticists were puzzled when they found that *C. elegans*, a tiny worm that lives in the soil and lacks a proper brain, has about 20,000 genes, which is almost the same number of genes found in a human being (~21,000). Of course, the genes in question are protein-coding genes. According to recent works on the subject, there is a large number of non-coding (nc) RNA molecules and they may help to explain why some creatures are more complex than others. There are micro RNAs (of about 20 nucleotides) that regulate gene expression. Large ncRNAs perform regulatory roles in certain chromosomes, meanwhile, guide RNAs are RNA genes that function in RNA editing. Another RNA, called XIST, has the power to turn off an entire chromosome [6]. Certain RNA molecules are involved in catalysis activities. In other words, genes are really RNA factories. There may be as high as 37,000 distinct types of RNAs.

As a consequence of the emerging studies on RNA, some groups have established new connections between it and illness. Small RNAs have been linked to many types of cancer [7] or to genetic diseases of the central nervous system. In another case, the expression of micro RNA (miR-107) decreases early in Alzheimer's disease and thus may accelerate the disease progression [8]. Other groups suggest that RNA molecules help the protein that causes Creutzfeldt-Jakob disease to recruit non-infections proteins to join its ranks. It is important to mention that the actual disease consists of misfolded proteins that form an aggregate called amyloid fibril (see Fig. 2).

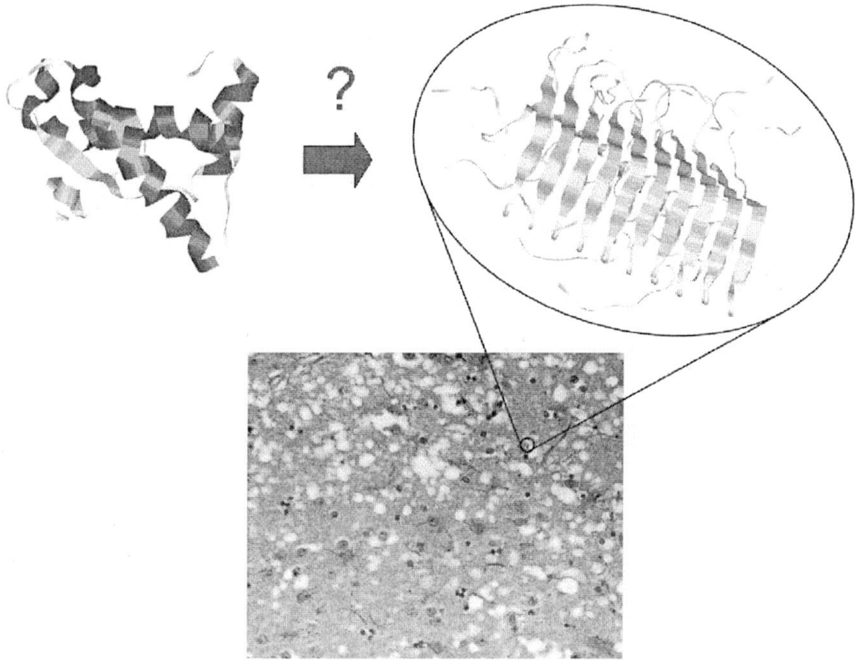

FIGURE 2. A prion (top left) in its native state. As a consequence of a still unknown process, prions sometimes fail to reach their native state and they form aggregates called amyloid fibrils (top right) which ultimately produced voids in the brain, as depicted in the photo.

Thus, it is not hard to deduce the relevance of the knowledge of the structural properties of RNA molecules. However, and in spite of its complexity, the structure of proteins is better understood. In the next sections, some of the existing models of biomolecular folding will be mentioned and our model will be discussed in more detail.

MODELS OF BIOMOLECULAR FOLDING

Proteins are composed of linked amino acids and there are 20 distinct types of amino acids or residues in a polypeptide chain. On the other hand, RNA molecules are made out of 4 different monomers, namely, adenine (A), cytosine (C), guanine (G), and uracil (U), called nucleic acids or nucleotides. In both cases, the composition of the polymer is known as the primary structure or sequence for short. From a linear sequence of residues or nucleotides, a relatively small biopolymer is believed to reach its conformation of minimum free energy, only with the interplay of interatomic forces and the thermal energy $k_B T$ [9, 10] with T being the absolute temperature and k_B is Boltzmann's constant.

The problem of biopolymer folding has been a challenge for several decades. In 1968, Cyrus Levinthal reported a simple calculation that shows the complexity of the problem. The relative position of two consecutive amino acids in a polypeptide chain can be described through the dihedral angles Φ and Ψ [5]. Let us suppose that each one of these angles has three stable positions. Therefore, there are about 10^N conformations for a protein of N residues. For a relatively small protein of 100 residues, a time of $\sim 10^{87}$ seconds would be required to explore all conformations (which is a time greater than the age of the universe). It is clear that proteins do not sample all possible configurations to reach a stable state. Moreover, all known proteins fold in shorter times than a few minutes.

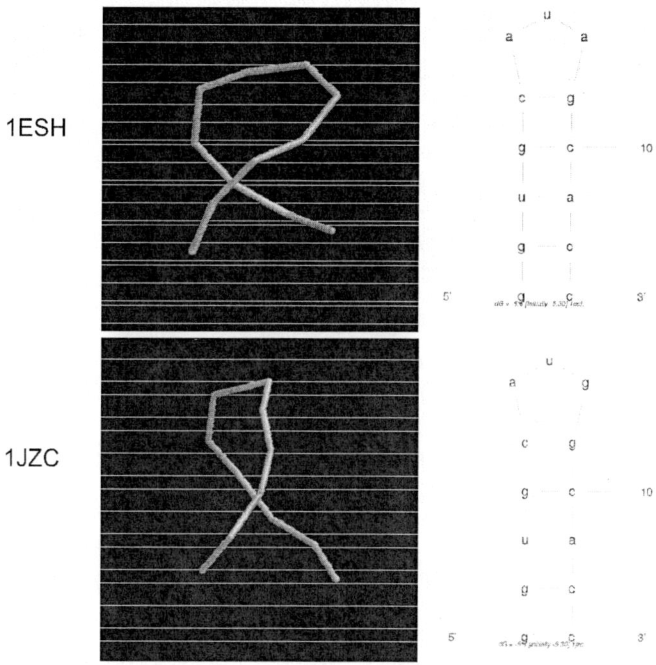

FIGURE 3. NMR experimental structures of two RNA molecules (left), from Refs. [15] (1ESH) and [16] (1JZC). Both molecules only differ by a single nucleotide. The corresponding secondary structures are also drawn [14] (right). It can be noticed that both secondary structures are identical.

Let us first mention that the problem of protein folding has been studied with rather different approaches. On the one hand, Ising-like models allow a full solution since it is possible to enumerate exhaustively all conformations [11]. These models can be enriched with pair contact potentials [12], however, the solution is not directly applicable to the description of actual proteins. On the other hand, *ab initio* methods, like Density Functional Theory, consist on solving the Schrödinger equation. Although these methods are highly accurate, quantum calculations are dominated by the computational cost of the diagonalization of the Hamiltonian matrix [10]. In a further simplification, all-atom methods based on semi-empirical force fields still remain accurate. These methods also require a vast computational effort [13]. Thus, a reasonable approach would be a combination of both simplicity and accuracy in a single model.

In contrast, the knowledge of the 3D structure of RNA is far from being complete. There are highly efficient algorithms that allow the prediction of the secondary structure of RNA [14]. Such algorithms are based on the formation of base pairs (bp). The original set consisting of the Watson-Crick bp (G-C and A-U) which are complemented by other bp like G-U 'wobble'. However, the knowledge of the secondary structure only provides a partial information about the spatial arrangement of RNA molecules, as illustrated in Fig. 3. Numerical simulations of the tertiary structure of RNA are still quite limited since nucleic acids are molecules with a considerable number of atoms in them [10]. To overcome this difficulty, we propose a coarse-graining model which is based on a series of effective pair potentials that are extracted from experimental data.

EFFECTIVE PAIR POTENTIALS

We define an effective pair potential (EPP) as the interaction energy between *particles* that reproduces the pair correlation function describing such *particles* [17]. The EPP can be mediated by other particles that can be observable or even non-observable like small ions or even water molecules. In the literature, similar EPPs have been used to describe other systems like confined colloidal suspensions [18], or the magnetic vortices of type-II superconductors [19]. In the case of proteins, EPPs were introduced in our previous papers [1, 2, 3].

The construction of a knowledge-based model relies on the different types of interactions that are extracted from the crystallographic structures of biomolecules. In the present model of RNA, we are interested in two types of interactions, namely, the bending energy between three consecutive nucleic acids and the distance dependent potential energy for the remaining nucleotides of the chain. In the last case, and considering the possible combinations of the 4 nucleotides, it is necessary to obtain 10 different pair potential functions. All of them are derived from their corresponding radial distribution function $g_{\gamma\mu}(r)$, where the subscripts γ and μ stand for A, C, G or U.

We analyzed a series of 10 distinct crystallographic structures of high molecular weight from the Protein Data Bank. All structures are assumed to be in thermodynamic equilibrium. For instance, in Fig. 4 we show one of them (1FFK), together with the corresponding positions of the centers of mass of guanines and cytosines. Let us first start by computing the bending energy from the averaged properties of the experimental data. The vector associated with the pairs of nucleotides $(i-1,i)$ is $\mathbf{a}_i = \mathbf{r}_i - \mathbf{r}_{i-1}$, and here $\mathbf{r}_i$ is the position of the nucleic acid i. The bending angle θ_i is obtained from the relation $cos(\theta_i) = -\mathbf{a}_i \cdot \mathbf{a}_{i+1}/(|\mathbf{a}_i||\mathbf{a}_{i+1}|)$. In an initial approximation, we computed the bending angle of the different combinations of nucleotides and we found a similar

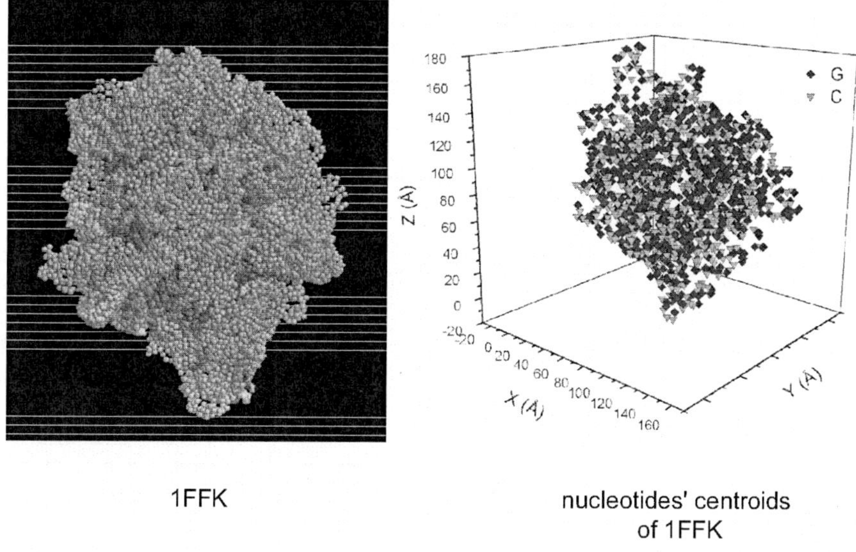

FIGURE 4. A ribosomal RNA (1FFK) is shown in the left figure. The right figure corresponds to the centroids' positions of the guanines (diamonds) and the cytosines (triangles) of 1FFK.

numerical value in all cases, up to experimental errors. The averaged angle of all combinations is then $<\theta>= 134.9°$, with a standard deviation $\sigma_\theta = 28.5°$. Taking into account these values, we propose a harmonic bending energy between nucleotides i and $i+2$, i.e.,

$$\beta v(\theta) = v_\theta \frac{(\theta - <\theta> + \sigma_\theta)(\theta - <\theta> - \sigma_\theta)}{\sigma_\theta^2}, \tag{1}$$

with $\beta = 1/k_B T$ and v_θ is a constant fixed to 0.4.

We also extracted the distance-dependent pair correlation functions between centroids of nucleotides according to the following definition [20],

$$g_{\gamma\mu}(r) = \frac{1}{\rho N \chi_\gamma \chi_\mu} \left\langle \sum_{i=1}^{N_\gamma} \sum_{j=i+3}^{N_\mu} \delta(\mathbf{r} - (\mathbf{r}_i - \mathbf{r}_j)) \right\rangle, \tag{2}$$

The number of particles of species γ (μ) is denoted as N_γ (N_μ) and $N = N_\gamma + N_\mu$. The angular parenthesis denote an ensemble average while $\mathbf{r}_i$ is the position of the centroid of nucleotide i and $\delta(r)$ is Dirac's delta. The number density is $\rho = N/V$ with V being the total volume and $\chi_\gamma = N_\gamma/N$. Eq. (2), however, is valid only in the case of infinite systems. For biomolecules large enough to extract a structural or thermodynamic property, it is possible to take into account the effect of finite size. In the case of nucleotides belonging to a molecule of RNA, let us first define a test sphere of volume $V = 4\pi r_{max}^3/3$ that contains one or two types of nucleic acids of species γ and μ, with at least 50 nucleotides of each type and homogeneously distributed inside the test sphere. For a given distance r, the radial distribution $g_{\gamma\mu}(r)$ is calculated through the equation [3],

$$g_{\gamma\mu}(r) = \frac{1}{\rho\chi_\gamma\chi_\mu}\left(\frac{h'(r)}{N4\pi r^2 dr - N'V_c(r)}\right), \tag{3}$$

with $h'(r)$ being the total number of nucleotides between two concentric spheres of radii r and $r+dr$, about a central one. If the central nucleotide is found inside the sphere of radius $r_{max} - r$ then $V_c(r) = 0$ (region I). For a central residue located outside region I and still inside the test sphere (region II),

$$V_c(r) = \pi\left[(r_{max}^2 - r^2)\ln\left(1 - \frac{r}{r_{max}}\right) + \frac{3}{2}r^2 + rr_{max}\right] dr, \tag{4}$$

where N' is the number of particles found in region II. The function $V_c(r)$ is thus a correction to the effect of finite size.

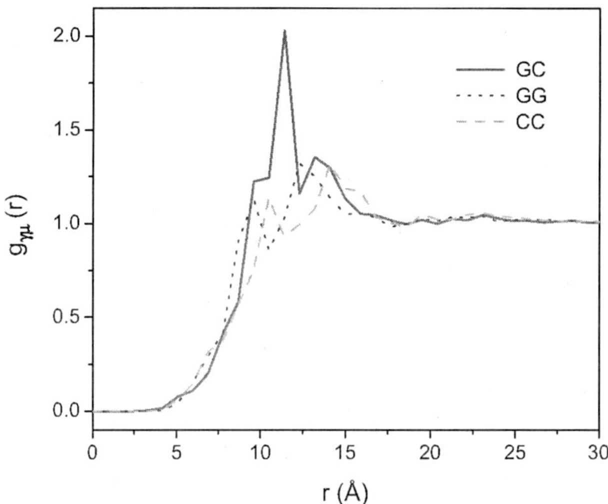

FIGURE 5. Radial distribution functions involving guanines and cytosines, i.e., $g_{GC}(r)$ (continuous line), $g_{CC}(r)$ (dashed line), and $g_{GG}(r)$ (dotted line). These correlation functions were obtained from an average of 10 RNA structures.

For instance, in Fig.5 we plot the three averaged radial distribution functions describing the correlations between guanines and cytosines. In comparison with the basically structureless curves of $g_{CC}(r)$ and $g_{GG}(r)$, the curve corresponding to $g_{GC}(r)$ has a sharp and well defined peak located at $r \sim 11.4$ Å. Let us recall that G and C are related to the formation of a Watson-Crick bp. In other words, the peak is originated by hydrogen bonds (i.e., electrostatic interactions) between the atoms of both nucleotides. Of course, and also in regard to Watson-Crick bp, $g_{AU}(r)$ (not shown) also presents a sharp peak. Otherwise, these pair correlation functions are the basis to estimate the distance-dependent interaction energies.

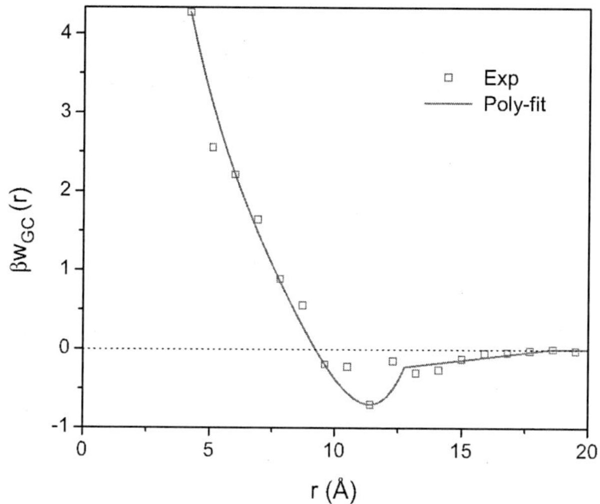

FIGURE 6. Experimental potential between guanines and cytosines (squares) that is obtained from its corresponding radial distribution through Eq. (5). The continuous line is a polynomial fit to the experimental data.

In the case of proteins, we have extracted effective potentials between pairs of amino acids of the same species by means of the Ornstein-Zernike (OZ) equation and a suitable closure relation like the hypernetted chain (HNC) approximation [1, 2]. This method can be extended to the case of amino acids of different species through the multicomponent OZ equations and the HNC closure [3]. It is important to mention that in all cases under study, the effective pair potentials determined in this way were basically identical to their corresponding potential of mean force which is defined as [20]:

$$\beta w_{\gamma\mu}(r) \equiv -ln[g_{\gamma\mu}(r)]. \tag{5}$$

It is well known that this behavior occurs only in the case of dilute systems. Although proteins are relatively dense objects, the density of individual residues is considerably lower.

On the other hand, RNA molecules are relatively extended structures and thus the approximation of Eq. (5) still remains valid. In Fig. 6, we present the interaction potential between guanines and cytosines obtained from Eq. (5) and using the experimental curve of $g_{GC}(r)$. In the same Fig., we also show a polynomial fit to the effective potential. A similar task was performed for the remaining potentials which are the main input of a numerical simulation.

DRESSED POLYMER MODEL

We performed Monte Carlo (MC) simulations of small RNA molecules. Our model of a biopolymer consists of freely rotating rigid sticks of variable sizes (which are determined from the average distances between the actual nucleotides). This polymer backbone is then dressed with both the bending potential energy $\beta v(\theta_{ij})$ (from Eq. (1)) describing the interactions between the pairs of nucleotides $(i, j = i+2)$ and also with the continuous version of the EPPs $\beta w_{\gamma\mu}(r_{ij})$, where $r_{ij} = |\mathbf{r}_i - \mathbf{r}_j|$, for the case $j > i+2$.

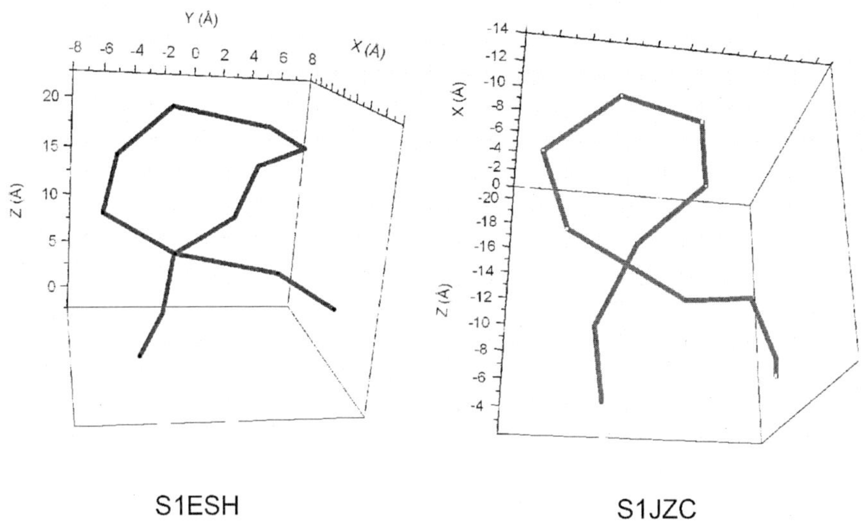

FIGURE 7. Simulated sequences S1ESH (left) and S1JZC (right) obtained within the dressed polymer model, as described in the text.

The basic algorithm is described in the following lines. Each segment is represented by a vector an thus the chain is just the sum of these individual vectors. A trial move consists of randomly rotating an aleatory segment and then reforming the chain. The trial move is accepted with a probability given by [21],

$$P_{acc} = min[1, exp(-\Delta\beta E_{RT})], \tag{6}$$

with $\Delta\beta E_{RT} = \beta E_T - \beta E_R$ is the change in potential energy between the conformations R (reference) and T (test). Our model requires an additional ingredient. We take into account chirality by defining a vector $\mathbf{c}_{i+1} = \mathbf{a}_i \times \mathbf{a}_{i+1}$ and a scalar $s = \mathbf{c}_{i+1} \cdot \mathbf{c}_{i+2}$. Therefore, configurations of the same chirality are distinguished by a positive value of s. It is important to mention that within this model, it is not possible to change the ionic strength or the temperature because these properties depend on the physical conditions of the original experimental data. A realistic change in temperature, for example, would require another set of EPPs.

In Fig. 7, we present the simulated structures S1ESH and S1JZC corresponding to the same sequences of the molecules 1ESH (GGUGCAUAGCACC) and 1JZC (GGUGCAUGGCACC). As it can be observed, although both figures have clear similarities in their general 3D shape, there are also noticeable differences between them. More important, the simulated molecules are at least qualitatively similar to their experimental counterparts (shown in Fig. 3). On the other hand, the internal energies $< \beta E >$ of the equilibrium structures are -16.45 and -16.43 for the sequences S1ESH and S1JZC, respectively.

CONCLUSIONS

Some of the most accurate models of biomolecular folding have an important number of variables which require expensive calculations to determine the native state. In this work, we presented a model that could be an alternative approach to the description of RNA folding, provided that a similar version of it has been tested in the case of proteins. The current version of our model has a reduced number of variables in the sense that there is only a potential curve between each pair of nucleotides. Our model is still able to capture the basic characteristics of such interactions. The resulting structures of Monte Carlo simulations based on EPPs reveal that it is possible to reproduce the most relevant spatial features of biopolymers with a relatively low computational cost. A more general version of our model also includes a variable separation between two consecutive nucleotides as well as dihedral angles [22].

ACKNOWLEDGMENTS

It is a pleasure to thank the organizers for their invitation to participate in the Advanced Summer School 2008. Enlightening discussions with O. Taxilaga-Zetina, P. Pliego-Pastrana, and I. Tinoco Jr. are acknowledged. This work was supported by CONACyT under Grant No. 49486.

REFERENCES

1. P. Pliego-Pastrana and M. D. Carbajal-Tinoco, Phys. Rev. E **68**, 011903 (2003).
2. P. Pliego-Pastrana and M. D. Carbajal-Tinoco, J. Chem. Phys. **122**, 244908 (2005).
3. P. Pliego-Pastrana and M. D. Carbajal-Tinoco, J. Phys. Chem. B, **110**, 24728 (2006).
4. F. H. C. Crick, Symp. Soc. Exp. Biol. XII, 139 (1958).
5. D. Voet and J. G. Voet, *Fundamentals of Biochemistry* (Wiley, New York, 1995).
6. J. Chow, Z. Zen, S. Ziesche, and C. Brown, Annu. Rev. Genomics Hum. Genet. **6**, 69 (2005).
7. L. He *et al.*, Nature **435**, 828 (2005).

8. W. X. Wang *et al.*, J. Neurosci. **28**, 1213 (2008).
9. *Protein Folding*, edited by T. E. Creighton (Freeman, New York, 1992).
10. T. Schlick, *Molecular Modeling and Simulation* (Springer, New York, 2002).
11. E. Shakhnovich and A. Gutin, J. Chem. Phys. **93**, 5967 (1990).
12. S. Miyazawa and R. L. Jernigan, J. Mol. Biol. **256**, 623 (1996).
13. G. Jayachandran, V. Vishal, and V. S. Pande, J. Chem. Phys. **124**, 164902 (2006).
14. M. Zuker, Curr. Opin. Struct. Biol. **10**, 303 (2000).
15. C. H. Kim, C. C. Kao, and I. Tinoco Jr., Nat. Struc. Biol. **7**, 415 (2000).
16. C. H. Kim and C. C. Kao, RNA **7**, 1476 (2001).
17. P. González-Mozuelos and M. D. Carbajal-Tinoco, J. Chem. Phys. **109**, 11074 (1998).
18. M. D. Carbajal-Tinoco, F. Castro-Román, and J. L. Arauz-Lara, Phys. Rev. E **53**, 3745 (1996).
19. C.-H. Sow, K. Harada, A. Tonomura, G. Crabtree, and D. G. Grier, Phys. Rev. Lett. **80**, 2693 (1998).
20. J.-P. Hansen and I. R. McDonald, *Theory of Simple Liquids* (Academic Press, London, 1986).
21. N. Metropolis, A. W. Rosenbluth, M. N. Rosenbluth, A. H. Teller, and E. Teller, J. Chem. Phys. **21**, 1087 (1953).
22. O. Taxilaga-Zetina, M. in Sci. Thesis, Cinvestav (2008).

Faraday waves on finite thickness smectic A liquid crystal and polymer gel materials

C. Ovando-Vazquez, O. Vazquez Rodriguez and M. Hernández-Contreras

Departamento de Física, Centro de Investigación y Estudios Avanzados del Instituto Politécnico Nacional
Apartado Postal 14-740, México Distrito Federal, México

Abstract. We studied with linear stability theory the Faraday waves on the surface of a smectic A liquid crystal and polymer gel-vapor systems of finite thicknesses. Model smectic A material exhibits alternating subharmonic-harmonic patterns of stability curves in a plot of driving acceleration versus wave number. For the case of highly viscoelastic gel media there are coexisting surface modes of harmonic and subharmonic types that correspond to peaks in the plot of the critical acceleration as a function of wave frequency. Larger frequencies lead to subsequent peaks of coexisting subharmonic waves only.

Keywords: complex fluids, fluid dynamics
PACS: 61.30.-v,47.50.Gj,47.20.-k

INTRODUCTION

Parametric waves develop on the surface of a liquid when the vessel that contains it experiences a vertical periodic motion. They constitute excellent prototype systems to study both linear and nonlinear phenomena on dynamical systems. Previous studies have investigated the dynamical evolution of these waves as a function of system material parameters in simple liquids [1], ferrofluids [2], polymer [3, 4] and micellar [5] solutions. Until now similar studies have not been performed in another important class of complex fluids, namely, smectic A liquid crystals. In this paper we demonstrate that there exist subharmonic and harmonic waves in model smectic A media. We also provide evidence for the formation of new surface wave modes on finite thickness polymer gels, for an important experimental range of material parameters. In the case of the gel system, we note that the predicted Faraday waves can be observed with present experimental surface dynamic light scattering techniques.

SMECTIC A LIQUID CRYSTAL

We consider a slab of smectic A media with thickness $h = 5$cm and whose surface is in contact with a vapor. The free energy of surface deformation is given by [6]

$$F = \frac{1}{2}\int dx\big[B(\partial_x\zeta)^2 + K[(\partial_y^2\zeta + \partial_z^2\zeta)^2]\big]. \tag{1}$$

CP1077, *Advanced Summer School in Physics 2008, Frontiers in Contemporary Physics—EAV08*
edited by L. M. Montaño Zetina, G. Torres Vega, M. García Rocha, L. F. Rojas Ochoa, and R. López Fernández

The smectic layers are considered to be parallel to the surface. $\zeta(x,y,t)$ is the deformation shape of the interface due to normal displacements from its equilibrium configuration. The first term in Eq. (1) represents the elastic energy of layer compression with $B = 10^5$dyn/cm^2 the compressibility modulus. The second term describes the layers curvature with $K = 10^{-5}$dyn being the Frank constant. Here $\partial_x \equiv \partial/\partial_x$. To determine the parametric waves we followed the approach given in Ref.[1]. This consists in solving within Floquet's method the Navier-Stokes equation with boundary conditions of vanishing waves at the bottom of the container, and not tangential surface forces, which were derived from the corresponding stress tensor component [6] together with Eq.(1). Elastic vertical restoring force obtained from the stress tensor plus Eq. (1) yields the balance of vertical forces on surface deformations. These were all boundary conditions required. The material layer is subjected to a vertical sinusoidal vibration force $g(t) = g - a\cos(wt)$, where a is the driven acceleration and w the frequency of oscillation with g being the gravitational acceleration in a frame of reference co-moving with the container.

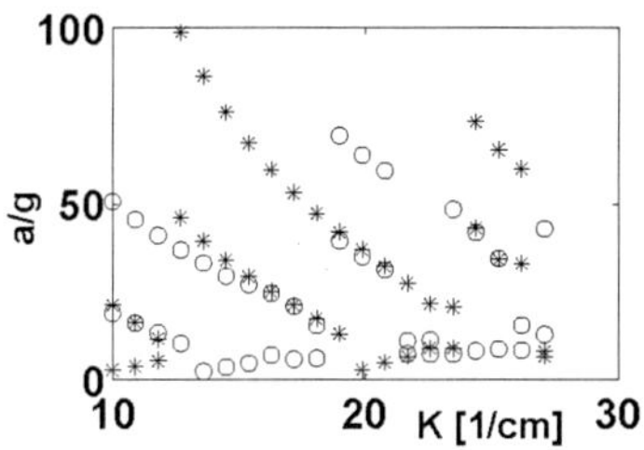

FIGURE 1. Neutral stability curves of Smectic A layer of thickness $h = 5$cm that result from the driving acceleration normalized to gravitational constant a/g, as a function of wave number k and $w = 6\pi/$s. Open circles correspond to the region of subharmonic waves, whereas star symbols to harmonic ones.

Figure 1 depicts the result of our numerical calculation [7] of the Faraday waves in a model of Smectic A liquid crystal of density $\rho = 1$gr/cm^3, permeation length $\lambda_p = 10^{-14}$cm^4/dyn s, viscocity $\eta_3 = 1$P, and in the limit of low frequency $w << B/\eta_3$, $w << k\sqrt{B/\rho}$ (we used $w = 6\pi/$s), and long wave length $\lambda k << 1$, with $\lambda = \sqrt{K/B}$. According to Fig. 1 there are tongue-like regions of subharmonic waves for low wave numbers k, followed by regions of harmonic waves at higher k which have the same driving critical acceleration. We are studying also this type of waves in a model nematic liquid crystal [7].

POLYMER GEL

We consider now a slab of gel media and parametrize the viscoelasticity of its interface by a surface tension γ. Surface adsorption of surfactant molecules provides an interfacial elasticity ε_0. A bending modulus κ determines the curvature elastic surface deformation, while coupling between transverse and lateral compression of the interface is taken into account through the parameter λ [8]. Thus, the elastic free energy of surface deformation is

$$F = \frac{1}{2}\int dx\left[\gamma(\partial_x\zeta)^2 + \varepsilon_0(\partial_x\xi)^2 + \kappa(\partial_x^2\zeta)^2 - 2\lambda(\partial_x\xi)(\partial_x^2\zeta)\right], \tag{2}$$

where the first term describes the coupling of a change in unit area of deformation with a corresponding change in volume and the second term is the curvature free energy. The third term is the free energy associated with fluctuations of the position of molecules in the plane of the interface and is related to the concentration of adsorbed material. $\zeta(x,y,t)$ and $\xi(x,y,t)$ are the deformation shapes of the interface due to normal and lateral displacements from its equilibrium configuration, respectively. The Navier-Stokes equation of a two component fluid model reads

$$\rho\frac{\partial \mathbf{v}}{\partial t} = \nabla\cdot\sigma - \rho g(t)\hat{\mathbf{e}}_{\mathbf{z}}, \tag{3}$$

where $\rho = \rho_s + \rho_p$, with ρ_s, ρ_p being the solvent and polymer densities. $\hat{\mathbf{e}}_{\mathbf{z}}$ is the unit vector in the z axis normal to the surface. Assuming a linear viscoelastic medium, the effective stress tensor is

$$\sigma_{ij} = \eta_s\dot{e}_{ij}(t) + \int_{-\infty}^{t} dt' E(t-t')\dot{e}_{ij}(t') - p\delta_{ij} + \rho g(t)z\delta_{iz}\delta_{jz}, \tag{4}$$

where p is the solvent hydrostatic pressure and η_s the solvent zero-shear viscosity, whereas $E(t)$ is the shear relaxation modulus of the polymer network assumed to be of Maxwell type.

For the boundary conditions at the material-vapor interface we impose normal and shear stress balance. Whereas at the bottom of the container the no-slip boundary condition applies. The surface elevation ζ and its lateral displacement ξ are related to the velocity field through a kinematic surface condition. The solution to Eq. (3) can be expressed according to Floquet's theory as superpositions of time-periodic functions

$$\zeta(t) = e^{(s+i\alpha w)t}\sum_{n=-\infty}^{\infty}\zeta_n e^{inwt}, \tag{5}$$

where s and α are real valued [1]. The surface wave will respond harmonically (H) to the driving frequency when $\alpha = 0$ and subharmonically (S) when $\alpha = 1/2$. Such an expansion was also performed on the other displacement variables, ζ, velocity, and pressure when boundary conditions were used as described in Ref.[1].

From the condition of balance of normal elastic forces we get the recursion relationship for ζ_n

$$S_n\zeta_n = a(\zeta_{n-1} + \zeta_{n+1}), \tag{6}$$

where

$$\begin{aligned} S_n \;=\; & \frac{2}{k}\left\{gk + \frac{\gamma}{\rho}k^3 + \frac{\kappa}{\rho}k^5 + \frac{1}{T_n}\Big[-k^2q_nP_n + q_nQ_n\cosh(q_nh)\cosh(kh)\right. \\ & -kR_n\sinh(q_nh)\sinh(kh) + k^2q_n\frac{\varepsilon_0}{\rho}(q_n\cosh(kh)\sinh(q_nh) \\ & \left.-k\sinh(kh)\cosh(q_nh))\Big]\right\}, \end{aligned} \tag{7}$$

with

$$P_n = 4\nu_n^2(q_n^2+k^2)+\frac{2k^2\lambda}{\rho\mu_n}\left[\frac{k^4\lambda}{\rho\mu_n}-\nu_n(q_n^2+3k^2)\right], \tag{8}$$

$$Q_n = \nu_n^2(q_n^4+2k^2q_n^2+5k^4)+\frac{2k^4\lambda}{\rho\mu_n}\left[\frac{k^4\lambda}{\rho\mu_n}-\nu_n(q_n^2+3k^2)\right], \tag{9}$$

$$R_n = \nu_n^2[q_n^4+6k^2q_n^2+k^4]+\frac{k^4\lambda}{\rho\mu_n}\times\left[\frac{k^2\lambda}{\rho\mu_n}(k^2+q_n^2)-2\nu_n(3q_n^2+k^2)\right], \tag{10}$$

and

$$\begin{aligned} T_n = {} & q_n\cosh(q_nh)\sinh(kh)-k\sinh(q_nh)\cosh(kh)+\frac{k^2\varepsilon_0}{\rho\mu_n^2}[2kq_n \\ & +(k^2+q_n^2)\sinh(q_nh)\sinh(kh)-2kq_n\cosh(q_nh)\cosh(kh)]. \end{aligned} \tag{11}$$

Equation (6) can be rewritten for the harmonic and subharmonic cases separately as

$$S\vec{\zeta} = aU_{\mathrm{H}}\vec{\zeta}, \qquad S\vec{\zeta} = aU_{\mathrm{S}}\vec{\zeta}, \tag{12}$$

where S, U_{H} and U_{S} are real matrices of order $2(N+1)\times 2(N+1)$ and $\vec{\zeta}$ is a vector of order $2(N+1)$.

$$S = \begin{pmatrix} S_0^r & -S_0^i & 0 & 0 & 0 & 0 & \cdots \\ S_0^i & S_0^r & 0 & 0 & 0 & 0 & \cdots \\ 0 & 0 & S_1^r & -S_1^i & 0 & 0 & \cdots \\ 0 & 0 & S_1^i & S_1^r & 0 & 0 & \cdots \\ 0 & 0 & 0 & 0 & S_2^r & -S_2^i & \cdots \\ 0 & 0 & 0 & 0 & S_2^i & S_2^r & \cdots \\ \vdots & \vdots & \vdots & \vdots & \vdots & \vdots & \ddots \end{pmatrix}, \tag{13}$$

and

$$\vec{\zeta} = \begin{pmatrix} \zeta_0^r \\ \zeta_0^i \\ \zeta_1^r \\ \zeta_1^i \\ \zeta_2^r \\ \zeta_2^i \\ \vdots \end{pmatrix}. \tag{14}$$

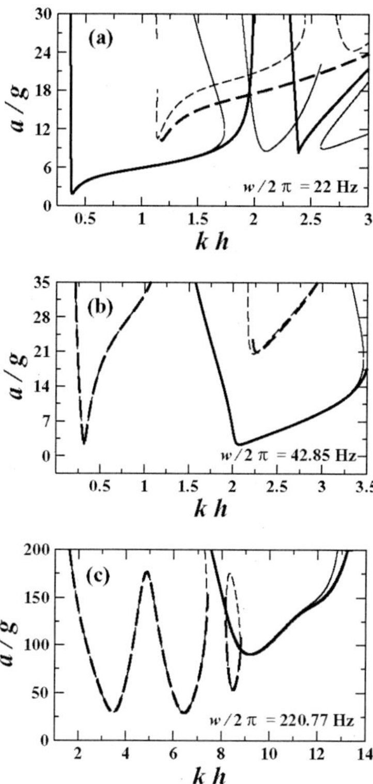

FIGURE 2. Calculated neutral stability curves of Faraday waves on polymer gel of depth $h = 0.42$cm, shear modulus $E = 110$Pa, and density $\rho = 1\text{gr/cm}^3$, at three driving frequencies $w/2\pi$. In each figure, thin continuous line is the full numerical solution of Eqs. (6-11) for harmonic waves, and thick black line is the corresponding analytical approximation given by Eqs.(19-20). Similarly, thin dashed line is a subharmonic wave predicted from the full numerical solution of Eqs. (6-11) while thick dashed line is the analytical result from Eqs.(17-18). Note that the calculated analytical values of a/g have a better agreement with those obtained from the full numerical solution for higher values of driving frequency, and from low up to intermediate values of wave number.

Here the complex numbers $S_n = S_n^r + iS_n^i$, $\zeta_n = \zeta_n^r + i\zeta_n^i$ and

$$U_{\mathrm{H}} = \begin{pmatrix} 0 & 0 & 2 & 0 & 0 & 0 & \dots \\ 0 & 0 & 0 & 0 & 0 & 0 & \dots \\ 1 & 0 & 0 & 0 & 1 & 0 & \dots \\ 0 & 1 & 0 & 0 & 0 & 1 & \dots \\ 0 & 0 & 1 & 0 & 0 & 0 & \dots \\ 0 & 0 & 0 & 1 & 0 & 0 & \dots \\ \vdots & \vdots & \vdots & \vdots & \vdots & \vdots & \ddots \end{pmatrix}, \tag{15}$$

with

$$U_S = \begin{pmatrix} 1 & 0 & 1 & 0 & 0 & 0 & \dots \\ 0 & -1 & 0 & 1 & 0 & 0 & \dots \\ 1 & 0 & 0 & 0 & 1 & 0 & \dots \\ 0 & 1 & 0 & 0 & 0 & 1 & \dots \\ 0 & 0 & 1 & 0 & 0 & 0 & \dots \\ 0 & 0 & 0 & 1 & 0 & 0 & \dots \\ \vdots & \vdots & \vdots & \vdots & \vdots & \vdots & \ddots \end{pmatrix}. \tag{16}$$

Solution of this eigenvalue problem for $s = 0$ yields the neutral stability curve $a(k,w)$ whose minimum provides the critical vibration amplitude a_c for a given wave number k_c corresponding to fixed material parameters of the gel [1].
In the case of layer thickness h larger than the viscous boundary layer $\sqrt{2\eta_s/(\rho w)}$ it is possible to truncate the matrices of Eqs. (12) in order to determine the stability threshold values a_c and k_c. We get the approximated analytical expression for the subharmonic eigenvalue

$$a_S = 2^{-1/2}\sqrt{\beta_S - \sqrt{\beta_S^2 - \sigma_S}}, \tag{17}$$

where

$$\beta_S = |S_1|^2 + 2\Re[S_0 S_1] \qquad , \sigma_S = 4|S_0 S_1|^2, \tag{18}$$

with $\Re[...]$ meaning the real part. While for the harmonic eigenvalue

$$a_H = \sqrt{\beta_H + \sqrt{\beta_H^2 - \sigma_H}}, \tag{19}$$

where

$$\beta_H = \frac{S_1^r |S_2|^2 + S_0^r \Re[S_1 S_2]}{2S_2^r + S_0^r} \quad , \quad \sigma_H = \frac{S_0^r |S_1 S_2|^2}{2S_2^r + S_0^r}. \tag{20}$$

We consider a gel with parameters similar to those of Ref.[9] $\gamma = 72 \times 10^{-3}$ N/m, $E \equiv \eta_p/\rho = 110$Pa, layer thickness $h = 0.42$cm, solvent viscosity $\eta_s = 0.01$Pa, and effective density $\rho = 1\text{gr/cm}^3$. For this high values of the shear modulus E the constitutive equation of the kinematic viscosity reads $\nu_n(w) = \eta_s/\rho + E/[\rho i(\alpha + n)w]$ [3].

In Figure (2) we show the neutral stability curves as obtained from Eqs. (6-11) with $\varepsilon_0 = \kappa = \lambda = 0$, for driving frequencies $w/2\pi = 22$, 42.85 and 220.77Hz, respectively. For $w/2\pi = 22$Hz the dominant surface waves are harmonic (see Fig. 2(a)) and is the first tongue-like stability curve that appears at small k followed by subharmonic waves at higher acceleration a and wave number k. At a higher driven frequency of strength $w/2\pi = 42.85$Hz we show in Fig. 2(b) that a reverse pattern occurs for the stability curve with respect to Fig. 2(a), since now the first waves to be excited are the subharmonic ones (note that $kh \approx 0.25$) followed by the harmonic waves at higher wave number $kh \approx 2$. However, both waves are found to occur at almost the same driving acceleration $a \approx 2.3g$. This occurrence of both types of waves at the same a is the first discontinuity peak displayed by the plot of the critical acceleration and wave number versus frequency curves shown below in Fig. 3.

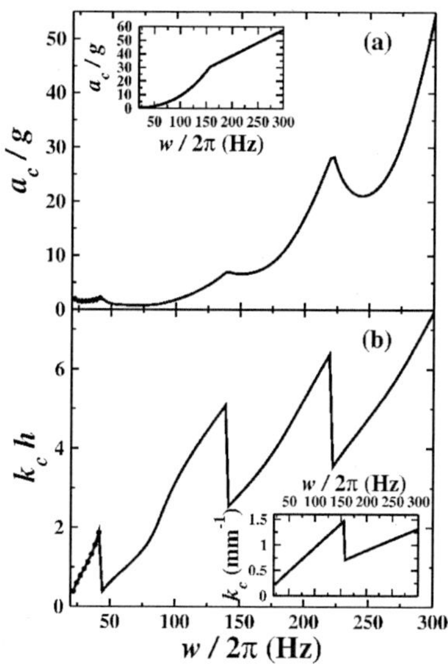

FIGURE 3. Film thickness effect on calculated (a) critical acceleration a_c and (b) critical wave number k_c of Faraday waves versus driving frequency $w/2\pi$ of polymer gel with same material parameters as in Fig. 2. Dot symbols and continuous lines are harmonic and subharmonic waves, respectively. At the first peak with $w/2\pi = 42.85$Hz there is coexistence of both types of waves as observed in Fig. 2(b), whereas the following peaks are subharmonic waves only (see Fig. 2(c)). Insets show subharmonic waves only that are found on a semi-infinite viscoelastic gel obtained at the limit $h \to \infty$.

Such discontinuity is the only one where both types of waves coincide for the same acceleration a. Since at higher frequencies, for instance, $w/2\pi = 220.77$Hz there is coincidence of two subharmonic waves at same acceleration, see Fig. 2(c).

ACKNOWLEDGMENTS

This work was supported by Conacyt Grants # 48794-F, and # 60595, México

REFERENCES

1. K. Kumar, *Proc. R. Soc. Lond. A* **452**, 1113 (1996).
2. H. Müller, *Phys.Rev.E* **58**, 6199 (1998).
3. S. Kumar, *Phys. Fluids* **11**, 1970 (1999).
4. H. Müller, and W. Zimmermann, *Europhys. Lett.* **45**, 169 (1999).
5. P. Ballesta, and S. Manneville, *Phys. Rev. E* **71**, 026308 (2005).
6. P. de Gennes, and J. Prost, *The Physics of Liquid Crystals*, Clarendon, Oxford, 1993.
7. M. Hernández-Contreras, *To be sent to Phys. Rev. E* (2008).
8. D. Buzza, J. Jones, T. Mcleish, and R. Richards, *J. Chem. Phys.* **109**, 5008 (1998).
9. F. Monroy, and D. Langevin, *Phys. Rev. Lett.* **81**, 3167 (1998).

IV. MEDICAL PHYSICS

Phase Contrast Imaging

Ralf Hendrik Menk

Sincrotrone Trieste & INFN Trieste, (Italy)

Abstract. All standard (medical) x-ray imaging technologies, rely primarily on the amplitude properties of the incident radiation, and do not depend on its phase. This is unchanged since the discovery by Röntgen that the intensity of an x-ray beam, as measured by the exposure on a film, was related to the relative transmission properties of an object. However, recently various imaging techniques have emerged which depend on the phase of the x-rays as well as the amplitude. Phase becomes important when the beam is coherent and the imaging system is sensitive to interference phenomena. Significant new advances have been made in coherent optic theory and techniques, which now promise phase information in medical imaging. The development of perfect crystal optics and the increasing availability of synchrotron radiation facilities have contributed to a significant increase in the application of phase based imaging in materials and life sciences. Unique source characteristics such as high intensity, monochromaticity, coherence and high collimating provide an ideal source for advanced imaging. Phase contrast imaging has been applied in both projection and computed tomography modes, and recent applications have been made in the field of medical imaging. Due to the underlying principle of X-ray detection conventional image receptors register only intensities of wave fields and not their phases. During the last decade basically five different methods were developed that translate the phase information into intensity variations. These methods are based on measuring the phase shift φ directly (using interference phenomena), the gradient $\nabla\varphi$, or the Laplacian $\nabla^2 \varphi$. All three methods can be applied to polychromatic X-ray sources keeping in mind that the native source is synchrotron radiation, featuring monochromatic and reasonable coherent X-ray beams. Due to the vast difference in the coefficients that are driven absorption and phase effects (factor 1,000 – 10,000 in the energy range suitable for medical imaging) phase based imaging techniques are inherently extremely sensitive.

INTRODUCTION

X-rays were used in biomedical applications almost immediately after their discovery by Wilhelm Conrad Roentgen in November 1895 [1]. Already in February 1896 the first clinical X-ray images were taken and still nowadays radiography with X-rays plays a major role in diagnostic imaging. Even though the window of visibility was constantly enlarged during the last 110 years, the present limitations of clinical imaging methods arise mostly from insufficient spatial resolution, contrast resolution and quantitative scaling. Synchrotron radiation (SR) sources added a new dimension in diagnosis and have

CP1077, *Advanced Summer School in Physics 2008, Frontiers in Contemporary Physics—EAV08*
edited by L. M. Montaño Zetina, G. Torres Vega, M. García Rocha, L. F. Rojas Ochoa, and R. López Fernández

the potential to provide new solutions and thus to overcome the current limitations [2]. SR offers a large variety of application in several fields such as material science, crystallography, micro-spectroscopy, X-ray diffraction and more recently a variety of medical imaging applications [3, 4].

The main feature of these SR sources is the wide and continuous energy spectrum providing a very high photon flux over an energy range up to some 50 KeV or even higher. Moreover, the beam comprises a high natural collimation at least in the vertical direction and a high degree of coherence in both, space and time. In the horizontal direction one can produce a large area, fan-like photon beam shape. The interested reader might refer to [5] in order to get a full overview of applications of SR.

All these unique features in combination with sophisticated optics make synchrotron sources well-suited instruments, e.g. for medical applications. Thus during the past years several synchrotron radiation facilities have developed dedicated medical beam lines. Those make not only use of the excellent source characteristics but also from fact that the high intensity SR spectrum allows to select and to tune monochromatic photon beams with a very narrow energy bandwidth (figure 1).

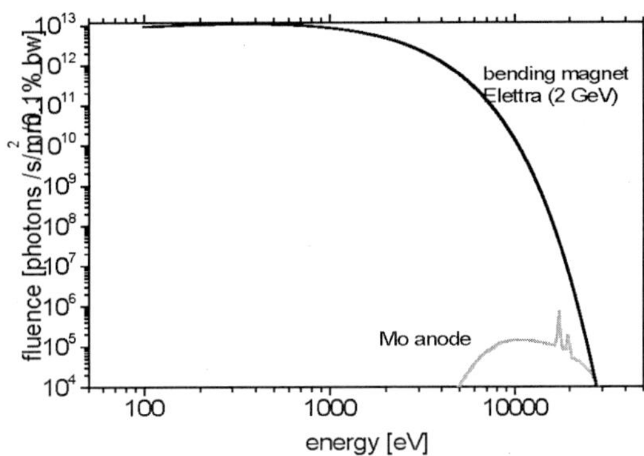

FIGURE 1: Spectral power distribution of the SYRMEP bending magnet beam line at Elettra operated at an energy of 2 GeV and a ring current of 100 mA. As a comparison the spectral power distribution of a mammographic X-ray tube featuring a Mo anode is shown as well [6].

Besides, the laminar geometry reduces scattered radiation and thus avoids image blurring. As a consequence an enhancement of the image quality is observed while the dose is conserved or is even reduced at the same time. This is simply due to the fact that for each application the right energy window can be chosen. Moreover, the monochromaticity can be utilized to implement multiple energy techniques. During the last two decades several systems for medical imaging based on synchrotron radiation have been developed. To mention are here synchrotron-based full field digital mammography, subtraction techniques at the k-edge of contrast agents implemented for intravenous coronary angiography and imaging of bronchi and lungs [7] and multi energy tomography [8].

Since the mid 1990s, new imaging modalities have been emerged providing further improvements in image quality. These techniques are based on the detection of phase information that occurs always when X-rays interact with matter. Phase becomes important to visualize small differences of low absorbing biological materials.

In conventional radiography the image contrast is due to variations in different X-rays absorption of the details within the considered object. This approach may not result in satisfactory image quality especially when weak absorbing (biological) samples are imaged. Taking advantage of the beam spatial coherence, it is possible to utilize imaging techniques, which involve the registration of the phase effects, which are related to a gradient of the real part of the refractive index inside the sample. This is particularly important for samples where small differences in the absorption coefficients result in low image contrast like for instance in mammography. Basically three phase-based methods have been developed: Interferometry, Analyzer Based Imaging (ABI) or Diffraction Enhanced Imaging (DEI) and Phase Contrast Imaging (PCI) also called in-line holography or free wave propagation imaging.

X-ray Interferometer based on perfect crystals can be used to directly measure the phase shifts of wave fields penetrating an object.

In the ABI or DEI approach phase perturbation effects are detected by the use of an analyzer crystal, which is placed between the sample and the detector. Phase perturbation results in scattering of the X-rays. Since the scattering angle is - to the first approximation - proportional to the gradient of the real part of the refractive index, the analyzer effectively converts the behavior of the refractive index inside the sample into intensity differences, which can be recorded by an X-ray detector

The PHC radiography is based on the observation of the interference patterns arising from the wave front phase shift produced by spatial variation of the refractive index in the sample. These variations are usually more relevant at the edges of structures within the sample resulting in an "edge enhancement" effect in the images.

Phase-based X-ray imaging techniques have been strongly pursued for more than ten years and they have been demonstrated to be extremely suitable for biomedical imaging applications; however, they are mainly restricted to synchrotron radiation laboratories.

In the following, we will discuss more in detail the associated effects involved in the interaction between x-rays and matter.

X-ray interaction with matter

From the absorption properties of the human body the interesting energy range of X-rays for diagnostic radiology is between few keV and about 100 KeV. The main contributions to the attenuation of the beam in this energy range are the photoelectric effect, Rayleigh and Compton effect. The former describes X-ray absorption by an atom in which the excess energy is transferred to a photoelectron, which is ejected from the atom itself. The latter refers to elastic and inelastic scattering, respectively.

Macroscopically the absorption is given by the linear attenuation coefficient μ(cm^{-1}); the attenuation of the a monoenergetic beam of intensity $I(z)$ through an infinitesimal thickness dz of material is:

$$-dI = I(z)\mu\, dz \rightarrow \frac{dI}{I(z)} = -\mu\, dz \qquad (1.1)$$

The solution of this differential equation is

$$I(z) = I_0 e^{-\mu z} \qquad (1.2)$$

where I_0 is the incident beam intensity at $z = 0$. For a pure material, the absorption coefficient is directly related to the total atomic absorption cross-section σ_a (cm^2 /atom)

$$\mu = \rho \cdot \frac{N_A}{A} \cdot \sigma_a = \rho_a \cdot \sigma_a \qquad (1.3)$$

where ρ (g/cm^3) is the material density, N_A is Avogadro's number, A is the atomic weight and ρ_a (1/cm^3) is the atomic number density.

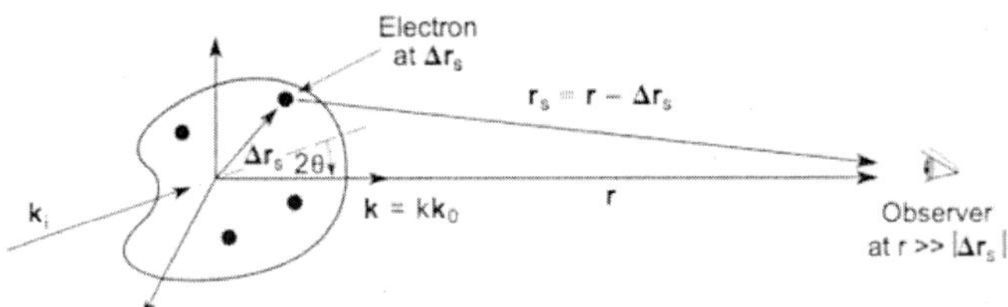

FIGURE 2: Multiple electrons bound to an atom comprising a nucleus with the charge Z at the position $r = 0$. Figure adapted by [9]

The total cross section is the sum of the cross section of all processes involved and it depends on the energy and on the material properties. The photoelectric effect is dominant below some tenth of keV for materials with low atomic number Z and its cross section in approximately proportional to λ^3 (λ radiation wavelength) and to Z^4. Eq. (1.2) is clearly an approximation since in the real case an object is rather a compound of several elements. If radiation crosses different materials along its path the equation can be generalized as:

$$I = I_0 \cdot e^{-\int \mu(z) \cdot dz} \tag{1.4}$$

Microscopically the interaction of X-rays with matter can be described as an oscillatory motion of multiple electrons bound to an atom comprising of a nucleus of the charge +Z at **r** = 0 (figure 2) which is due to the incident electro magnetic wave front described by the wave vector $\mathbf{k}_i$. This motion is caused by the Lorentz force $\boldsymbol{F}$

$$\vec{F} = -e\left(\vec{E}_i + \vec{v}_S \times \vec{B}_i\right) \sim \vec{F} = -e\left(\vec{E}_i\right) \tag{1.5}$$

where $\boldsymbol{E}_i$ is the electric field seen by the ith electron located at the positions $\Delta \boldsymbol{r}_s$. The electron distribution can be quoted as to

$$n\left(\vec{r},t\right) = \sum_{s=1}^{z} \delta\left(\vec{r} - \Delta\vec{r}_s\right) \tag{1.6}$$

where δ is the Dirac function. In a semi classical model we can now formulate the motion equation for each electron as

$$m\frac{d^2 x_S}{dt^2} + m\gamma\frac{dx_S}{dt} + m\omega_s^2 x_S = -e\left(\vec{E}_i\right) \tag{1.7}$$

The acceleration of the electrons is then given by

$$\vec{a_S} = \frac{-\omega_s^2}{\omega^2 - \omega_s^2 + i\gamma\omega}\frac{e\vec{E_i}}{m} e^{-i\left(\omega t - \vec{k}_i \Delta\vec{r}_s\right)} \tag{1.8}$$

and the associated scattered electric field at a distance $\boldsymbol{r}$ summed for all Z electrons, is then

$$E\left(\vec{r},t\right) = -\frac{r_e}{r}\underbrace{\left[\sum_{s=1}^{Z} \frac{\omega^2 e^{-i\Delta\vec{k}\Delta\vec{r_S}}}{\omega^2 - \omega_S^2 + i\gamma\omega}\right]}_{f\left(\Delta\vec{k},\omega\right)} E_i \sin(\Theta) \cdot e^{-i\omega(t - r_S/c)} \tag{1.9}$$

where the quantity $f(\Delta \boldsymbol{k}, \omega)$ is the complex atomic scattering factor, which tells us the scattered electric field due to a multi-electron atom, relative to that of a single free electron. Note the dependence on frequency ω (photon energy $h\omega$), the various resonant frequencies ω_S (resonant energies $h\omega_s$), and the phase terms due to the various positions of electrons within the atom, $\Delta \boldsymbol{k} \Delta \boldsymbol{r}_S$.

$$f\left(\Delta\vec{k},\omega\right)=\sum_{s=1}^{Z}\frac{\omega^2 e^{-i\Delta\vec{k}\Delta\vec{r_S}}}{\omega^2-\omega_S^2+i\gamma\omega} \tag{1.10}$$

In general the $\boldsymbol{\Delta k \Delta r_s}$ phase terms do not simplify, but in two cases they do. Noting that $|\boldsymbol{\Delta k}| = 2k_i \sin\theta = 4\pi a_0/\lambda \sin\theta$, and that the radius of the atom is of order the Bohr radius, a_0, the phase factor is then bounded by

$$\begin{aligned}&\left|\Delta\vec{k}\Delta\vec{r_S}\right|\rightarrow 0 \;\; for\;\; a_0/\lambda \ll 1\\&\left|\Delta\vec{k}\Delta\vec{r_S}\right|\rightarrow 0 \;\; for\;\; \Theta \ll 1\end{aligned} \tag{1.11}$$

In each of these two cases the atomic scattering factor $f(\boldsymbol{\Delta k},\omega)$ reduces to

$$f^0(\omega)=\sum_{s=1}^{Z}\frac{\omega^2}{\omega^2-\omega_S^2+i\gamma\omega}=f_1^0(\omega)-if_2^0(\omega) \tag{1.12}$$

The associated transversal wave equation to the electrical field in equation (1.9) is

$$\left(\frac{\partial^2}{\partial^2 t}-c^2\cdot\vec{\nabla}^2\right)\cdot\vec{E_T}\left(\vec{r},t\right)=-\frac{1}{\varepsilon_0}\left[\frac{\partial\vec{J}_T\left(\vec{r},t\right)}{\partial t}\right] \tag{1.13}$$

For the special case of forward scattering the positions of the electrons within the atom ($\boldsymbol{\Delta k \Delta r_s}$) are irrelevant, as are the positions of the atoms themselves, $n(\boldsymbol{r}, t)$ (equation (1.6)). The contributing current density is then

$$\vec{J}_0\left(\vec{r},t\right)=-e\cdot n_a\sum_{s=1}^{Z}g_S\vec{v}_S\left(\vec{r},t\right) \tag{1.14}$$

where n_a is the average density of atoms, and

$$Z=\sum_{s=1}^{Z}g_S \tag{1.15}$$

The oscillating electron velocities are driven by the incident field $\boldsymbol{E}$

$$\vec{v}\left(\vec{r},t\right)=-\frac{e}{m}\frac{1}{\omega^2-\omega_S^2+i\gamma\omega}\frac{\partial\vec{E}\left(\vec{r},t\right)}{\partial t} \tag{1.16}$$

such that the contributing current density is

$$\vec{J}_0(\vec{r},t) = -\frac{e \cdot n_a}{m} \sum_{s=1}^{Z} \frac{g_s}{\omega^2 - \omega_S^2 + i\gamma\omega} \frac{\partial \vec{E}(\vec{r},t)}{\partial t} \tag{1.17}$$

Substituting this into the transverse wave equation (1.13) one has combining terms with similar operators

$$\left[\left(1 - \frac{e^2 \cdot n_a}{\varepsilon_0 m} \sum_{s=1}^{Z} \frac{g_s}{\omega^2 - \omega_S^2 + i\gamma\omega}\right) \frac{\partial^2}{\partial^2 t} - c^2 \cdot \vec{\nabla}^2\right] \vec{E}_T(\vec{r},t) = 0 \Leftrightarrow \left(\frac{\partial^2}{\partial^2 t} - \frac{c^2}{n^2(\omega)} \cdot \vec{\nabla}^2\right) \cdot \vec{E}_T(\vec{r},t) = 0 \tag{1.18}$$

Comparing the left and right hand equations the frequency dependent refractive index $n(\omega)$ is now identified as to

$$n(\omega) = \sqrt{1 - \frac{e^2 \cdot n_a}{\varepsilon_0 m} \sum_{s=1}^{Z} \frac{g_s}{\omega^2 - \omega_S^2 + i\gamma\omega}} \sim n(\omega) = 1 - \frac{e^2 \cdot n_a}{2\varepsilon_0 m} \sum_{s=1}^{Z} \frac{g_s}{\omega^2 - \omega_S^2 + i\gamma\omega} \tag{1.19}$$

The latter approximation is valid to a high degree of accuracy if ω^2 is very large compared to the quantity $e^2 n_a / \varepsilon_0 m$. Substituting Equation (1.12) into Equation (1.19) we get

$$n(\omega) = 1 - \frac{n_a r_e \lambda^2}{2\pi} \left[f_1^0(\omega) - i f_2^0(\omega)\right] = 1 - \delta + i\beta \tag{1.20}$$

In the energy range for soft tissue imaging, like mammography, δ and β are very small (figure 3) δ being much larger than β, at least 3 order of magnitude. For a wave incident at normal angle to a surface the wave vector changes from k in vacuum to n_k in the medium and if n is a complex number the plane wave propagation in the medium in z direction is

$$E(z) = E_0(z) \cdot e^{i \cdot n \cdot k \cdot z} = E_0 \cdot e^{i \cdot (1 - \delta + i \cdot \beta) \cdot k \cdot z} = E_0 \cdot e^{i \cdot (1 - \delta) \cdot k \cdot z} e^{-\beta \cdot k \cdot z} \tag{1.21}$$

The real part δ of the refractive index results in a phase shift of the wave, while the imaginary part is called absorption index and it is linked to the attenuation coefficient μ. More precisely the intensity attenuation described by μ is related not to the amplitude of the electric field but to its intensity, which is the given by

$$I(z) = |E(z)|^2 = E_0^2 \cdot e^{-2\cdot\beta\cdot k\cdot x} = I_0 \cdot e^{-2\cdot\beta\cdot k\cdot x} \tag{1.22}$$

From equation (1.2) one can note the relationship between μ and β is

$$\beta = \frac{n_a r_e \lambda^2}{2\pi} f_1^0(\omega) \tag{1.23}$$

while δ is given by

$$\delta = \frac{n_a r_e \lambda^2}{2\pi} f_1^0(\omega) \tag{1.24}$$

These equations reveal that real and imaginary parts have very different dependence on the photon energy. In the regime where the photoelectric effect dominates absorption β is approximately proportional to λ^4 while δ is proportional to λ^2 except in the vicinity of absorption edges.

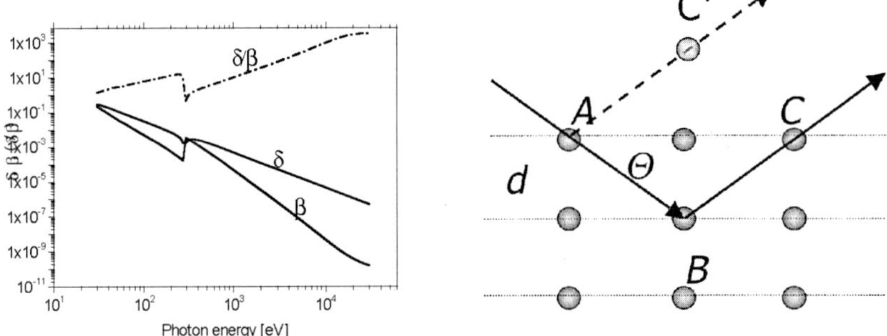

FIGURE 3: Left: Coefficients δ and β of the complex refractive index versus photon energy. Moreover the ration δ/β is shown as well. Right : Scattering of X-rays on multi electron atoms.

Bragg's Law

Bragg's Law derived by the English physicists Sir W.H. Bragg and his son Sir W.L. Bragg in 1913. It explains why the cleavage faces of crystals appear to reflect X-ray beams at certain angles of incidence Θ (figure 3). The variable d is the distance between atomic layers in a crystal, λ is the wavelength of the incident X-ray beam and n is an integer. With respect of the previous chapter Bragg's law can be derived calculating the scatter fields of multi electron atoms.

Braggs' observation was direct evidence for the periodic atomic structure of crystals postulated for several centuries. The Braggs were awarded the Nobel Prize in physics in 1915 for their work in determining crystal structures beginning with NaCl, ZnS and diamond. Although Bragg's law was used to explain the interference pattern of X-rays scattered by crystals, diffraction has been developed to study the structure of all states of matter with any beam, e.g., ions, electrons, neutrons, and protons, with a wavelength similar to the distance between the atomic or molecular structures of interest.

Under the Bragg condition two coherent waves are created inside the crystal resulting in a so-called standing wave field perpendicular to the reflecting atomic planes. There will be a path difference between the 'ray' that gets reflected along AC' and the ray that gets transmitted, then reflected along AB and BC paths respectively. This path difference is: (AB+BC). If this path difference is equal to any integer value of the wavelength then the two separate waves will arrive at a point with the same phase, and hence undergo constructive interference.

$$\begin{aligned}
&(AB+BC)-(AC')=n\lambda\\
&AC'=AC\cos(\Theta)=\frac{2d}{\tan(\Theta)}\cos(\Theta)\\
&n\lambda=\frac{2d}{\sin(\Theta)}-\frac{2d}{\tan(\Theta)}\cos(\Theta)\\
&n\lambda=\frac{2d}{\sin(\Theta)}\left(1-\cos^2(\Theta)\right)\\
&n\lambda=\frac{2d}{\sin(\Theta)}\sin^2(\Theta)\\
&n\lambda=2d\sin(\Theta)
\end{aligned} \tag{1.25}$$

Due to the interference between the incident and the reflected beam the amplitude of the standing wave inside the crystal falls off exponentially and eventually is extinct. In case of a symmetrical Bragg reflection the extinction thickness L_{ex} is given by

$$L_{ex}=\frac{V\cdot\sin(\Theta)}{r_e\cdot\lambda\cdot P\cdot F} \tag{1.26}$$

where V is the volume of the unit cell, P the polarization and F the structure factor describing dispersion effects. Since $sin(\Theta)/\lambda$ is constant for a given reflection L_{ex} can be considered constant at least as long F is small. Of note is that typically L_{ex} is much smaller than the absorption length and is observed only in a small band of incident angles of incident. This narrow angular range is called Darwin width of the reflection and can be expressed by:

$$w_D=\frac{\lambda}{\pi\cdot L_{ex}\cdot\cos(\Theta)} \tag{1.27}$$

From equation (1.27) it becomes obvious that the Darwin width is in the first order approximation is proportional to the wavelength of the incident radiation and subsequently inverse proportional to the energy (Table 1). Depicted in figure 4 is the Darwin curve of a single Si (111) crystal reflecting a monochromatic x-ray of 17 keV. Shown as well is the rocking curve of a second Si (111) crystal, which would reflect the monochromatic x-ray, generated by the first crystal. Note that the rocking curve is in this case a convolution of the two Darwin curves.

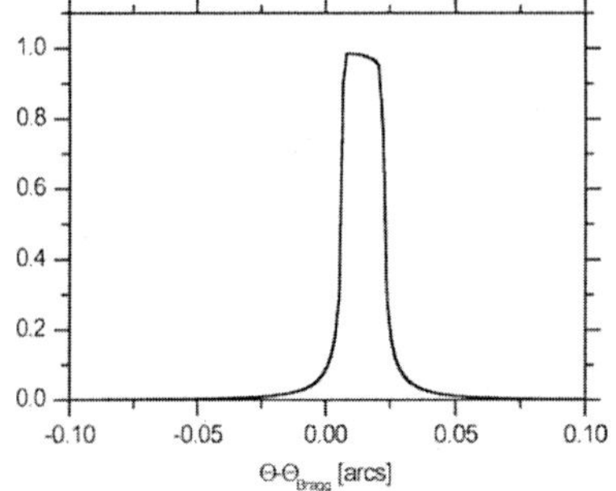

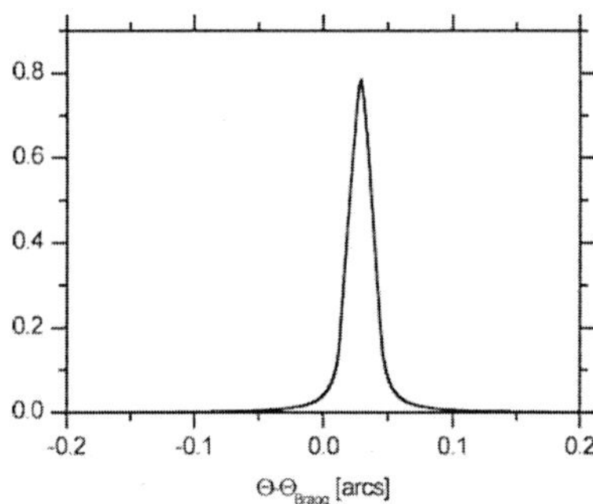

FIGURE 4: Darwin curve (left) and rocking curve (right) of a Si (111) crystal for a photon energy of 17 keV in arbitrary units.

In general perfect crystals are utilized to monochromize for example synchrotron radiation or x-rays generated by an x-ray tube. As discussed before and we will see later perfect crystals are also used in radiography to analyze scattered radiation emerging from the sample/patient.

Table 1: Extinction thickness and Darwin width for lower order reflections of Si at energies typically used in imaging with synchrotron radiation.

			W_d				
Hkl	L_{ex}	E(KeV)	15	20	25	30	35
111	1.543		17.20	12.85	10.26	8.54	7.32
220	2.201		12.24	9.09	7.23	6.01	5.15
311	3.951		6.88	5.09	4.04	3.36	2.87
400	3.751		7.37	5.40	4.28	3.55	3.03
331	6.113		4.56	3.33	2.64	2.18	1.86
422	5.337		5.31	3.85	3.04	2.51	2.14
333	8.441		3.39	2.48	1.93	1.59	1.36
440	7.102		4.11	2.94	2.30	1.90	1.62

Conventional absorption imaging and radiation dose

Let us consider in the following an ideal absorption measurement (figure 3) of monoenergetic X-rays of the energy E_γ, where the total attenuation of the direct beam is observed and all scattering is rejected (i.e. in the so called "good geometry condition). Main focus now is to visualize a detail of size w characterized by its absorption coefficient μ_W embedded in a background of thickness L and characterized by the absorption coefficient μ_T through an intensity modulation on a suitable detection device (see figure 5).

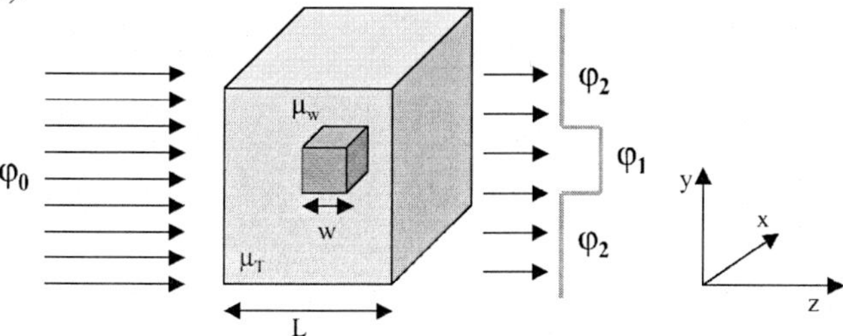

FIGURE 5: see text.

If φ_0 denotes the average photon fluence (photons/area) in front of the sample the detection device will record φ_1 photon fluence behind the detail, according to equation (1.28)

$$\varphi_1 = \varphi_0 \cdot e^{-\mu_T \cdot L} e^{-(\mu_w - \mu_T)\cdot w} \tag{1.29}$$

and φ_2 the average photon fluence in the region outside the detail.

$$\varphi_2 = \varphi_0 \cdot e^{-\mu_T \cdot L} \tag{1.30}$$

The contrast in the image is define as

$$C = \frac{N_2 - N_1}{N_2} = \frac{\varphi_2 - \varphi_1}{\varphi_2} = 1 - e^{-(\mu_W - \mu_T)\cdot w} \approx (\mu_W - \mu_T)\cdot w \tag{1.31}$$

if N_1 and N_2 denotes the total number of photons of the area w^2 behind the detail and the background, respectively. This approximation is valid only in cases of very small details or very small differences in the absorption coefficients. From equation (1.31) one can note that the object contrast results independent from the photon fluence, but depends only on the absorption properties of the sample.

The signal to noise ratio (SNR_{in}) for the detail of area w^2 in presence of Poisson noise only can be define as

$$SNR_{in} = \frac{(N_2 - N_1)}{\sqrt{N_2}} = w \cdot \frac{(\varphi_2 - \varphi_1)}{\sqrt{\varphi_2}} = w^2 \cdot \sqrt{\varphi_0} \cdot e^{-\frac{\mu_T}{2}\cdot L} (\mu_w - \mu_T) \tag{1.32}$$

It should be noted that the *SNR* depends on the photon statistics and scales with the square root of the photon number.

Equation (1.32) represents the inherent limit of the signal to noise due to Poisson statistics without considering the detection device. In the real case a detector with some inherent properties will be used to register the image. The signal to noise measured by the detection device (SNR_{out}) is coupled through the so-called detective quantum efficiency (*DQE*) as [10]

$$SNR^2_{out} = DQE \cdot SNR^2_{in} \tag{1.33}$$

Inserting equation (1.33) into equation (1.32) leads to

$$\varphi_0 = \frac{SNR^2_{out} \cdot e^{-\mu_T \cdot L}}{w^4 \cdot (\mu_w - \mu_T)^2 \cdot DQE} \tag{1.34}$$

The absorbed dose D is defined as the energy ΔE deposited by ionizing radiation per unit mass of material Δm. For monoenergetic X-ray photons with energy E the dose is related to the mass absorption coefficient μ_{en}/ρ as

$$D = \frac{\Delta E}{\Delta m} = \varphi_0 \cdot E \cdot \left(\frac{\mu_{en}}{\rho}\right) \tag{1.35}$$

The absorption coefficient describes the fraction the of the mass attenuation coefficient (μ/ρ) that contribute to the energy deposition in a given material. Typically in diagnostic radiology the entrance dose in air, i.e. the energy deposition in front of the tissue, is calculated considering in equation (1.35) the mass attenuation coefficient for air $(\mu_{en}/\rho)_{air}$. Inserting equation (1.34) into equation (1.35) one gets

$$D_{air} = \frac{SNR^2_{out} \cdot e^{\mu_T \cdot L} \cdot E}{w^4 \cdot (\mu_w - \mu_T)^2 \cdot DQE} \cdot \left(\frac{\mu_{en}}{\rho}\right)_{air} \tag{1.36}$$

Of note is that the entrance dose increases with the fourth power of the detail size w and increases exponentially with the thickness L of the tissue. Therefore in applications where dose restrictions are stringent such as mammography, compression of tissue during exposure is necessary.

As mentioned above the results so far are valid only in good geometry conditions and thus in absence of scattering. In the real case the intensity reaching the detector is the superposition of the transmitted radiation and secondary radiation i.e. scattering that can be consider in average a constant function S all over the image. In this case equation (1.31) can be rewritten as

$$C = \frac{(N_2 + S) - (N_1 + S)}{N_2 + S} = \frac{N_2 - N_1}{N_2 + S} = \frac{1 - e^{-(\mu_W - \mu_T)\cdot w}}{1 + R} \tag{1.37}$$

where R is the ratio S/N_2. As one can note the presence of scattering will decrease the contrast.

The expression found in equation (1.36) for the entrance dose limit is also valid for computed tomography (CT) with X-rays. CT refers to the cross-sectional imaging of an object from either transmission or reflection data collected by illuminating the object from many different directions. Thus it allows the two (or three-) dimensional reconstruction of the mass absorption coefficient $\mu(x,y)$ from measuring projections $P(\rho,\theta)$. The impact of this technique in diagnostic medicine has been revolutionary, since it has enabled doctors to view internal organs with unprecedented precision and safety to the patient. The first medical application utilized X-rays for forming images of tissues based on their X-ray attenuation coefficient. Although, from a purely mathematical standpoint, the solution to the problem of how to reconstruct a function from its projections dates back to the paper by Radon in 1917, the current excitement in tomographic imaging originated with Hounsfield's invention of the X-ray computed tomographic scanner [11] for which he received a Nobel Prize in 1972. He shared the prize with Allan Cormack who independently discovered some of the algorithms [12]. His invention showed that it is possible to compute high-quality cross-sectional images with an accuracy now reaching one part in a thousand in spite of the fact that the projection data do not strictly satisfy the theoretical models underlying the efficiently implementable reconstruction algorithms. His invention also showed that it is possible to process a very large number of measurements (now approaching a million for the case of X-ray tomography) with fairly complex mathematical operations, and still get an image that is incredibly accurate. In the following a brief mathematical description to CT will be presented. Let us consider the simplest case of a parallel beam of monoenergetic X-rays as depicted in figure 6.
According to equation (1.4) $P(\theta,\rho)$ can be expressed as the line integral along the ray S

$$\begin{aligned} P(\rho,\theta) &= \int_{-\infty}^{\infty} \mu(x,y)\cdot dS = \\ &= \int_{-\infty}^{\infty} \mu\big(\rho\cdot\cos(\theta) - S\cdot\sin(\theta), \rho\cdot\sin(\theta) + S\cdot\cos(\theta)\big)\cdot dS \end{aligned} \tag{1.38}$$

Each point of $P(\theta,\rho)$ is also called a ray-sum, while the resulting image is called a sinogram.

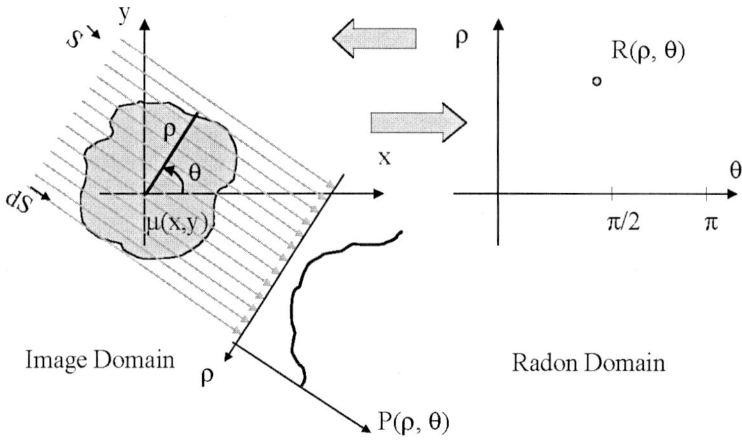

FIGURE 6: Basic concept of parallel beam tomography.

Now $\mu(x,y)$ can be reconstructed from its ray-sums using the so-called back projection operator:

$$B(x,y)=\int_0^\pi P\left(\theta, x\cdot\cos(\theta)+y\cdot\sin(\theta)\right)\cdot d\theta \tag{1.39}$$

where the output, $B(x,y)$, is a map of $\mu(x,y)$ blurred by the Radon transform.

This image blurring, which is due to an overestimation of each point $B(\theta,\rho)$ can be reduced by filtering $P(\theta,\rho)$ with appropriate filter function prior back projection (figure 7).

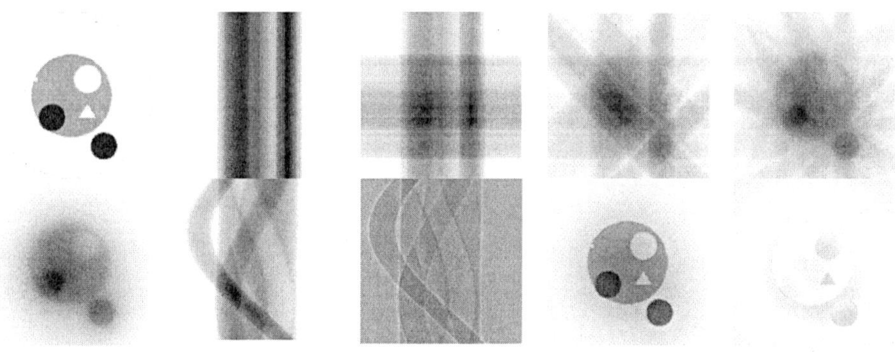

FIGURE 7: Simple back projection of a test object (most right) is shown increasing from the left to the right the numbers of projection data used (1,2,3,4 &128). The latter shows the characteristic 1/ρ blurring. With filtered back projection the raw projection data are convolved with a convolution kernel and the resulting projection data are used in the back projection process.

Phase Imaging

As seen before the absolute values of δ and β are very small for X-ray interacting with light materials, like for instance biological samples. However, δ is typically some order of magnitude larger than β; for instance for energies suitable for mammography (15-25 keV) for soft tissue β is in the order of 10^{-8} – 10^{-10}, while δ is in the order of 10^{-6} – 10^{-7}. The much greater magnitude of δ over β means that the phase shift of an X-ray beam propagating through a tissue can be much larger (2-3 order of magnitude) than the intensity change.

Assuming that the X-rays are propagating in the z-direction with wave vector k when an X-ray wave traverses a sample over a distance $\Delta z=(z_1-z_0)$ the beam is exponentially attenuated by $\mu\cdot\Delta z$ and its phase is changed by ϕ the electric field can be written as

$$E(z)=E_0(z)\cdot e^{-\mu\Delta z}\cdot e^{i\cdot(k\cdot z+\phi)} \tag{1.40}$$

where ϕ is given by

$$\phi(x,y)=-\left(\frac{2\cdot\pi}{\lambda}\right)\cdot\int_{z_0}^{z_1}\delta(x,y,z)\cdot dz=-r_e\cdot\lambda\cdot\int_{z_0}^{z_1}\rho_e(x,y,z)\cdot dz \tag{1.41}$$

where λ is the wavelength of the X-rays, r_e the classical electron radius, ρ_e the electron density. For example an organic sample of Δz = 50μm absorbs at 20 keV X-ray energy (δ = $3\cdot10^{-7}$ and $\beta = 1\cdot10^{-10}$) about 0.2% of the incident radiation but generates a phase shift of π which can produce complete extinction of the wave in case of interference. The local propagation vector $\vec{k}'(x,y,z)$, normal to the wavefront at the point (x,y,z), depends on the gradient of the phase and can be written in the paraxial approximation, $|\nabla_{xy}\phi| << k$, according to [13,14] as

$$\vec{k}'(x,y,z)\approx\left(\frac{\partial\phi}{\partial x},\frac{\partial\phi}{\partial y},k\right)=\nabla_{x,y}\phi(x,y)+k\hat{z} \tag{1.42}$$

The angular deviation of the wave front from the initial direction can be express as

$$\Delta\alpha(x,y)\approx\frac{1}{k}\left|\nabla_{x,y}\phi(x,y)\right|=\left|\nabla_{x,y}\int_{z_0}^{z_1}\delta(x,y,z)\cdot dz\right| \tag{1.43}$$

Note that the angle is proportional to the gradient of the real part of the refractive index and subsequently it is proportional to ρ_e. Due the principle of X-ray detection conventional image receptors register only intensities of wave fields and not their phases. Therefore, the phase information has to be translated into intensity variations. In the last decade several methods were developed to exploit phase information in this fashion. These methods are interferomerty [15], free propagation based imaging (PBI) [13, 16] and crystal analyzer based imaging methods (ABI) [17, 18] and they are based on measuring the phase shift ϕ directly, the gradient $\nabla\phi$, or the Laplacian $\nabla^2\phi$, respectively [19]. Both, PBI and ABI are routinely used in *in-vitro* medical applications practically on all organs [4] and very recently in clinical mammography [20].

Interferometer

X-ray interferometer retrieves information about the phase directly. A typical X-ray interferometer made of a perfect monolithic silicon crystal is shown in figure 8. It utilizes transmission-type (Laue-case) diffraction by crystal lamellae, which are used to divide and combine the X-ray wave field coherently. Such a configuration featuring three lamellae at regular intervals is called a triple Laue-case (LLL) interferometer. The first lamella divides the incident monochromatic synchrotron radiation into two beams, which are inclined by $2\cdot\theta_B$ where θ_B is the Bragg diffraction angle. Further downstream these two beams are spatially separated at the center lamella and subsequently divided again in two beams each. Those two beams that are propagating inside the X-ray interferometer overlap coherently at the third lamella. Each beam is again divided into two, and interference occurs in the outgoing beams. This configuration corresponds to the Mach - Zehnder interferometer. Rotating a slap of PMMA that intercept one of the diffracted beams it is possible to tune the phase shift (Phase shifter) und thus balance the phase shift introduced by the sample

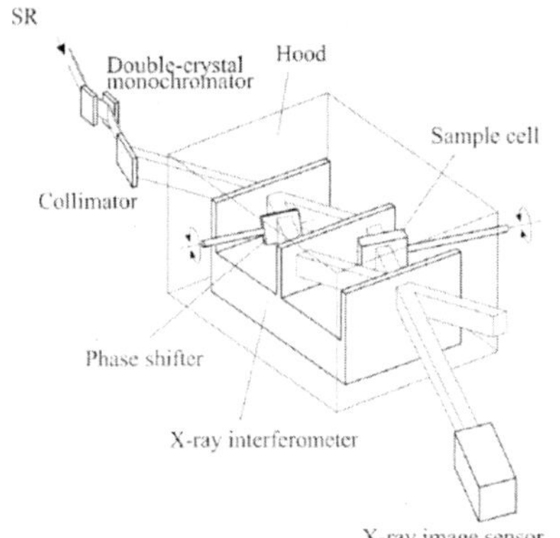

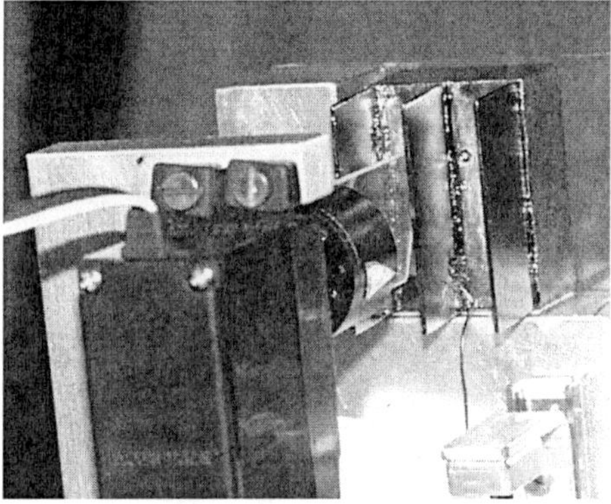

FIGURE 8: Left: Basic setup of a LLL X-ray interferometer used in synchrotron radiation experiments. Right: Instrument built by the University of Torino used at the medical beamline at Elettra.

An interference pattern $I(r_\perp)$ generated by the X-ray interferometer is given in general by

$$I(r_\perp) = a(r_\perp) + b(r_\perp) \cdot \cos(\varphi(r_\perp) + \Delta(r_\perp)) \tag{1.44}$$

where $a(r_\perp)$ and $b(r_\perp)$ are the average intensity and fringe amplitude, and $\varphi(r_\perp)$ and $\Delta(r_\perp)$ are the phase shift caused by a sample given by equation (1.41) and the inherent phase background owing to the imperfection of the X-ray interferometer. Thus, even when no sample is placed in the beam ($\varphi(r_\perp) = 0$), interference fringes (built-in Moire fringes) appear normally.

Utilizing the phase shifter at k different but equally spaces positions ($2\cdot\pi/k$) one obtains k different interference patterns given by

$$I_j(r_\perp) = a(r_\perp) + b(r_\perp) \cdot \cos\left(\varphi(r_\perp) + \Delta(r_\perp) + \frac{2\cdot\pi\cdot j}{k}\right) \tag{1.45}$$

($j = 1,2,..k$) which allow directly phase retrieval when summing with a weight of $exp(2\cdot\pi\cdot i\cdot j/k)$.

$$S(r_\perp) = \sum_{j=1}^{k} I_j(r_\perp) \cdot e^{\left(\frac{-2\cdot\pi\cdot i\cdot j}{k}\right)} = \frac{1}{2} \cdot k \cdot b(r_\perp) \cdot e^{i\cdot(\varphi(r_\perp)+\Delta(r_\perp))} \tag{1.46}$$

for $k > 3$. Thus $\varphi(r_\perp) + \Delta(r_\perp)$ is given by

$$\varphi(r_\perp) + \Delta(r_\perp) = \tan^{-1}\left(\frac{\mathrm{Im}[S(r_\perp)]}{\mathrm{Re}[S(r_\perp)]}\right) \tag{1.47}$$

Since $\Delta(r_\perp)$ can be determine by a separate measurement without sample present (a so-called flat field image) equation (1.47) can be solved. Amplitude information is represented by $b(r_\perp)$. The values of $\varphi(r_\perp)$ in equations (1.47) ranges from $[-\pi, \pi]$. Therefore, when $\varphi(r_\perp)$ exceeds jumps of 2π the grey values in the image restart from zero. Such an image is called a wrapped phase map (Figure 9) and phase unwrapping implies compensating the jumps to obtain a real phase map. When the phase shift is varying smoothly, the phase jump is easily detected because a phase difference near 2π is found between neighboring pixels. The jump can be removed by repeating addition or subtraction of 2π to or from one of the neighboring pixels. However, when the signal-to-noise ratio of an image is low or when abrupt phase variation exists, one fails in phase unwrapping occasionally because the phase jump becomes partially unclear. To overcome this problem, the so-called cut-line algorithm [21] is used. The cut-line algorithm detects unclear jumps automatically and sets cut-lines there to inhibit unwrapping across the cut-lines. Then, an almost consistent phase map is reproduced even if an image contains some defects. Equation (1.47) reveals what was said before, that in X-ray interferometry the image contrast is direct proportional to the phase distribution.

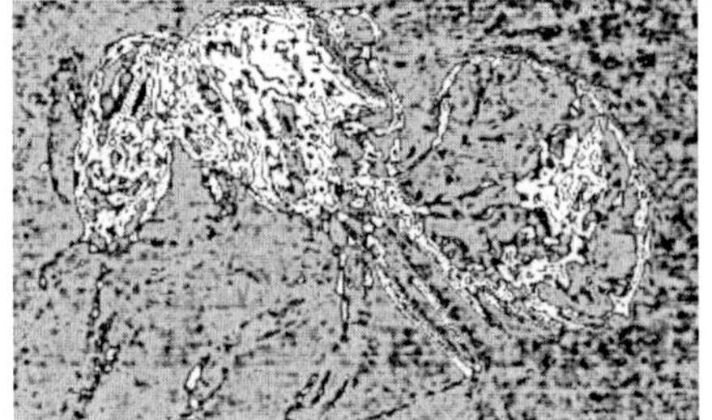

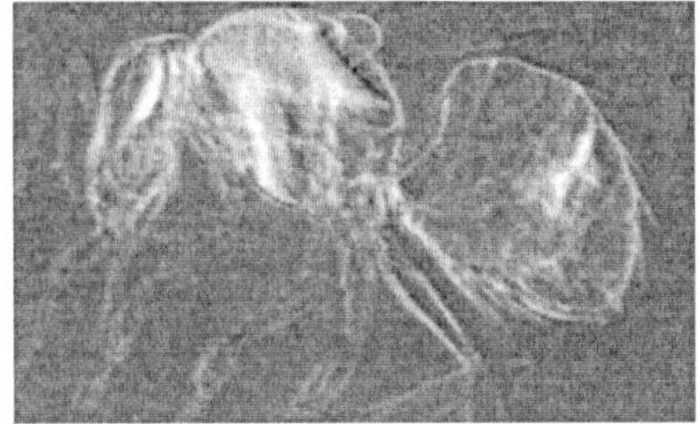

Figure 9: Left: Wrapped phase map of a bee recorded with the Torino LLL X-ray interferometer. Right: Unwrapped phase map of the bee.

Propagation based Phase Contrast

Free propagation phase imaging, often also referred as to Phase Contrast Imaging or "in line holography", is preferable in biomedical research because of its simple geometry and simple technical requirements because it does not imply the use of special optics or equipment. With the only requirement of a source with sufficient lateral (or spatial) coherence it is possible to detect the effects of phase perturbation, resulting in an enhancement of image contrast [22, 23].

For synchrotron radiation a typical vertical size s of the source can be about 100 μm and for long beamlines the source-sample distance r is some tenth of meters. Consequently the coherence length $L_s = \lambda r/(2\,s)$ for hard X-rays around 20 keV can be up to some tenth of μm. For conventional X-ray sources the typical distance r is about 50 cm and lateral coherence degree is too small to allow the visualization of phase effects. A solution is the use of micro focus X-ray tubes but typically the fluence reaching the detector requires quite long exposure time.

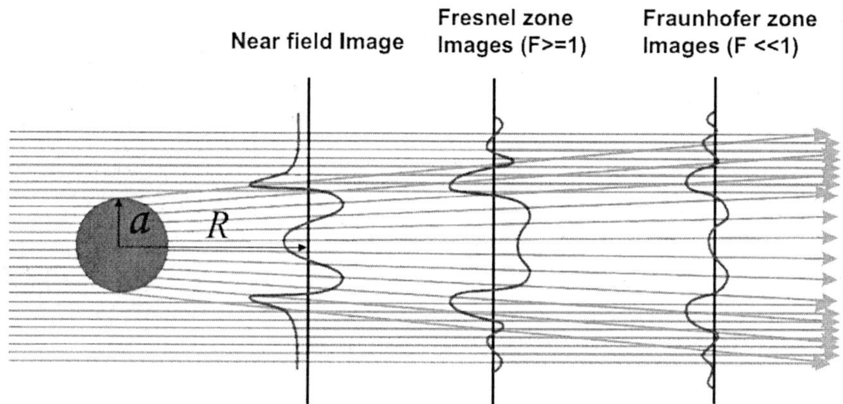

FIGURE 10: see text.

As discussed previously the X-ray wave field when it crosses an object is phase shifted. The waves refracted by a detail interfere with the not refracted waves traversing the background. This interference effect takes place along the border of the detail inside a narrow angular region (about 10 µradians), and it results in strong interference patterns inside this region that could be detected (figure 10).

The influence of detail size a, x-ray wavelength λ and sample to detector distance R on the interference pattern can be described qualitatively by the Fresnel number F.

$$F = \frac{a^2}{\lambda \cdot R} \tag{1.48}$$

For F << 1 Fraunhofer diffraction occurs and shows the far-field diffraction pattern of the detail, which is closely related to the spatial Fourier transform of the complex amplitude distribution of the wave field after the detail. Fresnel numbers around 1 or larger characterize the situation of Fresnel diffraction (or near-field diffraction), where the mathematical description is more complicated. For not too large Fresnel numbers and diffraction angles, the Fresnel approximation can be used. Most readily usable are configurations for $F >> 1$ (near field) where sharp edges seen as dark/light interference fringes, producing edge enhancement which can be observed with an appropriated imaging receptor. Obviously in order to appreciate these interference fringes detectors with high spatial resolution are necessary. The interference patterns are more pronounced at the edges of the details and they are revealed as strong intensity variation in the image. In particular this edge-enhancement effect enhances the visibility of the very thin and small details, usually almost invisible on the absorption image (figure 11).

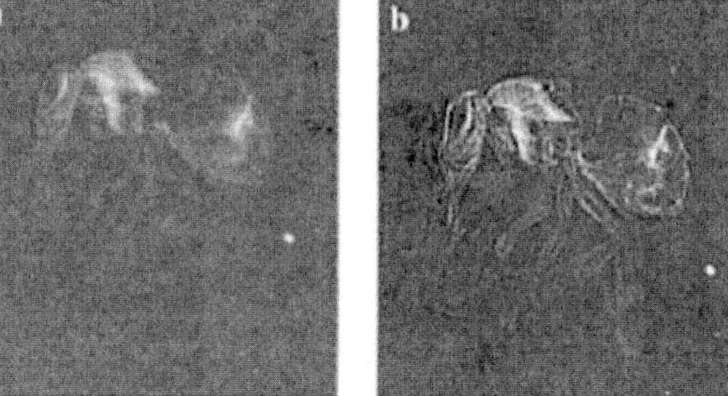

FIGURE 11: Absorption image (left) and PBI image (right) of a bee recorded at the SYRMEP beamline at Elettra. For the former the object was in contact with the detection device while for the latter the object detector distance was 50 cm.

Analyzer based imaging

Analyzer based imaging is an emerging imaging modality that has been showing promise for medical applications. This technique is based on an analyzer crystal placed between the object and the detection device, which due to its diffraction properties converts the angular distribution of refracted beams into intensity changes onto the detector. Analyzer Based Imaging (ABI) was pioneered by [21] but the recent widespread studies have brought significant technical as well as theoretical improvements [4]. At present the technique has been developed exclusively using synchrotron radiation sources for the high intensity and little divergence. Conventionally, ABI refers to an imaging process possessing high efficiency scattering rejection and/or forming images from X-rays deviated by only a few μradians (figure 12). The method is sensitive to the component of the refraction angle in the plane *yz*, that means, considering the equation (1.43).

$$\Delta\theta \approx \frac{\lambda}{2\cdot\pi}\cdot\frac{\partial\varphi(x,y)}{\partial y} \tag{1.49}$$

Illuminated with monochromatic radiation of wavelength λ, a perfect crystal 'reflects' radiation when Bragg's law is satisfied.

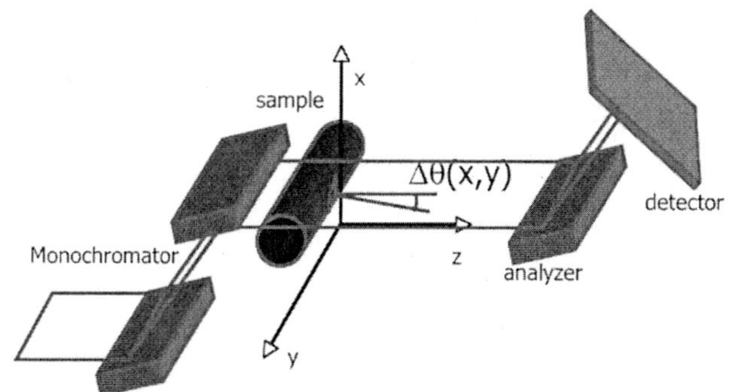

FIGURE 12: Sketch of an imaging setup with analyzer crystal.

As seen before a perfect crystal has a very narrow angular acceptance (around μrad) for X-rays of a certain energy or wavelength. For monoenergetic radiation only those beams are reflected whose angles with respect to the surface of the crystal are in a certain range around the Bragg angle. The reflectivity as function of the Bragg angle – also known as rocking curve- is shown in figure 4.

If the analyzer crystal is perfectly aligned with respect to the direction of the incident wave (peak position) then only the undeviated rays are reflected on the detector and subsequently form the image that are within the FWHM of the rocking curve. X-ray beams refracted under higher angles, however, are rejected. In this configuration the analyzer crystal serves as a perfect slit with almost no contributions from refracted X-rays.

Beyond simple scatter rejection a crystal analyzer reveals another more sophisticated feature when detuned (e.g. moved to the slope of the rocking curve). In this fashion well defined amounts of the refracted radiation contribute to the measured images recorded by the detector that can be now considered as a mixture of two components only - namely a pure absorption (I_R) and a pure refracted part ($\Delta\theta_z$). Initially the weight of both components is unknown.

A decade ago a geometrical optics based algorithm, called Diffraction Enhanced X-ray Imaging (DEI), was introduced by Chapman *et al.* [24]. The original was the combination of the two half-slope images according to simple equations based on a first order Taylor approximation of the RC, producing a couple of new images: the apparent absorption image and the refraction image.

If the analyzer is set to a given position $\bar{\theta}$, the intensity reaching the detector is given by:

$$I\left(\bar{\theta};x,y\right)=I_R\left(x,y\right)R\left(\bar{\theta}+\Delta\theta_Z\left(x,y\right)\right) \tag{1.50}$$

where I_R is the apparent absorption intensity and $\Delta\theta_R(x,y)$ is the refraction angle for photons exiting the object.

If two images I_L and I_H are recorded, one with the analyzer crystal detuned to smaller angles (θ_L) and the other with the analyzer detuned to higher angles (θ_H) as indicated in figure 13, then the equation with two unknowns can be solved by a simple matrix inversion on a pixel basis. At first the images are recorded at the steepest slope of the rocking curve $R(\theta)$. Here $R(\theta)$ is almost linear and can be developed in a first order approximation in a Taylor expansion and one obtains

$$
\begin{aligned}
I_L &= I_R\left(R(\theta_L) + \frac{\partial R}{\partial \theta}(\theta_L)\Delta\theta_z \right) \\
I_H &= I_R\left(R(\theta_H) + \frac{\partial R}{\partial \theta}(\theta_H)\Delta\theta_z \right)
\end{aligned}
\qquad (1.51)
$$

where $R(\theta_L)$, $R(\theta_H)$ are the values of the rocking curve at the angle θ_L and θ_H, respectively and $\partial R/\partial\theta\,(\theta_L)$and $\partial R/\partial\theta\,(\theta_H)$ are the values of the derivation at those angles. Since the rocking curve and its derivation are known equation (1.51) can be solved for the unknown I_R and $\Delta\theta_z$. One gets

$$
\begin{aligned}
I_R(x,y) &= \frac{I_L(x,y)\cdot \left.\frac{dR}{d\theta}\right|_{\theta_H} - I_H(x,y)\cdot \left.\frac{dR}{d\theta}\right|_{\theta_L}}{R(\theta_L)\cdot \left.\frac{dR}{d\theta}\right|_{\theta_H} - R(\theta_H)\cdot \left.\frac{dR}{d\theta}\right|_{\theta_L}} \\
\Delta\theta_Z(x,y) &= \frac{I_H(x,y)\cdot R(\theta_L) - I_L(x,y)\cdot R(\theta_H)}{I_L(x,y)\cdot \left.\frac{dR}{d\theta}\right|_{\theta_H} - I_H(x,y)\cdot \left.\frac{dR}{d\theta}\right|_{\theta_L}}
\end{aligned}
\qquad (1.52)
$$

I_R and $\Delta\theta_z$ are called *apparent absorption image* and *refraction image*. The name implies that the diffraction of a perfect analyzer crystal is utilizes to enhance the contrast of the images. Sources of contrast in the apparent absorption image are absorption and extinction /scatter rejection. Extinction means here missing intensities of those X-rays that are refracted under angles substantially higher than the acceptance of the analyzer. In contrast the refraction image gives for each pixel the angle of refraction in the object plane. Thus the white horizontal line in the refraction image indicates the upper boundary of the wire where the X-rays are refracted to positive angles. The black line represents the lower boundary of the wire.

As an example for the improved image quality a DEI radiography of bee is shown in figure 14. This image was taken at photon energy of 20 keV. Without the contributions of scattering even small details such as wings and joints can be appreciated as demonstrated in this image taken at the medical beam line at Elettra. The apparent absorption image is more pronounced than a normal transmission radiograph.

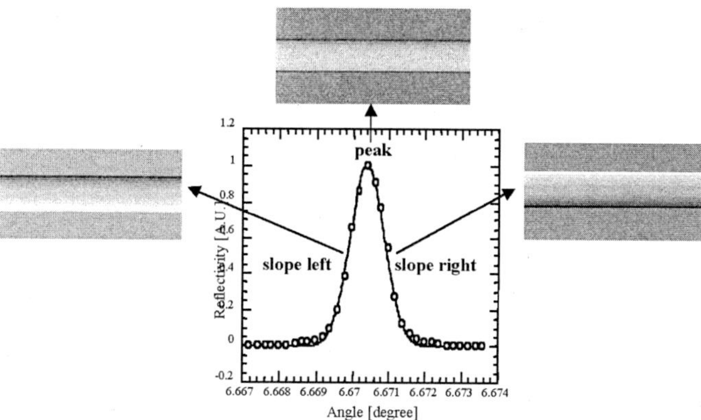

FIGURE 13: Experimental rocking curve of the Si [111] crystal at 17 keV at Elettra. Images of a 700μm Nylon wire were recorded: the upper image at the peak position of the rocking curve; images at the left slope and the right side image are also shown.

Additional information gives the refraction image. Here the overall attenuation differences between skeleton and soft tissue are not seen in the conventional fashion. Moreover, edges are enhancements. Since the physical effects upon which these images are based are sensitive to boundaries, one can also see inclusions in the bones that are not visible in the associated apparent absorption image.

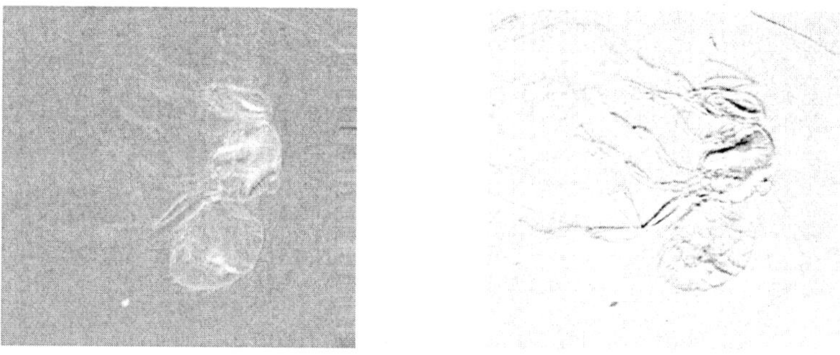

FIGURE 14: Apparent absorption image (left) and refraction image (right) of a bee.

There is an intrinsic limitation of the DEI algorithm since it is based on the hypothesis that the pixel size of the detector is small enough compared to the structure of the object, so that a specific part of the object, where a well-defined refraction angle $\theta(x,y)$ is expected, is mapped to a single pixel. However, this hypothesis is not fulfilled when the sample produces ultra-small angle scattering (USAXS) being characterized by a large amount of fine structures (i.e. smaller than the pixel size), not detectable separately. In this case, the signal recorded on the slopes may vanish [25] and the first order of the Taylor expansion disappears.

In absence of refraction, the intensity collected by the analyzer crystal set to an angle $\bar{\theta}$ is:

$$I(\bar{\theta};x,y) = I_R(x,y)\cdot\int R(\bar{\theta}+\Delta\theta_S)f(\Delta\theta_S;x,y)d(\Delta\theta_S) \qquad (1.53)$$

where $\Delta\theta_S$ is the stochastic scattering angle, governed by the probability density function $f(\Delta\theta_S;x,y)$ with the following the first three moments:

$$\begin{aligned} &I.\int f(\Delta\theta_S;x,y)d(\Delta\theta_S)=1\\ &II.\int \Delta\theta_S f(\Delta\theta_S;x,y)d(\Delta\theta_S)=0\\ &III.\int(\Delta\theta_S)^2 f(\Delta\theta_S;x,y)d(\Delta\theta_S)=\sigma^2_{\Delta\theta_S}(x,y)\end{aligned} \qquad (1.54)$$

One can expands to the second order and two images are taken on the peak and on the toe of the reflectivity curve obtaining:

$$\begin{aligned} I_{TOE}(x,y) &= I_R(x,y)\cdot\left(R(\theta_{TOE})+\frac{1}{2}\frac{\partial^2 R}{\partial\theta^2}|_{\theta_{TOE}}\cdot\sigma^2_{\Delta\theta}(x,y)\right)\\ I_{PEAK}(x,y) &= I_R(x,y)\cdot\left(R(\theta_{PEAK})+\frac{1}{2}\frac{\partial^2 R}{\partial\theta^2}|_{\theta_{PEAK}}\cdot\sigma^2_{\Delta\theta}(x,y)\right)\end{aligned} \qquad (1.55)$$

and subsequently

$$\begin{aligned} I_R(x,y) &= \frac{I_{TOE}(x,y)\cdot\frac{\partial^2 R}{\partial\theta^2}|\theta_{TOP}-I_{TOP}(x,y)\cdot\frac{\partial^2 R}{\partial\theta^2}|\theta_{TOE}}{R(\theta_{TOE})\cdot\frac{\partial^2 R}{\partial\theta^2}|\theta_{TOP}-R(\theta_{TOP})\cdot\frac{\partial^2 R}{\partial\theta^2}|\theta_{TOE}}\\ \sigma^2_{\Delta\Theta}(x,y) &= 2\frac{I_{TOP}(x,y)\cdot R(\theta_{TOE})-I_{TOE}(x,y)\cdot R(\theta_{TOP})}{I_{TOE}(x,y)\cdot\frac{\partial^2 R}{\partial\theta^2}|\theta_{TOP}-I_{TOP}(x,y)\cdot\frac{\partial^2 R}{\partial\theta^2}|\theta_{TOE}}\end{aligned} \qquad (1.56)$$

$I_R(x,y)$ can be called apparent absorption image, since it has the same physical meaning found in the original DEI algorithm. On the other hand, the image calculated according to equation (1.56) could be called a "refractive scattering image or USAXS image. It should be pointed out that in this case it does not represent a map of the refraction angle $\theta(x,y)$, but a map of the square width $\sigma^2_{\Delta\theta}(x,y)$ of the scattering distribution of the refraction angle.

Since in general a sample may produce a combination of absorption, refraction and scattering effects, multiple-image methods (i.e. based on the collection of several images) have been developed in order to separate this information. Numerical methods have been developed independently by Pagot *et al.* [26] and Wernick *et al.* [27] independently developed, that allow reconstructing the local RC combining several images acquired at different analyzer positions. Three parametric images, showing the area, the barycenter and the width of the local RC bring information on absorption, refraction and USAXS, respectively. A similar approach was followed by Oltulu *et al.*, where an analytical fitting function was utilized to extrapolate the local RC parameters [28].

A three-images generalized DEI (GDEI) algorithm has been also developed [29] that can be consider as a general model and offers a simple way to decouple the information relative to the three effects in three different parametric images.

$$I(\bar{\theta};x,y) \cong I_R(x,y)\cdot\left[R(\bar{\theta})+\left.\frac{\partial R}{\partial\theta}\right|_{\bar{\theta}}\Delta\theta_R(x,y)+\frac{1}{2}\left.\frac{\partial^2 R}{\partial\theta^2}\right|_{\bar{\theta}}\Delta\theta_R{}^2(x,y)+\frac{1}{2}\frac{\partial^2 R}{\partial\theta^2}\sigma^2_{\Delta\theta}(x,y)\right] \quad (1.57)$$

By recording three different images $I_i=I(\theta_i;x,y)$ at angular positions θ_i $(i=1,3)$ we obtain a three-equations system:

$$I(\theta_i)=I_R\cdot\left[R_i+\dot{R}_i\Delta\theta_R+\frac{1}{2}\ddot{R}_i(\Delta\theta_R)^2+\frac{1}{2}\ddot{R}_i\sigma^2_{\Delta\theta}\right] \quad (1.58)$$

[where a short-hand writing has been used, so that R_i, $\dot{R}_i$ and $\ddot{R}_i$ are respectively the rocking curve, first and second derivative values calculated in the angular position θ_i and spatial variables (x,y) are omitted]. The system in equation(1.58) can be solved to give the apparent absorption image $I_R(x,y)$, the refraction image $\Delta\theta_R(x,y)$ and the ultra-small-angle scattering image $\sigma^2_{\Delta\theta}$ (x,y):

$$I_R=\frac{I_1(\dot{R}_2\ddot{R}_3-\ddot{R}_2\dot{R}_3)-I_2(\dot{R}_1\ddot{R}_3-\ddot{R}_1\dot{R}_3)+I_3(\dot{R}_1\ddot{R}_2-\ddot{R}_1\dot{R}_2)}{R_1(\dot{R}_2\ddot{R}_3-\ddot{R}_2\dot{R}_3)-R_2(\dot{R}_1\ddot{R}_3-\ddot{R}_1\dot{R}_3)+R_3(\dot{R}_1\ddot{R}_2-\ddot{R}_1\dot{R}_2)} \quad (1.59)$$

$$\Delta\theta_R=-\frac{I_1(R_2\ddot{R}_3-\ddot{R}_2R_3)-I_2(R_1\ddot{R}_3-\ddot{R}_1R_3)+I_3(R_1\ddot{R}_2-\ddot{R}_1R_2)}{I_1(\dot{R}_2\ddot{R}_3-\ddot{R}_2\dot{R}_3)-I_2(\dot{R}_1\ddot{R}_3-\ddot{R}_1\dot{R}_3)+I_3(\dot{R}_1\ddot{R}_2-\ddot{R}_1\dot{R}_2)} \quad (1.60)$$

$$\sigma^2_{\Delta\theta}=2\frac{I_1(R_2\dot{R}_3-\dot{R}_2R_3)-I_2(R_1\dot{R}_3-\dot{R}_1R_3)+I_3(R_1\dot{R}_2-\dot{R}_1R_2)}{I_1(\dot{R}_2\ddot{R}_3-\ddot{R}_2\dot{R}_3)-I_2(\dot{R}_1\ddot{R}_3-\ddot{R}_1\dot{R}_3)+I_3(\dot{R}_1\ddot{R}_2-\ddot{R}_1\dot{R}_2)}-(\Delta\theta_R)^2 \quad (1.61)$$

Of note is that equations (1.59), (1.60) and (1.61) can be applied for any choice of the angular positions θ_i $(i=1,3)$. However, to ensure equation (1.58) provides a system of independent equations, the angular positions should be well separated from each other, thus providing a representative extract of the rocking curve. One particular case is, when θ_i $(i=1,3)$ are chosen to be respectively. Point 1 on one half-slope ($\ddot{R}_1=0$), point 2 on the peak ($\dot{R}_2=0$) and point 3 on the other half-slope ($\ddot{R}_3=0$) of the rocking curve. In this configuration equation (1.59) and equation (1.60) reduce to the conventional DEI (equation (1.52)) and equation (1.61) simplifies to: $\sigma^2_{\Delta\theta}=2\cdot(I_2/I_R-R_2)/\ddot{R}_2-(\Delta\theta_R)^2$.

Analyzer based tomography

For any given view angle the parametric images described in equations (1.59) - (1.61) can be written by means of a line integral along the path L [30]. Thus analyzer based CT can be implemented by simply applying conventional filtered back projection algorithms. The line integrals for apparent absorption, refraction and scattering are given by

$$I_R = I_0 \cdot \int_L dz \cdot \mu'$$

$$\Delta\theta_R = \int_L dz \cdot \frac{\partial \delta}{\partial x} \tag{1.62}$$

$$\sigma^2{}_{\Delta\theta s} = \int d\Delta\theta_S \cdot \Delta\theta_S{}^2 \cdot f = \int_L dz \cdot M^2$$

where μ' is the total x-ray attenuation coefficient including extinction effects, δ is the real part of the refractive index and f is the second moment of the local ultra small angle x-ray scattering distribution. Three tomographic datasets of a phantom (figure 15) have been acquired at 17 keV respectively at the peak of the rocking curve, at52% of the peak value on the low-angle side and at 48% on the high-angle side. Due to a slight instability of the beam-analyzer system (which is a well-known problem in DEI during the data acquisition these values drifted, respectively, to 97, 40 and 58% of the peak value. As can be seen in the following, this instability has not compromised the experimental test, at least not at the qualitative level. Each data set has consisted of 720 projections, the phantom turning around its axis, with the view angle increasingly rotated by 0.25 degrees. Each projection has been recorded in 0.5 s by a CCD utilizing a pixel size of 9 μm x 9 μm; the complete data acquisition required about 1h.

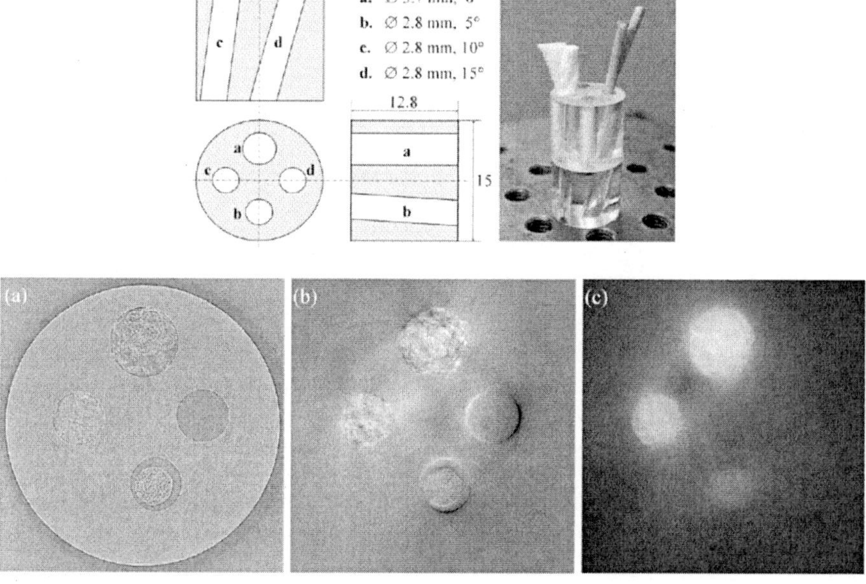

FIGURE 15: Upper panel. A custom made test object for analyzer based CT. Grooves have different diameters and have been filled with (a) soft paper, (b) wooden tooth sticks, (c) PMMA microspheres ~ 100μm in diameter and (d) air.
Lower panel: a slice of the phantom depicted ion the three different parametric axial views: (a) apparent absorption image, (b) refraction image and USAXS image. The images are oriented as indicated in the upper panel.

MEDICAL APPLICATION

Cell Tracking using phase contrast tomography

Cell therapies based on the implantation/transplantation of pluripotent cells of diverse origins in living hosts are currently of fundamental interest in the field of regenerative medicine [31] and encompass potential for treating diseases as demonstrated by numerous animal studies [32]. Small mammals are of special interest in cell tracking since gene targeting in pluripotent, mouse embryo-derived stem cells provides the means to generate mice of any desired genotype [33], which subsequently allows translation from the animal model to humans. However, many critical issues on cell implantation are still open and in order to develop effective cell therapies in animal models and subsequently in humans, the location, distribution, long-term viability and the preservation of original cell functionality of therapeutically implanted cells must be evaluated in a non-invasive fashion at any anatomic location over long time periods. Classically the assessment of the success of such cell therapies relies on stochastically methods involving a considerable number of animal trails. At different stages of tumor development larger numbers of animals are scarified and histological or histochemical analysis are conducted at those particular points in time. The result of such studies can be of high statistical value. However, studies of this type are not suited to the follow-up of pathological processes over time in individual subjects known as longitudinal studies.

In cancer research metastasis spread is an impediment to the development of effective cancer therapies. The understanding of metastasis is limited by the inability to follow this process *in vivo* again at any anatomic location [34]. Here animal models based on the implementation of tumour cells can help to understand the dynamic of cancer growth and metastasis spread.

In both areas *in vivo* imaging systems are needed to serially image and track implanted cells. Ideally, imaging technology used for cell tracking would have single-cell sensitivity and would permit quantification of exact cell numbers at any anatomic location. The ideal imaging technology would permit tracking of implanted cells for months to years because trials undoubtedly will require long-term follow-up of tissue function or host survival. At present, no imaging technology fulfils such requirements at a time[5] one challenge being the size of mammals, which does not necessarily match the spatial resolution provided by the imaging hardware.

In vivo imaging implies that the imaging system should be able to differentiate implanted cells from those of the target tissue. A useful approach is to specifically mark cells prior implantation. Successful *in vivo* cell tracking requires that a contrast agent associated with implanted cells exerts an "effect size" sufficient for detection by imaging hardware without altering the cell functionality [35]. Among possible contrast agent inorganic nano particles (NP) generally possess versatile properties suitable for cellular delivery and they appear biocompatible and noncytotoxic [36]. Colloidal gold NPs have been used successfully *in vivo* since 1950s[37]. Multiple functionality can be achieved by surface modification[38,39,40,41] allowing targeted drug delivery applications [42, 43]. Eventually their biological lifetime is limited by apoptosis only.

In cell tracking w in equation (1.36) is defined by the (multiple) size of single cells and subsequently $w << 1$ resulting in high and sometimes intolerable skin doses. Two parameters in equation (1.36) are free to compensate for small details and subsequently high skin doses: Firstly the choices of the most appropriate energy E_γ and secondly suitable high density and high atomic number material based contrast agents. While the former can be obtained with state of the art X-ray instrumentation such as synchrotron radiation sources the latter can be realized using gold nano particles (NPs). Gold NPs bound to serum proteins are taken up by cells in culture to be sequestered in lysosomes forming a permanent marker of cells [44].

For the experiments reported here (which were carried out in accordance with the Canadian laws for animal trials), we chose the C6 glioma cell line, a frequently used rodent model that mimics the malignant human brain tumor gliobalstoma multiforme. The C6 glioma cells were loaded with colloidal gold NPs before implantation into the brain of 6 rodents. The results of mass spectroscopy on those loaded C6 cells show that on average 341.68 μg of elemental gold were taken up per 10 Mio cells. No gold was detected by mass spectroscopy in the control sample, which was to be expected since gold is neither part of the normal growth medium nor is it part of the normal metabolic cell cycle. Since the gold loading protocol relies on cell metabolism only it can be translated to other non-malignant multipotent cell types. Following the C6 loading protocol we could also show that on average 332.43 μg / 10 Mio cells is taken up into human bone marrow stem cells, which translates into an effective gold density of ρ_{eff} = 0.026 g/cm^3 thus approximately 3% of the cell density. Moreover, about 27,000 NPs with an average diameter of ~ 50 nm are incorporated into a single cell. Growth assay using naïve cells and gold-loaded cells revealed that our loading protocol has no significant effects on cell proliferation.

Using equation (1.36) we can estimate entrance dose and photon fluence [photons /mm2] required for cell tracking. These values are depicted in Figure 16 versus the X-ray energy revealing a minimum of $D_{entrance}$ = 3 Gy and φ_{min} = 10^{12} (monoenergetic) photons /mm^2 (0.1% bandwidth) at a photon energy of 24 keV. Such fluence (within the mentioned bandwidth) is available at synchrotron sources. Also shown in figure 16 is the photon fluence corrected on the exposure time and the numbers of projections generated by the SYRMEP bending magnet at the Sincrotrone Trieste, Italy, verifying that cell tracking is possible with this type of radiation source.

All NP loaded tumors were prominent in the synchrotron phase contrast CT with a spatial resolution of about 20μm whilst tumors grown from non-labeled C6 cells were not resolvable. As an example depicted in figure 16 is a detail of the skull bone revealing sutures, burr hole through which the tumors were and implanted and diploe structure. In case of the NP loaded cells the infiltrating nature of the malignant brain tumour is evident, with finger-like extensions of gold-loaded cells extending from the main bulk of the tumor, which traced back to 100,000 implanted cells and an incubation of 14 days. It can be estimated from these images that the cell sensitivity is considerably below 10 cells. Tumors in the controls were not detectable and so far no brain soft tissue structures have been made visible in any of the renderings however, we believe that further refinement of the imaging technique will eventually visualize the anatomy.

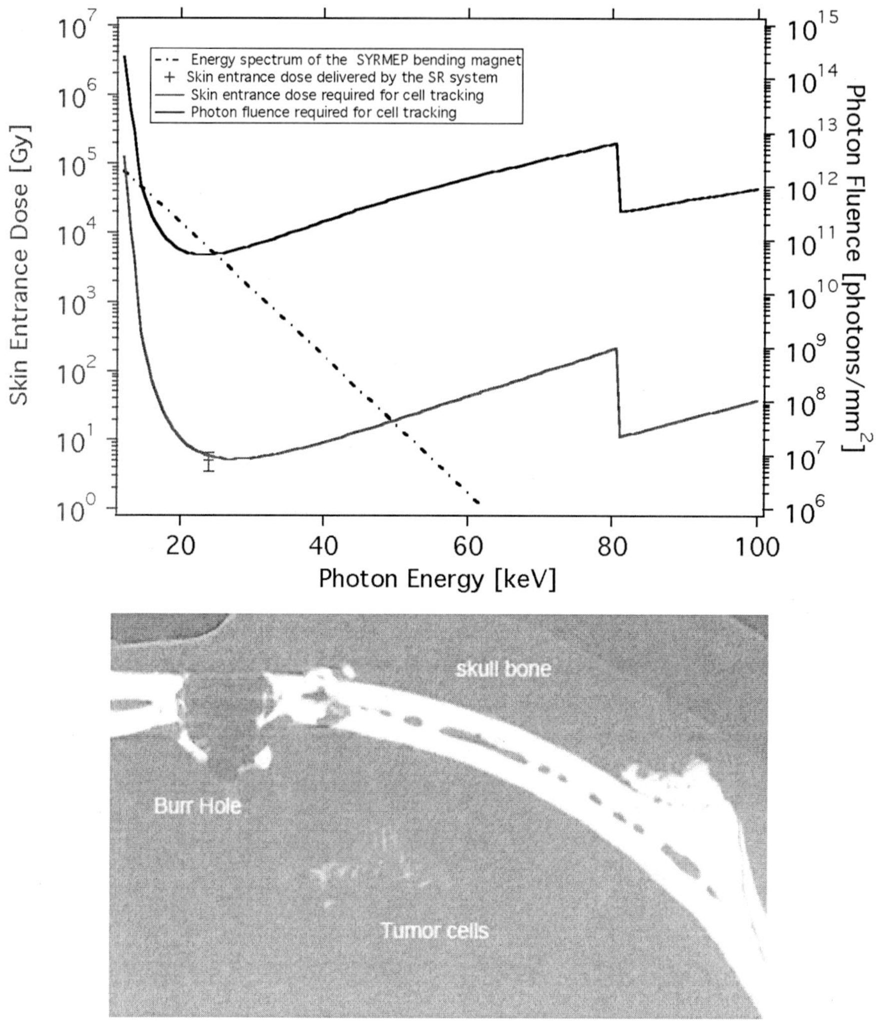

FIGURE 16: Upper panel. Skin entrance dose and photon fluence required and delivered by the synchrotron setup for cell tracking.
Lower panel: Full resolution (pixel ~ 14μm) slice reconstruction revealing burr hole, diploe and some clusters of gold loaded tumor cells.

Of note is that not only tumor cells can be loaded and visualized utilizing phase contrast CT but also different pluripotent cell lines that are used in novel cell therapies.

---

1 Röntgen, W. C. ,"*Über eine neue Art von Strahlen*". Sitzungsberichte der Physikalisch-medizinischen Gesellschaft zu Würzburg. 137 (1895)

2. H. Winick and S. Doniach, "*An Overview of Synchrotron Radiation Research,*" in Synchrotron Radiation Research, New York and London, Plenum Press, , 1980, pp. 1-10.

3 Thomlinson, W.C., *Medical applications of synchrotron radiation at the National synchrotron Light Source* in Synchrotron Radiaition in Biosciences, Oxford Science Publications, (1994)

4 Suortti, P., Thomlinson, W., *Physics in Medicine and Biology* **48** (13), R1-R35 (2003)

5 Winick H. and Doniach S., (eds.), "*An Overview of Synchrotron Radiation Research*" in Synchrotron Radiation Research, (1980)

6 Sánchez del Río M., Dejus R. J, *Synchrotron Radiation Instrumentation: Eighth International Conference*, edited by T. Warwick et al. American Institute of Physics, 784-787. (2004)

7 Giacomini J.C., Gordon H., O'Neil R., Van Kessel A., Carson B., Chapman D., Lavendar W., Gmür N., Menk R., Thomlinson W., Zhong Z., Rubenstein, E. *Nucl. Instr. and Meth. A* **406** *473,* (1998)

8 Nemoz, C., Bayat, S., Berruyer, G., Brochard, T., Coan, P., Le Duc, G., Keyrilainen, J., (...), Thomlinson, W. *AIP Conference Proceedings* **879**,*1887-1890,* (2007)

9 Attwood D. , Soft X-rays and Extreme Ultraviolet Radiation: Principles and Applications (Cambridge University Press), 1999

10 Menk, R.H., Thomlinson, W., Gmuer, N., Zhong, Z., Chapman, D., Arfelli, F., Dix, W.R., (...), Walenta, A.H. *Nuclear Instruments and Methods in Physics Research, Section A: Accelerators, Spectrometers, Detectors and Associated Equipment* **398** (2-3), 351-367(1997)

11 Hounsfield G.N. Computerized transverse axial scanning (tomography). 1. Description of system. Br J Radiol., 46(552):1016-22 (1973)

12 Cormack A., J. Applied Physics 34, 2722 (1963)

13 Wilkins S.W., Gureyev T.E., Gao D., Pogany A. and Stevenson A.W., *Nature* **384** (1996).

14 Snigirev A., Snigireva I., Kohn V., Kuznetsov S., Schelokov I., *Rev. Sci. Instrum.* **66** 5486. (1995)

15 Bonse U., Hart M. *Appl. Phys. Lett.* **6** 155 (1965)

16 S.W. Wilkins et al. , *Nature* **384**, 335-338 (1996)

17 L.D. Chapman, W. Thomlinson, R.E. Johnson., D. Washburn, E. Pisano, N. Gmür, Z. Zhong, R.H. Menk, F. Arfelli and D. Sayers, *Phys.Med. Biol.* **42**, (1997).

18 Foerster E., Goetz K., Zaunseil P. *Kristall und Technik* **15** 973 (1980)

19 Fitzgerald, R. *Physics Today* **53** (7), 23-26 (2000)

20 Menk R.H. *Medical Imaging Technology* **24**, 373-379 (2006)

21 Huntley J.M. *Appl. Opt.* ***28***, 3268-3270 (1989)

22 Cloetens P., Barrett, Baruchel, Guigay, Schelokov, *J.Phys. D: Appl. Phys.***29**, 133, 1996

23 Snigirev A., Snigireva I., Kohn V., Kuznetsov S.and Schelokov I., *Rev. Sci. Instrum.* **66**, 5486, (1995).

24 Chapman L.D., Thomlinson W., Johnson R.E., Washburn D., Pisano E., Gmür N., Zhong Z., Menk R.H., Arfelli F. and Sayers D., *Phys. Med. Biol.* 42, (1997).

25 Rigon L. et al *J. Phys. D: Appl. Phys.* **36**, A107–A112,(2003)

26 Pagot E. et al *Applied Physics Letters* Vol. **82**, No 20, (2003)

27 Wernick M.N.et al., *Phys.Med.Biol.***48**, 3875, (2003)

28 Oltulu O. et al., *J.Phys.D:Appl.Phys.***36**, 2152, (2003)

29 Rigon L., Arfelli F., Menk R.-H., *Journal of Physics D: Applied Physics* **40** (10),. 3077-3089 (2007)

30 Rigon L., Astolfo A., Arfelli F., Menk R.H., *European Journal of Radiology,* Article in Press (2008)

31 Rogers W.J., Meyer C.H., Kramer C.M., Technology Insight: in vivo cell tracking by use of MRI, *Nature Clinical Practice Cardiovascular Medicine* **3**, 554-562 (2006)

32 Zandonella C., Stem-cell therapies: The first wave, *Nature* **435**, 877-878 (2005)

33 Capecchi, M.R., The new mouse genetics: Altering the genome by gene targeting, *Trends in Genetics* Volume **5**, Issue 3, 70-77 (1989)

34 Voura E.B., Jaiswal J.K., Mattoussi H., Simon S.M., Tracking metastatic tumor cell extravasation with quantum dot nanocrystals and fluorescence emission-scanning microscopy, *Nature Medicine* **10**, 993 - 998 (2004)

35 Frangioni J.V., Hajjar R.J., In Vivo Tracking of Stem Cells for Clinical Trials in Cardiovascular Disease, *Circulation* **110**; 3378-3383 (2004)

36 Connor E.E., Mwamuka J., Gole A., Murphy C.J., Wyatt M.D. Gold NPs Are Taken Up by Human Cells but Do Not Cause Acute Cytotoxicity. *Small*, **1**, 325-327 (2005)

37 Hultborn K.A., Larsonn L.G., Ragnhult I. The lymph drainage from the breast to the axillary and parasternal lymph nodes studied with the help of colloidal Au 198. *Acta Radiol.*, **43**, 52-63 (1955)

38 Sokolov K., Follen M., Aaron J., Pavlova I., Malpica A., Lotan R., Richards-Kortum R. Real-Time Vital Optical Imaging of Precancer Using Anti-Epidermal Growth Factor Receptor Antibodies Conjugated to Gold NPs. *Cancer Res.*, **63**, 1999-2004. (2003)

39 Katz E., Willner I. Integrated NP-Biomolecule Hybrid Systems: Synthesis, Properties, and Applications. *Angew. chem. Int. Ed.*, **43**, 6042-6108 (2004)

40 El-Sayed I.H., Huang X., El-Sayed M. A. Surface Plasmon Resonance Scattering and Absorption of anti-EGFR Antibody Conjugated Gold NPs in Cancer Diagnostics: Applications in Oral Cancer. *Nano Lett.*, **5**, 829-834 (2005)

41 Dixit V., Van den Bossche J., Sherman D.M., Thompson D.H., Andres R.P. Synthesis and Grafting of Thioctic Acid-PEG-Folate Conjugates onto Au NPs for Selective Targeting of Folate Receptor-Positive Tumor Cells. *Bioconjug. Chem.*, **17**, 603-609 (2006)

42 Paciotti G.F., Myer L., Weinreich D., Goia D., Pavel N., McLaughlin R.E., Tamarkin L. Colloidal Gold: A Novel NP Vector for Tumor Directed Drug Delivery. *Drug Delivery*, **11**, 169-183 (2004)

43 Jain P.K., El-Sayed I.H., El-Sayed M.A. Au NPs target cancer. *Nanotoday*, **2**, 18-29 (2007)

44 Juurlink, B.H. and Devon, R.M., Colloidal gold as permanent marker of cells. *Experientia* **47(1)**: 75-77(1991)

Deep Brain Electrical Stimulation in Epilepsy

Luisa L. Rocha

Depto. Farmacobiología. Centro de Investigación y de Estudios Avanzados del IPN. Calz. Tenorios 235. Col. Granjas Coapa. México, D.F. C.P. 14330

Abstract. The deep brain electrical stimulation has been used for the treatment of neurological disorders such as Parkinson's disease, chronic pain, depression and epilepsy. Studies carried out in human brain indicate that the application of high frequency electrical stimulation (HFS) at 130 Hz in limbic structures of patients with intractable temporal lobe epilepsy abolished clinical seizures and significantly decreased the number of interictal spikes at focus. The anticonvulsant effects of HFS seem to be more effective in patients with less severe epilepsy, an effect associated with a high GABA tissue content and a low rate of cell loss. In addition, experiments using models of epilepsy indicate that HFS (pulses of 60 µs width at 130 Hz at subthreshold current intensity) of specific brain areas avoids the acquisition of generalized seizures and enhances the postictal seizure suppression. HFS is also able to modify the status epilepticus. It is concluded that the effects of HFS may be a good strategy to reduce or avoid the epileptic activity.

Keywords: Epilepsy, Pharmacoresistance, Electrical Stimulation.

DRUG-RESISTANT EPILEPSY

The word epilepsy is derived from the Greek verb epilamvanein ("to be seized," "to be taken hold of," or "to be attacked"). Epilepsy refers to recurrent paroxysmal episodes of brain dysfunction manifested by stereotyped alterations in behavior and electrographic activity. It represents a broad category of syndromes arising from any number of brain disorders that themselves may be secondary to a variety of pathologic processes [1].

The incidence of epilepsy is about 2% and approximately 60–80% of patients can be controlled with antiepileptic drugs [2]. In more than 60% of all cases, seizures remit permanently. Nevertheless, a substantial proportion of patients (30%) do not respond to antiepileptic drug medication, despite administration in an optimally monitored regimen. Such cases are often loosely termed intractable [3].

Medically intractable epilepsy implies a long-standing epilepsy in which years of therapeutic efforts with single or combined drugs have failed for a sufficient period of time. Intractable epilepsy is a major risk for personal injury, poor quality of life, and in some cases mortality. Furthermore, the socioeconomic cost of recurrent seizures is significant, both to the individual and society [3].

Pharmacoresistance appears to correlate with certain features of the epileptic condition, febrile seizures prior to treatment, early onset of seizures, or the presence of certain types of structural brain lesions. Intractability is more common in those with

CP1077, *Advanced Summer School in Physics 2008, Frontiers in Contemporary Physics—EAV08*
edited by L. M. Montaño Zetina, G. Torres Vega, M. García Rocha, L. F. Rojas Ochoa, and R. López Fernández

mental retardation, neurologic deficits, or both and generally in patients with detectable structural brain damage and gross cerebral malformations [4]. A high frequency of seizures (especially of grand mal attacks) is also predictive of difficulties of control [5]. The duration of active epilepsy is a very important factor for the prediction of subsequent course; the longer the history of active epilepsy, the lower the chance of ultimate remission. Indeed, considerable evidence supports that most patients who do not respond to correct monotherapy for 2 years have only a limited chance of responding to further trials [6].

Since a wrong pharmacological treatment is considered as a major condition to develop drug-resistant epilepsy in many patients, an adequate medical therapy is the first step to control seizures, and an essential point is to set a long-term treatment plan. This depends on the previous treatment. The history of previous pharmacological treatment should begin with the onset of treatment and includes doses, duration of administration, blood determinations with dates performed, and evaluation of the effects of drugs previously taken on seizures [7].

Surgical therapy is considered the treatment of choice for certain well-defined surgically remediable syndromes for which (a) the pathophysiology is understood; (b) the natural history is reasonably well known to be medically refractory or even progressive; (c) presurgical evaluation requires only noninvasive studies; and (d) surgery offers a 70% or greater chance of abolishing disabling seizures. The following are surgically remediable syndromes: a) mesial temporal lobe epilepsy, where the underlying pathophysiological substrate is hippocampal sclerosis; b) partial epilepsy produced by discrete structural cerebral lesions that can be resected without introducing additional neurologic deficit; and c) catastrophic unilateral or secondary generalized epilepsies of infants and young children that result from disturbances confined to one hemisphere (hemimegalencephaly, Sturge-Weber syndrome, Rasmussen encephalitis, and cortical dysplasias and porencephalic cysts) [8]. Unfortunately, there are patients in which surgery cannot be applied, such as with several epileptic foci. In these patients, other therapeutic strategies, such as brain electrical stimulation, are applied to control the seizure activity.

BRAIN ELECTRICAL STIMULATION

Electrical stimulation has been used since ancient times to modulate the nervous system. The electric ray *Torpedo nobiliana* was so named by the Romans for its ability to cause torpor. Scribonius Largus suggested applying the live ray to the head of a patient suffering from a headache. Electrical stimulation was later used for hemorrhoids, gout, depression, and epilepsy. During the 18th Century the electrical stimulation was used for therapeutic purposes. For example, sparks were used to treat arm paresias; electrical shocks were applied in an attempt to reverse blindness or to revive the drowned; electric fishes were used for pain control. Peripheral applications of electrical current had become common in the 19th Century for anesthesia and a variety of medical ailments. In 1870, Fritsch and Hitzig performed a series of experiments on dogs at Hitzig's home and they found that they could elicit graded responses, from small arm movements to generalized tonic clonic seizures, based on the amount of current applied. Thereafter, there was an explosion of experiments applying brain electrical stimulation some of them using stereotactic functional neurosurgery in the late 1950s [9].

At present, electrical stimulation of discrete brain structures has been considered as potential treatment of neurological disorders such as Parkinson disease and chronic pain [10, 11, 12]. Stimulation of the nervous system has also been performed in patients with medical-resistant epilepsy in whom epilepsy surgery in not an option. Stimulation sites, such as cerebellar cortex [13], centromedian [14] and ventralis anterior [15] thalamic nuclei, subthalamic nucleus [16], hippocampal epileptic foci [17, 18] and vagus nerve [19] have been claimed effective to control seizures.

Cerebellar Stimulation

The efficacy and safety of cerebellar stimulation was first documented to decrease seizure activity in several experimental models of epilepsy, from which it was proposed as an alternative treatment for patients with intractable epilepsy [20, 21]. For the human cerebellum, the stimulation frequency effective for seizure reduction is 10-20 Hz [13]. Velasco et al. [20] found that the superomedial cerebellar cortex appears to be a significantly effective and safe target site for electrical stimulation for reduction of intractable motor seizures. These authors reported that patients receiving cerebellar stimulation present a reduction of generalized tonic-clonic seizures after 1-2 months and continue to decrease over the first 6 months and then maintains this effectiveness over 2 years.

It is important to consider some consequences of the surgery carried out to apply the cerebellar stimulation such as infection of the tissue or migration of the electrodes. Although studies support the efficacy of cerebellar stimulation to control motor seizures, future studies should be carried out to determine its precise indication.

High frequency electrical stimulation of deep brain areas

The application of high frequency electrical stimulation (HFS) at 130 Hz in limbic structures of patients with intractable temporal lobe epilepsy abolishes clinical seizures and significantly decreases the number of interictal spikes at focus [14, 18]. Subacute HFS of the parahippocampal cortex (PHC), which has been suggested to play an important role in the intractable epilepsy [22], significantly decreases the number of interictal spikes at focus and abolished clinical seizures in patients with intractable mesial temporal lobe epilepsy [14]. Blockage of temporal lobe epilepsy by PHC stimulation seems to be due, at least in part, to an inhibitory process of the stimulated area according with the following effects observed in the patients receiving HFS: a) an increased threshold and decreased duration of the PHC afterdischarge and depression of the PHC evoked response recovery cycles, comparing the same patients before vs. after PHC stimulation; and b) a single photon emission computed tomography (SPECT) hypoperfusion of the hippocampal region comparing the epileptic and stimulated side vs. the normal and non-stimulated side in the same patient [14, 23]. Subacute HFS of PHC is more effective in producing antiepileptic effects in patients with less severe epilepsy, an effect associated with higher GABA (the most important inhibitory neurotransmitter) tissue content and a low rate of cell loss [24].

Electrical stimulation of the hippocampus (a brain area involved in memory and learning processes) has been considered a safe, nonlesional alternative in patients with mesial temporal lobe epilepsy who are not candidates for resective surgery. The long-term (18 months to 7 years) HFS of the hippocampal epileptic foci of patients with refractory partial complex seizures provides a nonlesional method that improves seizure outcome without memory deterioration. HFS of hippocampus effectively reduces seizures without a negative effect on memory performance [25]

HFS of the thalamic centromedian nucleus has been proposed as a minimally invasive alternative for the treatment of difficult-to-control seizures of multifocal origin and seizures that are generalized from the onset. This strategy intends to interfere with seizure propagation in a non-specific manner through the thalamic system. The HFS of thalamic centromedian nucleus significantly decreases generalized seizures of cortical origin and focal motor seizures. Best results are obtained in non-focal generalized tonic clonic seizures and atypical absences of the Lennox-Gastaut syndrome. Experience has indicated that the most effective target for seizure control is the thalamic parvocellular centromedian subnucleus [26].

Vagus nerve stimualtion

The vagus nerve is a cranial nerve raising at the level of the brain stem that innervates most of the organs in the body. The vagal nerve electrical stimulation (VNS) therapy was approved by the US Food and Drug Administration (FDA) in 1997, as an adjunctive therapy in reducing the frequency of seizures in patients with partial seizures that are refractory to antiepileptic medications. At present it is also suggested to be effective on adults with pharmacoresistant generalized epilepsy syndromes [19]. The VNS is delivered by a pulse generator implanted in the chest and connected to the left vagus nerve with a lead ending with electrodes. The patient who senses an oncoming seizure may activate an extra burst of stimulation with a hand-held magnet in an attempt to abort or diminish the seizure. The patients with generalized seizures experience 43 to 46% median reduction in seizure frequency as consequence of VNS [19, 27].

The restrictions to apply VNS are as follows: progressive neurologic disease other than epilepsy; a history of active cardiac or pulmonary disease; peptic ulcer; cervical vagotomy; gastric surgery; addictions within the previous year; self-mutilation; suicidal attempts; general anesthesia within the previous 3 months;

previous VNS therapy or brain stimulation; swallowing dysfunction or pneumonia; or being on ketogenic diet. Adverse events associated with VNS therapy are voice change or hoarseness, throat pain, swallowing complaints and cough [19, 27].

HIGH FREQUENCY ELECTRICAL STIMULATION IN EXPERIMENTAL MODELS

The HFS consisted of pulses of 60 μs width at 130 Hz and the subthreshold current intensity of stimulation that is determined for each animal as follows: HFS is delivered starting at 100 μA and it is gradually increased by 100 μA until motor behaviors (head nodding, forelimb muscle twitches, or backing up) occur during short (15 s) stimulation trains with 2 min pauses. The current intensity is then decreased until no motor effects occurred, i.e., it is set at the highest level which did not produce motor behaviors. The HFS is continuously applied and is generated with a Grass S-48 stimulator. Under these conditions, HFS at 130 Hz applied in both ventral hippocampi is able to modify the epileptogenesis induced by experimental models and the refractoriness to subsequent seizures during the postictal depression in rats [28].

Wet dog shakes are an interesting phenomenon that accompanies the hippocampal kindling [29] and represent neuronal hyperactivity in limbic structures that spread to midbrain areas and motor stem [30]. The ventral hippocampus and especially the dentate granule cells are essential for WDS expression and seizure spreading [31]. Animals receiving HFS during the epileptogenesis process show decreased incidence of wet dog shakes. This effect may denote hippocampal hypoactivity probably induced by this type of electrical stimulation.

The predominant protective effect of HFS during the epileptogenesis is to avoid the establishment of generalized seizures, an effect associated with HFS applied at low current intensity (range 120-300 μA). The higher efficacy of HFS in avoiding generalized seizures extends to other types of electrical stimulations such as vagal nerve stimulation [32] and quenching [33]. In contrast, higher intensities did not modify the progression of epileptogenesis. It is possible that HFS at higher current intensity results in spread of electrically altered neuronal activity to nearby structures, which function in part to influence seizure expression [34].

The status epilepticus, considered a major neurologic and medical emergency, is characterized by continuous or rapidly repetitive seizures [35]. SE is associated with one of the highest morbidities and mortalities in epilepsy [36]. Evidences support that the electrical stimulation of deep brain areas can be effective in reducing the SE. In a preliminary report, Benabid et al. [37] described that the SE induced by kainic acid injection into the amygdala was interrupted for several dozen seconds by HFS at 130 Hz applied in hippocampus. Similarly, the electrical stimulation of the anterior nucleus of the thalamus significantly increased the latency for seizures and pilocarpine-induced SE. This effect is more effective at 500 μA, regardless of the stimulation frequency used (20 Hz and 130 Hz) [38].

The protective effects of HFS in status epilepticus is enhanced when is associated with subeffective doses of diazepam, a drug enhancing the GABAergic neurotransmission [39]. Similarly, studies have demonstrated that unsatisfactory effects in neuropathic pain induced by electrical stimulation in spinal cord may be considerably improved by administration of low doses of analgesic drugs [40].

MECHANISMS OF THE BRAIN ELECTRICAL STIMULATION

The reduction in epileptic activity as well the enhanced refractoriness to subsequent seizures during the postictal period induced by HFS could be explained by different mechanisms. A) A rise in extracellular potassium concentrations and "depolarization block" of neurons induced by HFS [41]. The increased potassium can depolarize neurons sufficiently to tonically inactivate Na^+ channels such that action potentials cannot be initiated [42]. B) Neuronal hyperpolarization by activation of postsynaptic receptors in the hippocampus [43]. C) Enhanced activation or release of inhibitory neurotransmitters such as GABA [44]. In fact, we found that HFS of parahippocampal cortex is more effective in patients with less severe epilepsy, an effect associated with a high GABA tissue content [24]. D) Decreased synchronization and propagation of paroxysmal activity, an effect mediated by desynchronization [45]. The suppression of firing rate and abnormal synchronized oscillatory activity during HFS involve the activation of GABAergic terminals and subsequent GABA release [44], as well as the depression of subliminal voltage-gated currents underlying spontaneous spikes (persistent Na^+ currents) [46].

The clinical efficacy of HFS has been proposed to result from a decrease in the firing rate of neurons and the abnormal synchronized oscillatory activity [47]. It is described that HFS induces dual effects, i.e., it suppresses the spontaneous activity and imposes a new activity pattern [48]. In addition, HFS increases output from the stimulated site and changes the firing pattern and mean discharge rate of neurons at the projection sites [49]. It is possible that HFS alters the synchronous neuronal activity in hippocampus, and disrupts the propagation of ictal activity from the epileptic focus to other brain areas [50].

CONCLUSIONS

Epilepsy represents a complex disorder and the prevention of the development of resistance to medical therapy has been recognized as a major challenge. An important issue to consider is that many factors are believed to play a key role in the generation of the drug-resistant seizures, a situation that makes difficult the adequate treatment. Although several therapeutic strategies have been designed to control the drug-resistant epilepsy, the actual mechanisms and consequences of many of them are at present unknown. There are already promising results with certain strategies, but additional studies are mandatory. In addition, pharmacogenomic studies are necessary to design new therapeutic strategies to avoid or control drug-resistant epilepsy.

ACKNOWLEDGEMENTS

The authors thank Ms. Leticia Neri Bazan and Mr. Hector Vazquez Espinosa for their excellent technical assistance. This study was supported by the National Council for Sciences and Technology of Mexico (grant 45943-M).

REFERENCES

1. J.Jr. Engel and TA Pedly. Introduction: what is epilepsy. In *Epilepsy: A comprehensive textbook*, edited by J Engel Jr. and TA Pedley. Philadelphia: Lippincott-Raven Press, 1997, pp. 1-10.

2. WA Hauser, JF Annegers, LT Kurland. *Epilepsia* **32**, 429–445 (1991).

3. G Regesta and P Tanganelli. *Epilepsy Res* **34**, 109–122 (1999).

4. PR Huttenlocher, RJ Hapke. *Ann Neurol* **28**, 699–705 (1990).

5. R Emerson, BJ D'Souza, EP Vining, KR Holden, ED Mellits, JM Freeman. *N Engl J Med* **304**, 1125–1129 (1981).

6. J Aicardi and SD Shorvon. Intractable epilepsy. In *Epilepsy: A comprehensive textbook*, edited by J Engel Jr. and TA Pedley. Philadelphia: Lippincott-Raven Press, 1997, pp.1325-1332.

7. JT Gilman. *J Epilepsy* **3**(Suppl 1), 21–34 (1990).

8. MS Duchowny, AS Harvey, MR Sperling, PD Williamson. Indications and criteria for surgical intervention. In *Epilepsy: A comprehensive textbook*, edited by J Engel Jr. and TA Pedley. Philadelphia: Lippincott-Raven Press,1997, pp.1677-86.

9. JM Schwalb and C Hamani. *Neurotherapeutics* **5**:3-13 (2008).

10. M Kitagawa, J Murata, S Kikuchi, Y Sawamura, H Saito, H Sasaki, K Tashiro. *Neurology* **55**, 114-116 (2000).

11. AE Lang and H Widner. *Mov. Disord*, **17** (Suppl. 3) S94-S101 (2002).

12. JP Nguyen, JP Lefaucher, C Le Guerinel, JF Eizenbaum, N Nakano, P Carpentier, P Brugieres, B Pollin, S Rostaing, Y Keravel. *Arch. Med. Res.* **31**, 263-265 (2000).

13. IS Cooper, I Amin, M Ricklan. *Arch Neurol*, **33,** 559– 70 (1976).

14. M Velasco, F Velasco, AL Velasco, B Boleaga, F Jiménez, F Brito, I Marquez.

Epilepsia **41**, 158-169 (2000).

15. JF Kerrigan, B Litt, RS Fisher, S Cranstoun, JA French, DE Blum, M Dichter, A Shetter, G Baltuch, J Jaggi, S Krone, M Brodie, M Rise, N Graves. *Epilepsia* **45,** 346–54 (2004).

16. S Chabardès, P Kahane, L Minotti, .A Koudsie, E Hirsch, AL Benabid. *Epileptic Disord* **4(** suppl**),** 583– 593 (2004).

17. F Velasco, AL Velasco, M Velasco, L Rocha, D Menes. Electrical neuromodulation of the epileptic focus in cases of temporal lobe seizures. In *Deep brain stimulation and epilepsy*, edited by HO Lüders HO. London: Taylor & Francis, 2004 pp. 285– 97.

18. K Vonck, P Boon, E Achten, J De Reuck, J Caemaert.. *Ann Neurol* **52**, 556– 565 (2002).

19. MD Holmes, DL Silbergeld, D Drouhard, AJ Wilensky, LM Ojemann. *Seizure* **13**, 340-345 (2004).

20. F Velasco, JD Carrillo-Ruiz, F Brito, M Velasco, AL Velasco, I Marquez, R Davis. *Epilepsia.* **46**, 1071-81 (2005).

21. IS Cooper, I Amin, A Upton, M Riklan, S Watkins, L McLellan. *Appl Neurophysiol.* **40**, 124-34 (1977-1978).

22. EH Scharfman. *Ann New York Acad Sci* **911**, 305-327 (2000).

23. AL Velasco, M Velasco, F Velasco, D Menes, F Gordon, L Rocha, M Briones, I Márquez. *Arch Med Res* **31**, 316-328 (2000).

24. M Cuellar-Herrera, M Velasco, F Velasco, AL Velasco, F Jiménez, S Orozco, M Briones, L Rocha. *Epilepsia* **45**, 459-466 (2004).

25. AL Velasco, F Velasco, M Velasco, D Trejo, G Castro, JD Carrillo-Ruiz. *Epilepsia* **48** 1895-1903 (2007).

26. F Velasco, AL Velasco, M Velasco, F Jiménez, JD Carrillo-Ruiz, G Castro. *Acta Neurochir* Suppl. **97**(Pt 2), 337-42 (2007).

27. D Labar, J Murphy, E Tecoma. *Neurology* **52**, 1510-1522 (1999)

28. M Cuellar-Herrera, L Neri-Bazan, L Rocha. *Epilepsy Res* **72**, 10-7 (2006).

29. M Lerner-Natoli, A Hashizume, G Rondouin, M Baldy-Moulinier. C.R. Seances Soc Biol Fil **177** 93-101 (1983).

30 L Velíšek, P Mares. *Physiol Res* **53** 453-461 (2004).

31. G Rondouin, M Lerner-Natoli, A Hashizume. *Exp Neurol* **95** 500-5 (1987).

32. M Magdaleno-Madrigal, A Valdés-Cruz, D Martínez-Vargas, A Martínez, S Almazán, R Fernández-Mas, A Fernández-Guardiola. *Epilepsia* **43** 964-969 (2002).

33. ML Lopez-Meraz, L Neri-Bazan, L Rocha. *Epilepsy Res* **59** 95-105 (2004).

34. JB Ranck. *Brain Res* **98** 417-440 (1975).

35. RJ DeLorenzo. *Semin. Neurol* **2** 396-405 (1990).

36. NJ Aminoff, RP Simon. *Am. J. Med* 69:657-666 (1980).

37. AL Benabid, L Vercueil, K Bressand, M Dematteis, A Benazzouz, L Minotti, P Kahane. 2004 Absence seizures in the GAERS model: subthalamic nucleus stimulation. In *Deep brain stimulation and epilepsy*, edited by HO Lüders HO. London: Taylor & Francis, 2004: pp. 335-348.

38. C Hamani, FIS Ewerton, SM Bonilla, G Ballester, LEAM Mello, AM Lozano. *Neurosurgery* **54** 191-197 (2004).

39. M Cuellar-Herrera, JF Peña-Ortega, L Neri-Bazan, L Rocha. *Epilepsia* (submitted).

40. Z Song, BA Meyerson, B Linderoth. *Neurosci Lett* **436** 7-12 (2008).

41. M Bikson, J Lian, PJ Hahn, WC Stacey, C Sciortino, DM Durand DM. *J Physiol* **531** 181-191 (2001).

42. B Hille. Ionic channels of excitable membranes, 3rd Ed. Sinauer Associates, Sunderland, MA, U.S.A. 1992

43. KT Lu, PW Gean. *Neuroscience* **86** 729-37 (1988).

44. F Windels, N Bruet, A Poupard, N Urbain, G Chouvet, C Feuerstein, M Savasta. *J Neurosci* **12** 4141-4146 (2000).

45. MA Mirski, RS Fisher. *Epilepsia* **35** 1309-1316 (1994).

46. C Beurrier, B Bioulac, J Audin, C Hammond. *J Neurophysiol* **85** 1351-1356 (2001).

47. W Meissner, A Leblois, D Hansel, B Bioulac, CE Gross, A Benazzouz, T Boraud. *Brain* **128** 2372-2382 (2005).

48. L Garcia, J Audin, G D'Alessandro, B Bioulac, C Hammond. *J Neurosci* **23** 8743-51 (2003).

49. T Hashimoto, CM Elder, MS Okun, SK Patrick, JL Vitek. *J Neurosci* **5** 1916-1923 (2003).

50. EJ Hadar, Y Yang, U Sayin, PA Rutecki PA. *Epilepsy Res* **49** 61-71 (2002).

Clustering microcalcifications techniques in digital mammograms

Claudia. C. Díaz[†], Paolo Bosco[‡], Piergiorgio Cerello[‡]

[†]*Physics Department, Centro de Investigación y de Estudios Avanzados del IPN, A. P. 14-740, 07000 Mexico City, Mexico.*
[‡]*INFN Sezione di Torino, Via Pietro Giuria 1, 10125 Turin, Italy*

Abstract. Breast cancer has become a serious public health problem around the world. However, this pathology can be treated if it is detected in early stages. This task is achieved by a radiologist, who should read a large amount of mammograms per day, either for a screening or diagnostic purpose in mammography. However human factors could affect the diagnosis. Computer Aided Detection is an automatic system, which can help to specialists in the detection of possible signs of malignancy in mammograms. Microcalcifications play an important role in early detection, so we focused on their study. The two mammographic features that indicate the microcalcifications could be probably malignant are small size and clustered distribution. We worked with density techniques for automatic clustering, and we applied them on a mammography CAD prototype developed at INFN-Turin, Italy. An improvement of performance is achieved analyzing images from a Perugia-Assisi Hospital, in Italy.

INTRODUCTION

Breast cancer has become one of the major public health problems in Mexico. In 2006 the mortality rate was of 16 deceases for 100 thousand of women older than 25 years old, becoming the second cause of death due to cancer in women [1]. Many non-palpable breast cancers are detected by the mammographic demonstration of clustered calcifications, which are the smallest structures identified on mammograms. However, the great majority of them are benign, therefore causing a problem in diagnosis. Because both benign and malignant lesions can have such a mammographic appearance, most radiologists encourage biopsy in this situation, even though only about 20-30 per cent of cases prove to be cancer [2,3,4,5].

To differentiate benign from malignant clustered calcifications is considered the linear, curvilinear and branching shapes suggesting malignancy, and round or oval shapes indicating benign lesions.

Many radiologist suspect malignancy in any group of calcifications within a 1 cm^3 volume comprising at least five discrete particles less than 0.5 mm, although some would broaden the definition to encompass at least three particles [4].

Our work focused on automatic clustering of potential microcalcifications using this information, considering the density distribution or them. We briefly discuss the algorithm DBSCAN and its implementation on a prototype of CAD for

CP1077, *Advanced Summer School in Physics 2008, Frontiers in Contemporary Physics—EAV08*
edited by L. M. Montaño Zetina, G. Torres Vega, M. García Rocha, L. F. Rojas Ochoa, and R. López Fernández

mammography named GPCALMA. We report its performance, confronting with the previous diagnosis given by a specialist.

DBSCAN ALGORITHM

The algorithm DBSCAN (Density-Based Spatial Clustering of Applications with Noise) is designed to find clusters considering the density distribution of points in a database DB. The main difference among this algorithm and others (like k-means or k-medoids) is that DBSCAN discovers clusters of arbitrary shape and doesn't require to know a priori how many clusters must be created. It's very efficient even for large spatial DB's and it works even if they are noisy, because the density within the areas of noise is lower than the density in any of the clusters to be created.

The algorithm requires two global parameters: epsilon (ε) and minimum points (MinPts), which must be previously determined according to the characteristics of the clusters to find in the DB.

The algorithm is based on density distribution of points in the DB, so we define the ε-neighborhood of a point p, denoted by $N_\varepsilon(p)$ by the set of points whose distance to de point p is lower or equal to a radius ε, that is, $N_\varepsilon(p) = \{q \in D | \text{dist}(p,q) \leq \varepsilon\}$.

In order to differentiate between a border point and a core point in the DB, it's necessary to define a point p as directly density reachable from a point q wrt parameters ε and MinPts if p satisfies the following conditions:

1) $p \in N_\varepsilon(q)$ and
2) $|N_\varepsilon(q)| \geq$ MinPts (core point condition)

An extension of direct density reachability es the concept of density reachable, which is applied in the asymmetric case:

Definition 1. A point p is density-reachable from a point q wrt parameters ε and MinPts if there's a chain of points $p_1, \ldots, p_n$, $p_1=q$, $p_n=p$, such that p_{i+1} is directly density-reachable from p_i.

In the case that two border points are not density-reachable from each other, we apply the definition of density-connectivity to cover this relation of border points.

Definition 2. A point p is density-connected to a point q wrt parameters ε and MinPts if there is a point o such that both, p and q are density-reachable from o wrt ε and MinPts.

Intuitively, a cluster is defined to be a set of density-connected points which is maximal wrt density-reachability.

Definition 3. Let DB be a database of points. A cluster C wrt ε and MinPts is a non-empty subset of DB satisfying the following conditions:

1) For all p,q: if $p \in C$ and q is density-reachable from p wrt ε and MinPts then $q \in C$
2) For all $p,q \in C$: p is density-connected to q wrt ε and MinPts.

Once we have defined what a cluster is, in terms of density-connectivity and density-reachability, we are ready to describe the DBSCAN algorithm, which is visualized in Figure 1. First we assigned each point a label that indicates it doesn't belong to a cluster. The algorithm begins with an arbitrary starting point p. Then it retrieves all points density-reachable from p wrt ε and MinPts, according to definition 1, and if p is a core point, a cluster is created. We add all density-reachable objects or density-connected objects to this cluster. If p is a border point, no points are density-reachable from p, so we visit the next point in the database. The algorithm repeats recursively the evaluation process until all points have been processed [6].

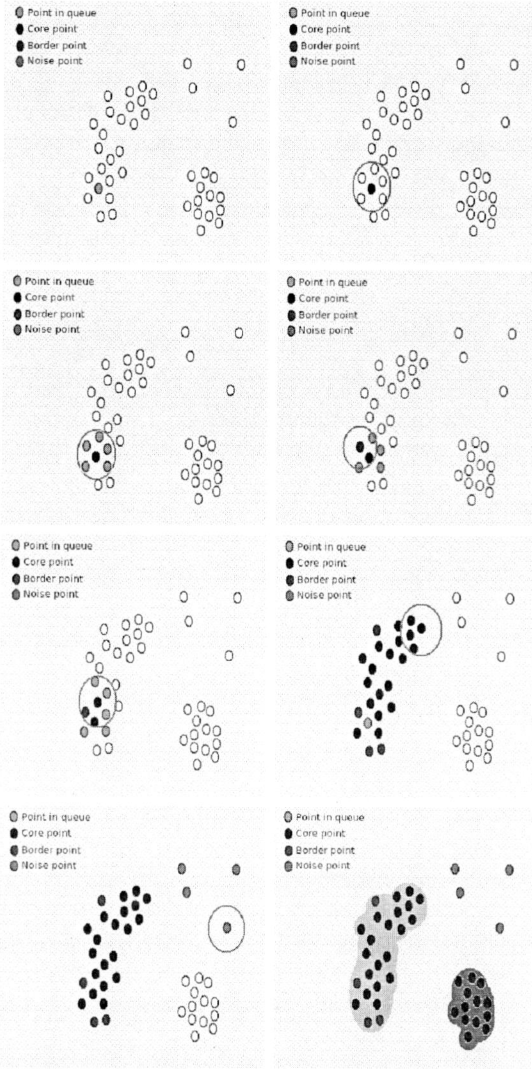

Figure 1. Stages of DBSCAN algorithm.

PROCEDURE

We analyzed a group of 26 digital mammograms of both craniocaudal and oblique views from 13 patients, obtained from a Perugia-Assisi Hospital, in Italy. The pixel size of the images was 75 microns and a 10 bit grey depth. They were visualized through the graphical interface (GUI) for mammography called GPCALMA (Grid Platform Computer Assisted Library for Mammography), which is a CAD prototype designed at INFN-Turin [7]. The GUI includes the diagnosis of benign, malign or no microcalcifications, which was previously made by a radiologist.
The algorithm DBSCAN was programmed in C++ using the available libraries in ROOT version 5.18/00. In our case, the DB consists of spatial coordinates of regions previously considered as potential microcalcifications on the mammograms. The optimal density parameters of the DBSCAN algorithm, ε and MinPts, were determined by analyzing this group of images. The results of the algorithm were the possible clusters of microcalcifications, which were confronted with the diagnosis made by a specialist.

RESULTS AND DISCUSSION

From the group of 26 images we found the optimal parameters for DBSCAN. In a radius ε equals to 50 pixels we search, at least, 10 potential microcalcifications. The CAD prototype correctly diagnosed 23 images using this clustering algorithm. We found a mean value of 5 false-positives per image. The possible reasons were the intensity variations of gray levels on skin line in the stage of image preprocessing. In order to improve the performance, we have to work in the binarization of images. In addition, a review of the diagnosis made by the radiologist is necessary, due to different clusters found by the CAD, and not marked by the specialist.
Disadvantages of the algorithm DBSCAN are that it doesn't work well for datasets with varying densities and the performance is very sensitive to changes in the parameters.

CONCLUSIONS

In this work we presented the algorithm DBSCAN applied to the detection of potential clusters of microcalcifications in digital mammograms. The main advantage is that two global parameters are necessary to be determined. However, the algorithm is very sensitive to changes in these parameters. The main disadvantage is the dependence of these optimal parameters of the group of images that are being analyzed.

ACKNOWLEDGMENTS

The authors wish to thank to the organizers of the 2008 Advanced Summer School for the invitation.

REFERENCES

1. http://www.dgepi.salud.gob.mx/diveent/RHNM.htm#situacion
2 J.N. Wolfe, *Am. J. Roentgenol.* **121,** 846-853 (1974).
3. E.A. Sickles. *Am. J. Roentgenol.* **143,** 461-464 (1984).
4. E.A. Sickles.*Radiology.* **160,** 289-293 (1986).
5. H.P. Chan, et. al. *Med. Phys.* **14(4),** 538-548, 1987.
6. M. Ester. "A density-based algorithm for discovering clusters in large spatial databases with noise" in The Second Internation Conference on knowledge Discovery and Data Mining 1996, edited by Evangelos Simoudis, et. al. AAAI Conference Proceedings, USA 1996.
7. Bottigli U. et. al. *International Congress Series,* 1256, 944-949 (2003).

V. SOLID STATE PHYSICS

ZnCdMgSe-Based Semiconductors for Intersubband Devices

Maria C. Tamargo

Department of Chemistry
The City College of New York
Convent Avenue and 138th Street
New York, NY 10031

Abstract. This paper presents a review of recent results on the application of ZnCdMgSe-based wide bandgap II-VI compounds to intersubband devices such as quantum cascade lasers and quantum well infrared photodetectors operating in the mid-infrared region. The conduction band offset of ZnCdSe/ZnCdMgSe quantum well structures was determined from contactless electroreflectance measurements to be as high as 1.12 eV. FT-IR was used to measure intersubband absorption in multi-quantum well structures in the mid-IR range. Electroluminescence at 4.8 μm was observed from a quantum cascade emitter structure made from these materials. Preliminary results are also presented on self assembled quantum dots of CdSe on ZnCdMgSe, and novel quantum well structures with metastable binary MgSe barriers.

Keywords: II -VI compounds, quantum wells, quantum cascade lasers, intersubband devices
PACS: 71.20.Nr, 73.21.Fg, 73.21.La, 78.30.Fs, 78.66.Hf, 78.67.De, 78.67.Hc, 78.67.Pt

Intersubband devices are devices that rely on electronic transitions within the quantum well energy levels in one band (the conduction band or the valence band) of the semiconductor. [1] This is in contrast to the more typical mechanism observed in semiconductor devices, which involves transitions between the conduction band and the valence band (interband or band-to-band transitions). Besides several improved performance features, one of the principal advantages of these devices is the relative independence of the device properties, such as emission wavelength, on the materials from which the structure is made. Rather, their properties depend on the design of the structure itself (through band structure engineering). These devices, which are no longer just a laboratory curiosity, but are being deployed in actual commercial applications, have evolved primarily due to the convergence of two technologies: advanced epitaxial growth techniques and band structure computation techniques. However, in some cases, there are some limits in the device properties that do depend on the materials used. An example of these materials dependent properties is the short wavelength limit of the intersubband (ISB) transitions, which is determined by the conduction band offset (CBO) of the two materials combined in the quantum well (QW) structure. [2]

In this paper we will discuss the potential advantages that wide bandgap II-VI

CP1077, *Advanced Summer School in Physics 2008, Frontiers in Contemporary Physics—EAV08*
edited by L. M. Montaño Zetina, G. Torres Vega, M. García Rocha, L. F. Rojas Ochoa, and R. López Fernández

compounds may offer for ISB device applications. In particular, we will present results recently obtained in the implementation of a particular family of wide bandgap II-VI materials in quantum cascade lasers and other ISB device applications, for the purpose of achieving shorter wavelength operation than that offered by the materials that are currently being employed. The paper will be divided into four sections. First, we will provide general background, including a brief description of the ISB device concepts and an introduction to the wide bandgap II-VI ZnCdMgSe materials. Secondly, we will discuss conduction band offsets and their measurement by a novel approach that uses modulation spectroscopic technique known as contactless electroreflectance. Third, we will present the recent results on quantum cascade (QC) emitters made from these II-VI compounds, and lastly, we will present several new directions in related ISB devices and materials that further advance the field.

BACKGROUND

Intersubband (ISB) devices exhibit ultra fast responses and very narrow linewidths, among other improved performance properties. They also rely almost entirely on the structural design parameters (QW layer thicknesses) rather than on the actual materials selected. [1] Figure 1 illustrates the concept of ISB transitions and compares them to the more frequently considered interband transitions in a semiconductor QW structure. Most well known semiconductor devices rely on transitions between the e1 and h1 levels in the conduction and valence bands, respectively (interband transitions). Although quantum confinement effects affect the energy of these transitions, they are largely determined by the bandgap of the QW layer material selected for the device, and thus when significantly different properties are desired, new materials must be utilized. Alternatively, in ISB devices, transitions between levels E2 and E1 in the conduction band (or H2 and H1, if the valence band ISB transitions are being considered) are the transitions of interest.

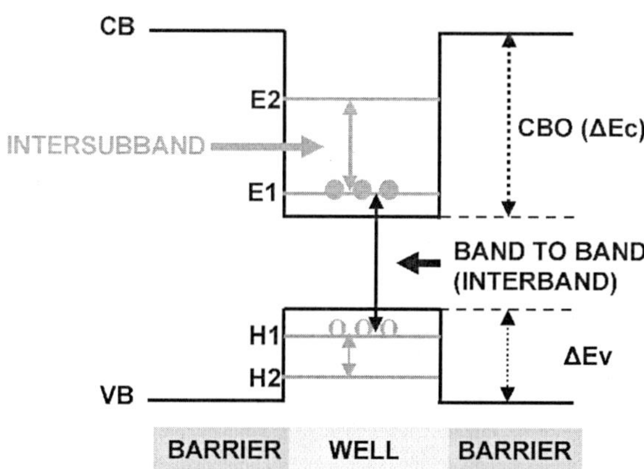

FIGURE 1. Interband and intersubband quantum well transitions

In this case, the transition energies are largely independent of the materials selected and can be tuned over a very large range by the selection of the appropriate layer thicknesses. This tunability is possible by the presence of advanced crystal growth techniques, such as molecular beam epitaxy, which enable the reproducible growth of ultrathin layers with extreme accuracy and reproducibility. The evolution of bandstructure engineering concepts and modeling capabilities has also enabled this very promising technology.

Two types of ISB devices have been extensively investigated over the last several years: quantum well infrared photodetectors (QWIPs) [3] and quantum cascade lasers (QCLs). [1, 2] Figure 2 shows a schematic representation of the basic operating principles of the QCL. This device is composed of alternating injector and active regions designed precisely such that when an external field is applied, the lowest energy level of one active region is aligned with the higher level of the next. This alignment allows electrons injected into the first active region to tunnel through the injector region into the next active region, after decaying through the emission of a photon. Once in the second active region, another photon emission can take place, followed by another tunneling event, and so on, as the electron "cascades" through the sequential active region stages. One can immediately appreciate an important advantage of such a device: quantum efficiencies much greater than one are possible, determined by the number of "cascading" stages in the structure. It should be noted that each active and injection region actually consists of many layers having precise thicknesses and compositions. These multilayered structures must meet several other important requirements that we will address at some length in a later section. In the case of QWIPs the process is one of ISB absorption instead of emission, and again, high quantum efficiencies are achieved by the multiple quantum well (MQW) nature of this device.

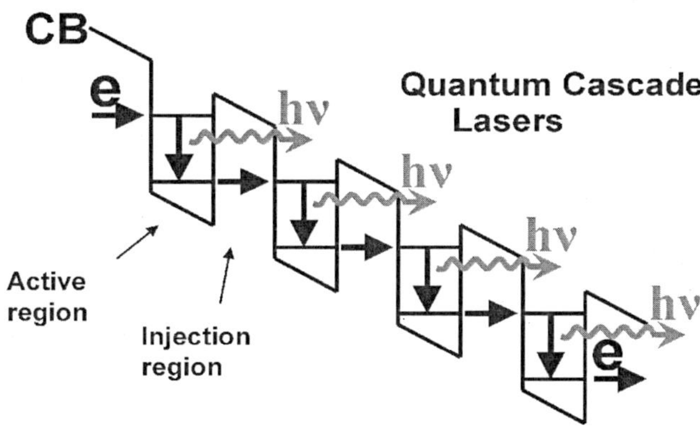

FIGURE 2: Schematic diagram of the QCL principle

The materials that have been used extensively with great success in the development of these devices to date have been InGaAs/InAlAs MQW grown on InP substrates, [4] and GaAs/AlGaAs MQWs on GaAs substrates. [5] The structures have been largely grown by molecular beam epitaxy, a growth technique that provides maximum control of layer thickness and precision control over complex multi-layered structures. The devices have been very successful in applications requiring transitions in the mid to far infrared region of the spectrum, namely in the 7-24 μm range. However, many applications, such as chemical and atmospheric sensing require operation at shorter wavelengths. It has now become rather clear that, although a certain degree of independence of materials parameters is a characteristic of these devices, the short wavelength response limit is given by the QW energy barrier height, that is, the conduction band offset (CBO), of the materials combined in making the QW. Thus, most recent research in these devices has focused on the identification and implementation of alternative materials for short wavelength applications.

TABLE 1. Heterostructure materials properties

Materials	CBO (eV)	Heterostructure Type	Intervalley Scattering	Effective CBO (eV)	Short Wavelength Limit
GaAs/AlGaAs	0.360	Type I	X-valley	0.300	8 μm
InGaAs/InAlAs	0.600	Type I	L-valley	0.520	6 μm
InGaAs/InAlAs (strained)		Type I	L-valley	0.700	4.5 μm
InAs/AlSb	2.1	Type II	L-valley	0.800	3 μm
ZnCdSe/ZnCdMgSe	1.12	Type I	none	1.12	

Table 1 summarizes the properties of the materials that are being explored most successfully to fabricate ISB devices. It also includes the II-VI materials that we are describing here. From this table it is seen that the shortest wavelength limit with the more well-developed materials is ~ 4.5 μm. To achieve shorter wavelengths, InAs/AlSb is being pursued very vigorously by several groups. [6] These materials have already demonstrated room temperature operation at wavelengths as short as 3 μm. [7] In spite of this significant success, several issues, such as the presence of intervalley scattering processes, limit the short wavelength reach in this system, and difficulties involved in the growth of these structures, whose layers have no common ions, may ultimately diminish their success. We propose that wide bandgap II-VI materials: in particular, ZnCdSe/ZnCdMgSe heterostructures, are a promising alternative. [8] These materials provide a CBO of 1.12 eV, [9] and have no intervalley scattering problems. Thus the entire CBO is available for device design, which should allow for devices operating well below the 3 μm limit thus far demonstrated. Furthermore, the incorporation of strain may further reduce the wavelength that can be attained to numbers as low as 1.55 μm, which is of interest for some new device concepts, such as all optical switches [10].

ZnCdSe/ZnCdMgSe

ZnCdSe/ZnCdMgSe materials have been investigated extensively for interband device applications, such as red-green-blue semiconductor lasers and light emitting diodes. [8] Their bandgap vs lattice constant characteristics are summarized in Figure 3. The circles indicate experimental data points of samples previously grown by our group. The family of interest is the subset of these alloys that can be grown lattice matched to InP substrates, indicated by the vertical line in Figure 3. These materials offer a large range of bandgaps, from 2.1 eV to 3.6 eV. Their structural, optical and electronic properties have been optimized to device quality levels. [8]

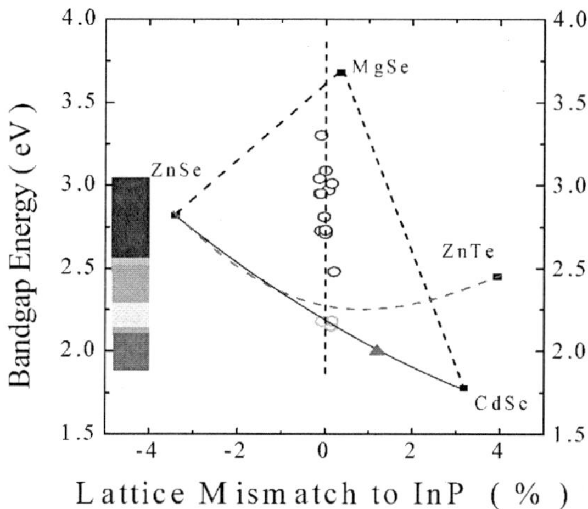

FIGURE 3: Relationship between bandgap and lattice mismatch to InP for the ZnCdMgSe quaternary alloy system

Molecular beam epitaxy has been the technique of choice in the growth of these materials. Its low temperature and non-equilibrium conditions favor the growth of dissimilar materials, allowing the optimization of conditions for the growth of these materials on well-established high-quality III-V substrates, such as GaAs and InP. The heterostructures presented in this work were grown using a Riber 2300 system composed of a chamber for the growth of As-based III–V materials and another for II–VI materials. The two growth chambers are interconnected by ultra-high vacuum transfer modules.

Before the growth of the II-VI layers, the InP substrate is heated in the III-V chamber under an arsenic overpressure until desorption of the oxide surface layer is achieved. The oxide desorption is assured by observing the transition from a (2x4) As-stabilized InP surface to the (4x2) In-stabilized InP surface. On observing this transition, the substrate temperature is rapidly lowered by 30° C before commencing growth of a lattice matched $In_{0.53}Ga_{0.47}As$ buffer layer. The InGaAs buffer with a growth rate of 1μm/h is grown for 15 minutes under As-rich growth conditions at a substrate temperature that is 30° below the oxide layer desorption temperature. This 0.25 μm InGaAs buffer layer is grown to ensure that a smooth III-V surface with group-V termination is present for the initiation of the II-VI layer growth.

A sequence of steps was developed to initiate ZnCdSe growth on an InGaAs surface. [11] Upon completion of the InGaAs layer growth, the sample is transferred to the II-VI chamber where the As-terminated InGaAs surface is exposed to a zinc flux for 40 s at a temperature of 200° C, a step referred to as "Zn-irradiation". The Zn-irradiation step is performed to suppress the formation of In_2Se_3 and Ga_2Se_3 compounds at the III-V/II-VI interface. The presence of these selenium compounds must be minimized to prevent the formation of defects such as stacking faults which are known to degrade the material. This is followed by the growth of a low-temperature (100° C below the optimal growth temperature) $Zn_{0.43}Cd_{0.57}Se$ buffer layer (~90 Å) to promote two-dimensional nucleation. It was previously shown that this sequence of steps, including the low temperature $Zn_{0.43}Cd_{0.57}Se$ interfacial layer results in high quality epitaxy and reduces the defect densities to ~ $5 \times 10^4 cm^{-3}$ [11].

After the low temperature layer, the growth is interrupted and the substrate temperature is raised and stabilized at the growth temperature (~300° C) under a selenium flux before carrying out the growth of the II-VI structure. The growth rates of ZnCdSe and ZnCdMgSe were typically ~ 0.5 and 1.5 μm/hour, respectively, and the Se/group-II flux ratio was maintained at above 4, ensuring Se-stabilized growth conditions. Reflection high energy electron diffraction (RHEED) is used routinely to monitor and control the growth. RHEED intensity oscillations were often measured after the growth interruption, at the initiation of the growth at 300 °C. The RHEED intensity oscillations indicate a clear layer-by-layer growth mechanism during the growth of these materials, and enable us to establish the growth rates with high accuracy. We also previously demonstrated the growth of the high bandgap ternary end point material, ZnMgSe, lattice matched to InP. [12] Excellent structural and optical properties were obtained following the growth procedure described above.

DETERMINATION OF THE CONDUCTION BAND OFFSET

The short wavelength limit of the ISB devices is ultimately set by the CBO of the materials selected for the structure. Thus, measurement of the CBO is of great interest in the development of short wavelength devices. We have implemented a method for measurement of the CBO using a modulation spectroscopy technique known as contactless electroreflectance (CER). CER provides an accurate determination of the optical transitions and is relatively easy to implement. CER [13] measures the changes in the optical reflectance of the material with respect to a modulating electric field. This gives rise to sharp, differential-like spectra in the region of the transitions, and allows the observation of higher energy transitions.

Our experimental setup [14] is composed of a 150 W xenon-arc lamp, 0.2m focal-length monochromator with a 1200-line/mm diffraction grating blazed at 500nm, silicon detector, and silica lenses. The sample holder utilizes a condenser-like system consisting of a front wire grid electrode with a second metal electrode separated from the first electrode by insulating spacers, which are ~0.1 mm larger than the sample dimension. The sample is placed between these two capacitor plates. The electromodulation is achieved by applying an ac voltage of 1.2kV, 200 Hz across the electrodes.

Figure 4 shows a CER spectrum of a single ZnCdSe/ZnCdMgSe QW sample. [15] The solid line is the measured RT CER spectrum. The energies corresponding to the transitions observed were obtained using a fit, shown by the dashed line, based on the first derivative of a Gaussian lineshape [13, 16]. The values of the measured transitions are indicated by arrows in the figure and are summarized in Table 2. The notation EnH(L)m indicates that the transitions are from the nth conduction subband to the mth valence subband of heavy (H) or light (L) hole character, respectively.

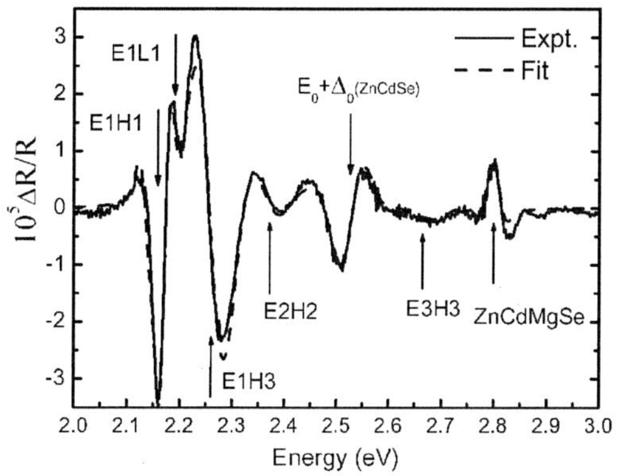

FIGURE 4: Contacless electroreflectance spectrum of a ZnCdSe/ZnCdMgSe QW

To identify the transitions, we first looked for several of them that could be established through other methods. For example, the signal from the barrier at 2.8 eV was straightforward when we considered the 77K PL signal at 2.88eV and its thermal energy shift. The intensities of the transitions at 2.159 eV and 2.192 eV exhibit a ratio close to three suggesting that they are associated with the heavy and light hole transitions, respectively [11, 13]. The band gap of the bulk $Zn_{0.53}Cd_{0.47}Se$ was

TABLE 2. Experimental and calculated interband energies.

Transition	Experiment (eV)	Theory (eV)
E1H1	2.159±0.005	2.158
E1L1	2.192±0.005	2.185
E1H3	2.259±0.005	2.261
E2H2	2.369±0.005	2.371
E3H3	2.660±0.005	2.664
$E_0+\Delta_0$	2.525±0.005	2.522
E_0(Barrier)	2.800±0.005	

determined by CER measurements on a thick sample grown during the same run. The value that we found was E_0=2.080±0.005eV, in good agreement with reference [17], which reported E_0=2.078±0.002 and spin-orbit splitting Δ_0=0.442±0.02eV. Using our value for the band gap and this value for Δ_0 we obtain $E_0+\Delta_0$=2.522eV, which agrees well with the experimental value of Table 2.

In order to calculate the expected energies corresponding to the observed transitions we performed a calculation based on the envelope approximation [18, 19], which considers the nonparabolicity of the bands. Using the effective masses and the spin-orbit splitting from the literature, we calculated the energies of the different transitions as a function of the parameter $Q_c=\Delta E_c/\Delta E_0$. Figure 5 shows the results of this calculation in solid lines for the transitions that most closely fit the experimental values of the transitions. The experimental values are represented by the horizontal dashed lines. The transitions correspond to the symmetry allowed (n=m) and symmetry forbidden but parity allowed (n=m±2,4...) transitions. As indicated in this figure by the dotted vertical line, the best agreement between the calculated and the experimental values for all the transitions was found for Q_c=0.82 (ΔE_c=590meV).

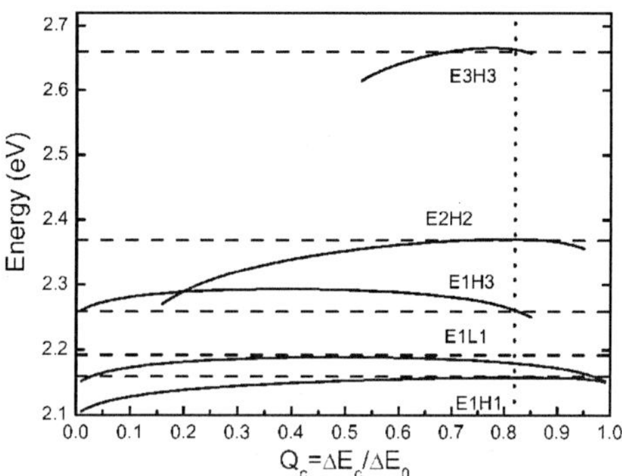

FIGURE 5: Energies of the transitions determined by the envelope function approximation vs. Q_c (=$\Delta E_c/\Delta E_0$). The dashed lines indicate the experimental values measured by CER.

Since we want to explore the full range of CBO that could be achieved with these materials, the high bandgap limit of the lattice matched compositions, ZnMgSe lattice-matched to InP, was also grown and characterized [9]. We performed a similar study with QW structures of $Zn_{0.53}Cd_{0.47}Se$, using the lattice-matched composition $Zn_{0.13}Mg_{0.87}Se$ for the barriers. [12] Figure 6 shows the room-temperature CER measurement of a $Zn_{0.53}Cd_{0.47}Se/Zn_{0.13}Mg_{0.87}Se$ single QW with a nominal thickness of ~35 Å. A CdSe cap layer was used for this structure in order to avoid overlap between the CER signal of the $Zn_{0.53}Cd_{0.47}Se$ QW and the cap layer. As before, the energies corresponding to the observed transitions were obtained using a fit, shown by the dashed line, based on the first derivative of a Gaussian line shape [13, 16]. The signal at 1.696 eV corresponds to the CdSe cap layer. The transitions of the

$Zn_{0.13}Mg_{0.87}Se$ barrier layer, determined by RT PL to be at 3.52 eV cannot be seen in our spectrum due to the limit of the CER apparatus at high energies. The transitions in the range of 2.2–2.8 eV correspond to five QW transitions. Excellent agreement was obtained between the experimental and calculated transitions for ΔE_c=80% of the band-gap discontinuity (ΔE_0), which yields a very large value for the CBO of 1.12 eV. This value is much larger than that of the InGaAs/InAlAs III-V materials typically used in ISB devices. It is also larger than the effective bandgap of the antimony-based systems being considered for short wavelength applications.

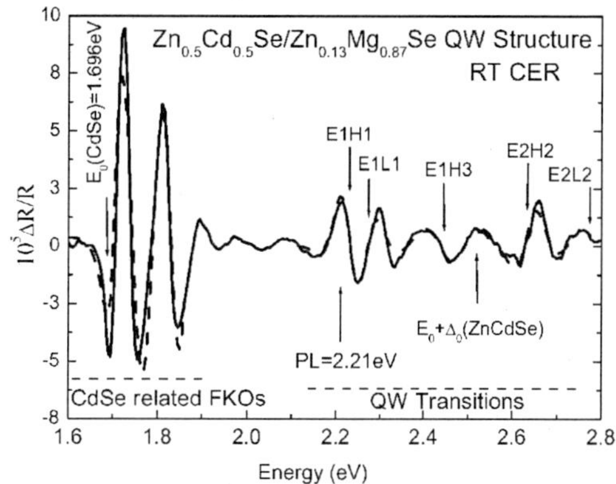

FIGURE 6: CER spectrum of a $Zn_{0.13}Mg_{0.87}Se/Zn_{0.53}Cd_{0.47}Se$ QW structure (solid line). Dashed lines are fits yielding the energies indicated by the arrows.

We also investigated MQW samples using CER. [20] Two samples were grown. Sample A had ten periods of $Zn_{0.5}Cd_{0.5}Se/Zn_{0.2}Cd_{0.2}Mg_{0.6}Se$ QWs sandwiched between two quaternary $Zn_{0.2}Cd_{0.2}Mg_{0.6}Se$ layers. Sample B had 60 periods of $Zn_{0.5}Cd_{0.5}Se/ Zn_{0.2}Cd_{0.3}Mg_{0.5}Se$ QWs, in this case, sandwiched between two ternary $Zn_{0.5}Cd_{0.5}Se$ layers. The nominal thicknesses of the wells were 50 and 40 Å in samples A and B, respectively, while the barriers are nominally 140 Å in both samples. In both

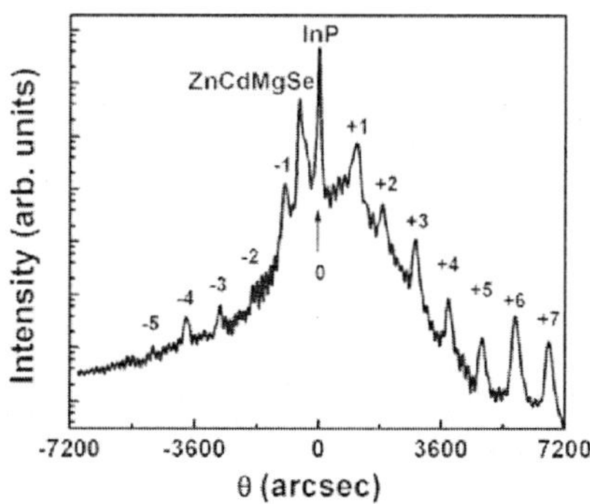

FIGURE 7: High resolution X-ray diffraction of the ZnCdSe/ZnCdMgSe MQW structure of sampleA.

samples, the QW layers are doped *n*-type using $ZnCl_2$ as the dopant source. Doping was needed in order to measure the ISB absorption. The structure was capped by a 90 Å thick CdSe layer. XRD measurements showed that for both samples the ZnCdMgSe barrier and the ZnCdSe QW layers were nearly lattice matched to the InP substrate. The high resolution XRD shown in Figure 7 confirms the excellent crystalline quality.

The interband transitions in the samples were determined using CER. The *E1-E2* ISB transition was estimated from the CER results at 6.9 and 5.3 μm for samples A and B, respectively. The predicted values of the *E1-E2* transitions were confirmed by FTIR measurements. Absorption peaks at 180 meV (6.89 μm) and 231 meV(5.37 μm) are clearly observed, which they are strongly polarization dependent. The full width at half-maxima (FWHM) are 21 and 30 meV, indicating a ratio of $\Delta E/E$peak of 10% for both samples. This ratio supports the assertion that the absorption is due to the bound-to-bound ISB transition *E1-E2*.[21] The excellent agreement between the FTIR results and the CER predictions confirm that CER can be used effectively to predict the energy of the ISB transitions.

QUANTUM CASCADE EMITTERS

Quantum cascade lasers (CQLs) are unipolar intersubband devices with a gain medium that is composed of repeated periods of active and injector regions. [1, 2] Lasing results from the radiative emission and resonant tunneling that occurs as injected electrons traverse a series of lasing stages. The gain is directly proportional to the number of active regions since one electron could generate a photon at each stage. The active region typically consists of a two or three coupled-QW structure that gives rise to a three-level system. An example of a three coupled-QW active region is shown in Figure 8. Other designs will be discussed shortly.

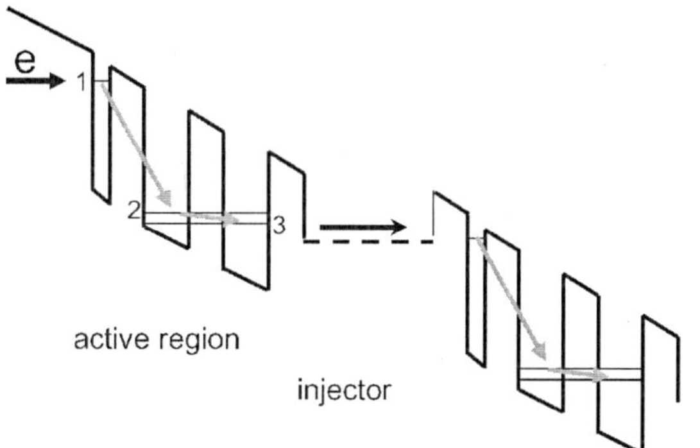

FIGURE 8: Three level active region for a quantum cascade laser

One of the attractive features of the QCL is that many of the device properties are controlled solely by the structure design (layer thicknesses), and not by the choice of the materials. However, as was stated before, the maximum separation between the energy levels, which gives rise to the emission energy, is limited by the value of the CBO for the two materials selected for the structure. Shorter wavelengths can only be achieved for large CBO. At higher temperatures, electrons can also escape to the continuum if the higher level is weakly confined. This escape gives rise to leakage currents that can be reduced by increasing the size of the CBO. The CBO also affects the barrier "transparency". For high bandgap barrier materials the barrier layers must be kept thin in order for coupling between the wells to exist. Another example of a material parameter that plays an important role in the QCL design and operation is the effective electron mass, which has an influence on the device gain.

The most well established material in the production of mid-infrared QC lasers is the InGaAs/InAlAs on InP materials system, which exhibits emission in the 4.5 to 24 μm region [4]. Within this wavelength range, the technology has progressed to the point where QC lasers are now finding commercial applications. These lasers can be made to operate in the absence of cryogenic cooling, allowing them to be part of very portable gas sensing instruments and infrared transmitters. Due to the high output power, tunability, and narrow linewidth of their emission, these devices are ideally suited for commercial applications in the areas of chemical spectroscopy and high-speed free-space telecommunication [1].

Research for QC structures operating below 4 μm using InGaAs/InAlAs has not had similar success because of the limited CBO of this material system. This limitation has led to the exploration of alternative materials. With CBOs of 2.1 and 1.6 eV, several authors have reported QCL emission as low as 2.75 μm using InAs/AlSb QWs grown on InAs substrates and InGaAs/AlAsSb grown on InP [6, 22]. However, the performance of Sb-based devices is still not comparable to their InGaAs/InAlAs counterparts at longer wavelengths. This is in part due to the limitation of the effective CBO of InAs/AlSb and InGaAs/AlAsSb to about 0.8 and 0.5 eV, respectively, due to inter-valley electron scattering. Other problems with the antimony-based materials have to do with difficulties in their growth due to the absence of common ions between the two materials. We propose that the ZnCdMgSe/ZnCdSe system offers a large CBO (as high as 1.12 eV) with no intervalley scattering, and a single common anion (Se) thus representing a very promising alternative.

Device Fundamentals

The active region of a QCL can be modeled by a general three level system, [1, 22] illustrated in the schematic of Figure 9. The population in levels 2 and three are given by n_2 and n_3 respectively:

$$n_3 = (\eta_i \, J/e)\, \tau_3$$

where: η_i is the injection efficiency ($\eta_I = J_3/J \sim 1$), and

$$n_2 = n_3 \, \tau_2$$

The condition for population inversion, a requirement for lasing, is given by the equation:

$$n_3 - n_2 = \frac{\eta_i J}{q}(\tau_3 - \tau_2) \succ 0$$

Thus, for population inversion to occur the lifetime of electrons in level 3, τ_3 (which may include both radiative and nonradiative processes), must be longer than the lifetime of electrons in level 2, τ_2. In order to ensure that relationship, the active region of a QCL is designed such that that the separation of levels 1 and 2 is one LO phonon, ensuring the need for $\tau_3 > \tau_2$; and the separation of levels 2 and 3 is the desired emission energy.

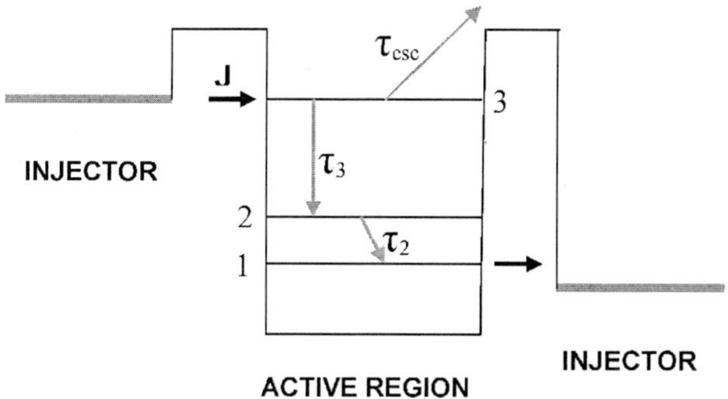

FIGURE 9: Schematic representation of a three level QCL active region

The active region illustrated in Figure 8 is an example of a design that relies on a diagonal transition (that is a transition between states that are localized in adjacent wells). Another type of active region, such as the asymmetric coupled QW (ACQW) active region of the structure in Figure 10, relies on vertical transitions. Each of these has its advantages and disadvantages. Diagonal transitions, have larger τ3 values, thus potentially leading to larger gains, but have broader linewidths due to the effects of interface roughness. Vertical transitions lead to sharper emission lines, but more complex active region designs are needed to further reduce the value of τ2.

The injector region is also composed of a complex multi-layered structure and its design is meant to facilitate the tunneling of electrons between active regions and their effective channeling into level 3 of the subsequent active region. It consists of digitally graded alloys of the well and barrier material (typically 50-100 layers), with precise thicknesses of each layer in order to optimize the transfer of electrons into the next active region and minimize the probability of the electrons repopulating the level 1 in the previous active region ("thermal backfilling").

$Zn_xCd_{(1-x)}Se/Zn_{x'}Cd_{y'}Mg_{(1-x'-y')}Se$-InP QC emitters

With these basic design concepts in mind, and in order to investigate the $Zn_xCd_{(1-x)}Se/ Zn_{x'}Cd_{y'}Mg_{(1-x'-y')}Se$-InP suitability for QCL fabrication, a QC emitter structure was designed to emit light at a wavelength of 4.5 μm. [23] The structure consists of several hundred layers of alternating thin layers of $Zn_xCd_{(1-x)}Se$ and $Zn_{x'}Cd_{y'}Mg_{(1-x'-y')}Se$ having precise thicknesses and compositions. Initially, calibration layers of the ZnCdSe and ZnCdMgSe materials were grown. Based on these calibration runs, the material composition for the II-VI epilayers in the QC structure was selected to be $Zn_{0.43}Cd_{0.57}Se$ and $Zn_{0.20}Cd_{0.19}Mg_{0.61}Se$. These compositions produce ternary and quaternary layers lattice-matched to InP with bandgaps of 2.08 and 3.03 eV at 300 K, respectively, and are expected to have a CBO of 0.78 eV. The QC emitter structure was designed using a program based on the envelope function approximation assuming an electron effective mass of m_0 x 0.128 and m_0 x 0.181 (where m_0 is the free electron mass) for $Zn_{0.43}Cd_{0.57}Se$ and $Zn_{0.20}Cd_{0.19}Mg_{0.61}Se$, respectively [24]. The simulation results, showing the squared moduli of the wave functions in the conduction band are displayed in Figure 10. The proposed structure is composed of 20-30 repeats of asymmetric coupled quantum well (ACQW) active regions with alternating (multi-layer) injector regions. One period (one active region and one injector region) of the designed structure starting from the injection barrier is made up of the following layers in angstroms: ***30****/34/****10****/28/****20***/24/**10**/<u>22/**12**/20/**16**/20/**18**/18/**18**/18/**20**/16/**20**/16/**20**/14</u>/**22**/14/**24**/12/**26**/12 with the $Zn_{0.20}Cd_{0.19}Mg_{0.61}Se$ barrier indicated in bold and the Cl-doped layers underlined. The active region layers are shown in italics.

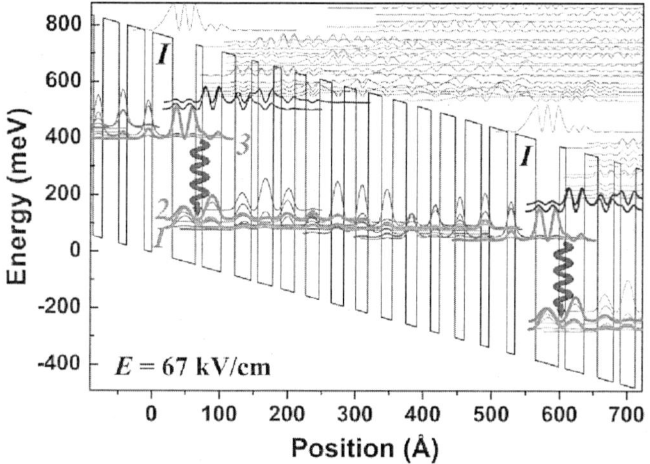

FIGURE 10: Proposed bandstructure of a QC emitter based on $Zn_xCd_{(1-x)}Se/Zn_{x'}Cd_{y'}Mg_{(1-x'-y')}Se$-InP with an asymmetric coupled QW active region.

In order to further optimize the accurate growth parameters that would lead to the growth of this complex multi-layered structure, another set of calibration samples composed of several repeats of the ACQW layers separated by simple quaternary barrier layers, rather than the multi-layered injector regions, were grown. The alternating ACQWs and quaternary barrier layers were repeated 25 times and the wells were doped with chlorine (n ~ $4x10^{18} cm^{-3}$) in order to observe intersubband (ISB) absorption through multi-pass transmission experiments using FTIR spectroscopy. Prior to the growth of the ACQW test structures, individual layers were grown to determine the exact growth rate and composition of the well and barrier materials used. In some of the samples, RHEED intensity oscillation measurements were made to accurately determine the growth rates.

The room temperature FTIR absorption for the first ACQW structure grown is shown in Figure 11. The two absorption peaks seen at 0.240 and 0.301 eV were assigned to the $\mathbf{e_2} - \mathbf{e_3}$ and $\mathbf{e_1} - \mathbf{e_3}$ transitions. Simulations based on the envelope function approximation, predict $\mathbf{e_2} - \mathbf{e_3}$ and $\mathbf{e_1} - \mathbf{e_3}$ absorption at 0.266 and 0.313 eV, respectively. Furthermore, a separation of $\mathbf{e_2} - \mathbf{e_1}$ of 0.061 eV is deduced from the two FTIR peaks, while the predicted value based the simulation was 0.047 eV [24]. The combination of these energy discrepancies is consistent with the ACQW barrier being slightly thinner than intended. At this point, a second ACQW structure was grown and characterized after recalibrating the growth conditions. For the re-calibration, we used RHEED intensity oscillations to improve the accuracy of the measured growth rates.

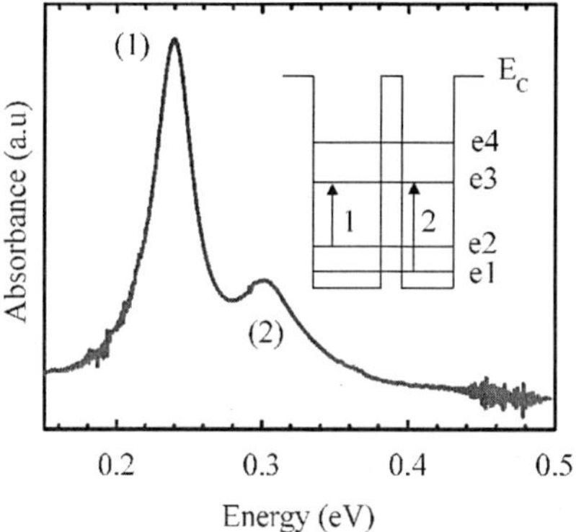

FIGURE 11: FTIR absorption measurement of a calibration structure consisting of 25 repeats of ACQWs separated by quaternary layers

A QC structure was grown according to these improved growth parameters [25, 26]. Two peaks, a sharp one at ~380 meV and a broad one centered around 420 mev were observed. The lower energy transition was assigned to the $\mathbf{e_1} - \mathbf{e_3}$ transition, due to its agreement with the simulation results. We also identified the broad, higher energy absorption peak observed in the FTIR spectrum as originating from the $\mathbf{e_2} - \mathbf{e_4}$

and $e_1 - e_4$ transitions present in the un-biased QC emitter structure. In addition to the dominant peak due to emission from the active region, the PL spectrum of the QC emitter exhibits emissions originating from the injector region and the contact layer.

QC electroluminescent device structures were fabricated in the form of semicircular cleaved mesas. [24] Circles of 400 μm diameter were etched into mesa structures using a photolithography and wet chemical etching, followed by a second lithography step to apply the top contact metallization consisting of 150 Å Ti followed by 2500 Å Au. Contacts of Ge/Au were deposited on the InP side. The mesas were cleaved into semicircular QC emitter structures. Electroluminescence was collected for temperatures between 78 and 300K for several applied currents. We observed electroluminescence (EL) emission centered near 4.8 μm. The emission polarization characteristics were examined to confirm the fact that it was originating from ISB transitions since ISB optical transitions in quantum wells are TM polarized. Results of the EL measured at 78K are shown in Figure 12. The emission peak grows with increasing pumping current, as anticipated. Current-voltage curves showed typical QC behavior, with a characteristic current "turn-on" once sufficient voltage has been applied. The turn-on voltage at 78 K occurs at 5.40 V. Although a somewhat lower turn on would ultimately be desired, this value is reasonable given the bandgap of the materials.

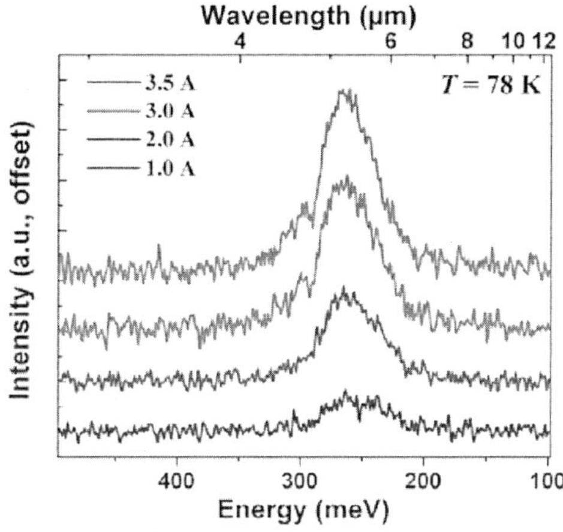

FIGURE 12: QC electroluminescence from a ZnCdSe/ZnCdMgSe QC structure.

OTHER INTERSUBBAND DEVICES AND MATERIALS

Quantum Well Infrared Photodetectors

Quantum well infrared photodetectors or (QWIPs) are being used to replace narrow bandgap IR and far-IR detectors such as HgCdTe-based detectors. [3] Unlike those devices, based on conventional interband optical absorption, QWIPs use ISB absorption in quantum wells. Advantages of the QWIPs are a rapid carrier relaxation time, a large tunability of transition energies (detection wavelengths) by precise structural design, and the fact that they are made from well established III-V materials, thus overcoming the growth difficulties associated with small bandgap semiconductors. However, as in the discussion above, for short wavelength detection (in the 2-5 micron range) alternative materials such as the ZnCdMgSe materials are of interest.

To demonstrate the feasibility of this material in QWIP applications we have grown MQW structures and characterized them using FT-IR and time resolved PL. The 10 periods MQW structures, with different QW widths, were sandwiched between a 3000 Å (bottom) and a 1000 Å (top) n-type $Zn_{0.5}Cd_{0.5}Se$ contact layers. [26] The nominal thickness of the $Zn_{0.2}Cd_{0.2}Mg_{0.6}Se$ barrier is 5 nm. In the investigation described here, four samples with $Zn_{0.5}Cd_{0.5}Se$ well widths of 4 nm, 3 nm, 2 nm and 1 nm (samples A, B, C and D, respectively) were grown. All the QW layers and the contact layers were doped by chlorine (Cl), obtained with a $ZnCl_2$ source.

Photoluminescence and high resolution X-ray diffraction (HRXRD) were used to establish the quality of the materials. HRXRD shows numerous satellite peaks as well as interference fringes both indicators of high structural quality. The PL exhibits sharp peaks with no evidence of deep level emission, also indicative of high materials quality. Time-resolved PL spectra of the four samples were measured at different temperatures in the range of 77 K ~ 295 K. All the PL traces decay exponentially with time and can be well described by a first order exponential equation,

$$I(t) = I_0 e^{-t/\tau_{PL}} + C$$

where I is the PL intensity at time t, I_0 is the maximum PL intensity at $t = 0$, τ_{PL} is the PL decay time and C represents noise level. Since the investigated QWIP structures were n-type and heavily doped, the PL decay time represents the lifetime of the minority carriers, holes. The linear dependence of the decay process with temperature indicates that it is dominated by radiative recombination, as desired for ISB devices.

Fourier transform infrared (FT-IR) spectroscopy was used to measure the ISB absorption. Figure 13 shows the normalized absorbance of the samples at room temperature, obtained by taking the ratio of P-polarized spectra over S-polarized spectra. For samples (A, B and C) with different well widths from 5 to 3 nm, peak absorption λ_p were observed at 6.88 μm (0.18 eV), 5.35 μm (0.23 eV) and 3.99 μm (0.31 eV), respectively. Fitting by a Gaussian line shape gives the FWHM (ΔE) values of 21 meV, 30 meV and 46 meV for the three samples. The value of $\Delta E/E_{peak}$ is of the order 10% to 20%, which is typical for a bound-to-bound transition and comparable to the values obtained for the well-studied III-V semiconductors. [27]

The solid orange line shown in Figure 13 is the absorption of the thinnest well sample D. Instead of shifting the absorption peak to a shorter wavelength, the peak was observed at 4.43 μm (0.28 eV). Also, the absorption peak is much wider ($\Delta E/E_{peak}$ = 32%) than those of the other three samples and has an asymmetric shape with a tail

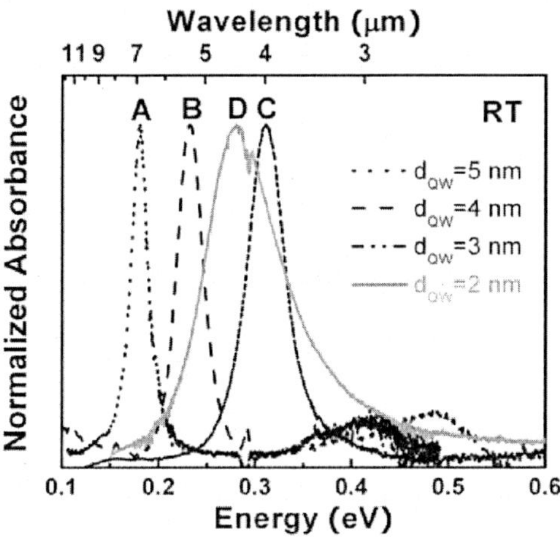

FIGURE 13: Normalized FTIR absorption spectra for MQW QWIP structures.

at the higher energy side. These characteristics are typical of bound-to-continuum transitions, and so we attribute the unexpected shift to the fact that the second level in this well is no longer confined, but is in the continuum. [27] With a bandgap of 2.9 eV, the barrier is not high enough to confine the second level in the QW with a width of 2 nm. The results suggest that for the barrier composition used in this study it would not be possible to achieve shorter wavelength by simply reducing the QW width. In order to achieve further reduction of the absorption wavelength a wider bandgap barrier layer must be used. The use of strain compensated structures may further result in even shorter wavelength devices.

Self Assembled Quantum Dots (SAQDs)

So far we have confined our discussion to the use of two dimensional structures or QWs. Lower dimensional structures, such as quantum dots (QDs) may offer additional advantages in ISB device applications. For example, quantum dot infrared photodetectors (QDIPs) are expected to allow normal incidence detection, since the selection rules that make normal incidence absorption forbidden for QWs, will be relaxed. [3] Also, QDs in QCLs are predicted to add benefits in terms of reducing the non-radiative decay processes in those devices. Thus, it is of interest to investigate the ISB transition properties in SAQDs in the II-VI materials.

We have previously demonstrated that SAQDs are formed under the appropriate growth conditions, when CdSe is deposited by MBE on ZnCdMgSe surfaces. [28] The formation of these structures is driven by strain. However, several mechanisms are possible to control the size, and thus the electronic and optical properties of these zero dimensional structures. In the case of CdSe on ZnCdMgSe, we have shown that good control of QD size can be achieved by either changing the CdSe deposition time or by changing the Mg content of the quaternary barrier layer. This latter phenomenon is

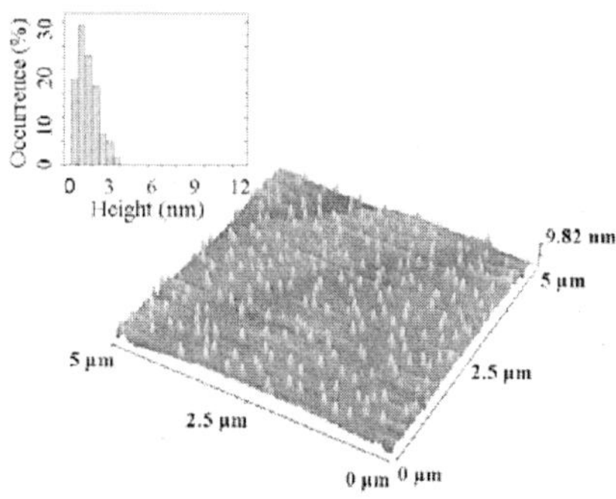

FIGURE 14: Atomic force micrograph of uncapped CdSe QDs on a ZnCdMgSe surface.

ascribed to surface energy effects. Figure 14 shows an atomic force micrograph of a surface where CdSe self assembled QDs have formed on the ZnCdMgSe surface. When the QD structures are capped with another layer of ZnCdMgSe they exhibit a shift of the PL emission energy as a function of QD size. Emission from these capped SAQDs can be observed over the visible spectrum range.

To explore the use of these QD structures in ISB devices, we have investigated the ISB absorption of three SAQD structures. [29] The samples were grown by MBE on (001) semi-insulating InP substrates in our dual chamber Riber system. After the removal of the oxide layer under an As flux and the growth of a 0.15 µm InGaAs buffer layer in the III-V growth chamber, the samples were transferred through the vacuum modules to the II-VI growth chamber. There, the InGaAs surface was exposed to a Zn flux for 20 s followed by the growth of a 10 nm low-temperature ZnxCd1−xSe buffer layer at 200 °C. After these steps, which are needed for adjusting the III-V and II-VI interfaces to improve the material quality of the epitaxial layers, the substrate temperature was raised to 300 °C to grow the subsequent layers. A 60 nm $Zn_xCd_{1-x}Se$ buffer layer was grown before the growth of the multilayer stacks of $CdSe/Zn_xCd_yMg_{1-x-y}Se$ QDs. Three samples with 50 QD layers were grown for this study. In all three samples, the $Zn_xCd_yMg_{1-x-y}Se$ barrier layer thickness vertically separating the dots is 18 nm. The band gap of the $Zn_xCd_yMg_{1-x-y}Se$ barrier is about 3.0 eV at 77 K as determined from separate calibration layers. Samples A and B have identical structures except that sample A is undoped while in sample B the CdSe QD layers were doped with Cl. The nominal CdSe thickness deposited for samples A and B is 1.56 nm, corresponding to 5.2 ML. For sample C, CdSe QD layers with a nominal deposited thickness of 6.9 ML were also doped with Cl. From AFM measurement results on uncapped QDs, the height of the dots can be inferred to be in the range between 2.5 and 4.5 nm while the lateral size is between 30 and 60 nm. The overall structures are capped with 25 nm $Zn_xCd_{1-x}Se$ layers.

Figure 15 shows the absorbance of the three samples at room temperature. The spectra were obtained by taking the ratio of p-polarized spectra over s-polarized spectra. For sample A, which is undoped, no absorption features were observed. Absorption with a peak at 2.54 µm (0.49 eV) with a FWHM of 46 meV was observed for sample B. The narrow linewidth ($\Delta E / E_{peak}$ =9.4%) is an indication of a relatively uniform distribution of dot sizes. For sample C, two absorption peaks were observed: at 2.69 μm (0.46 eV) with a FWHM of 29 meV ($\Delta E / E_{peak}$ = 6.3%) and at 3.51 μm (0.35 eV) with a FWHM of 20 meV ($\Delta E / E_{peak}$ =5.7%).

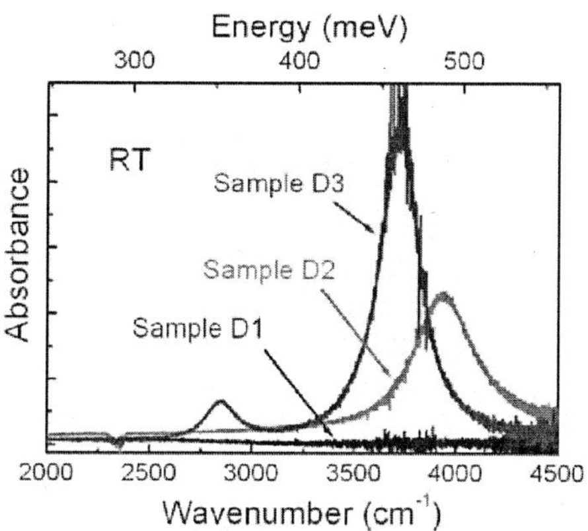

FIGURE 15: ISB absorption spectra for three SSAQD samples.

There are several possible explanations for the origin of the two peaks observed in sample C, including: 1) the presence of a bimodal size distribution of the QDs and 2) the existence of transitions involving more than two energy levels in the QDs. If we assume that the two peaks originate from a bimodal size distribution, we would expect to see two emission peaks in the PL spectra. PL measurements were made to help clarify the possible origin of the two absorption peaks. The PL spectrum for sample C did not exhibit a second peak that would be present in the case of a bimodal QD size distribution; thus, we attribute the presence of the two absorption peaks in sample C to the presence of more than two levels within the dots. We tentatively assign the absorption peak at 0.35 eV (Figure 15) to the transition from the ground state (n=1) to the first excited state (n=2) and the peak at 0.46 eV to the transition from the first excited state (n=2) to the second excited state (n=3). Further experiments, such as temperature-dependent ISB absorption measurements and photoconductivity measurements, may be needed to further confirm this interpretation.

New Materials and Structures

In order to obtain the highest possible CBO in these II-VI materials we have explored the use of binary MgSe as the barrier material. [30] This structure has several advantages over the one using the lattice matched ternary or quaternary layers discussed above. These are: 1) the increased CBO of ~1.2 eV enables the tuning of ISB transition energies over a wider range; 2) the structure can be grown lattice matched to InP substrates with a net strain compensation; 3) the binary nature of the barrier material makes easier to control the growth and may improve its transparency.

Three samples consisting of 15 periods of MgSe/$Zn_xCd_{1-x}Se$ double quantum wells separated by $Zn_xCd_{1-x}Se$ spacer layers (100 MLs) are reported here. The thicknesses of the $Zn_xCd_{1-x}Se$ wells in each sample are nominally 11, 13, and 15 MLs, respectively. The well layer composition is x = 0.46, which is the composition that is lattice-matched to InP. The well was uniformly doped with $ZnCl_2$ to an electron density of about (1-2) x 10^{18} cm^{-3}. The MgSe barrier thickness is 20 MLs for all the samples. Photoluminescence (PL) measurements were carried out at 77 K and room temperature using the 325 nm line of a He-Cd laser as the excited source. The ISB absorption was measured with a multiple-pass waveguide geometry using a Fourier-transform IR (FTIR) spectrometer with a wire-grid polarizer and a liquid-nitrogen-cool HgCdTe detector.

The lattice mismatch between zincblende (ZB) MgSe and the InP substrate is only 0.2%. However, theoretical calculations and experiments indicate that MgSe prefers the rocksalt (RS) structure over the ZB structure. [31, 32] The critical thickness of ZB MgSe grown on InP is affected not only by the lattice mismatch but also by the structural phase transition. RHEED intensity oscillations were observed when the growth of MgSe was initiated on the $Zn_xCd_{1-x}Se$ surface, indicating that the growth proceeds by a layer-by-layer two-dimensional growth in the initial growth stages. Furthermore, the surface reconstruction of MgSe shows a (2 x 1) pattern, which is similar to the Se-terminated $Zn_xCd_{1-x}Se$ surface, suggesting that MgSe retains the ZB structure. However, after the growth of 5 periods of MgSe/$Zn_xCd_{1-x}Se$ QWs (without the thick spacers), the RHEED pattern becomes spotty, possibly due to the onset of the MgSe phase transition from ZB to RS. To improve the structural quality, thick $Zn_xCd_{1-x}Se$ spacer layers (~100 MLs) were introduced for every three MgSe barriers. With the insertion of the spacer layers, sharp streaky RHEED patterns persist until the end of the growth of the entire structure (30 QWs).

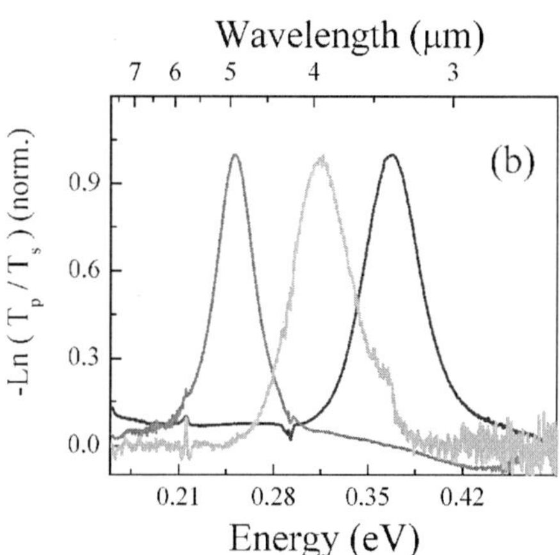

FIGURE 16: FTIR Absorption spectra for ZnCdSe QWs grown with metastable MgSe barriers.

The PL spectra of the three samples measured at 77 K exhibit strong emission peaks at 526 .5nm, 533.6 nm, and 540.6 nm for three samples with well layer thickness of 11 ML, 13 MLs, and 15 MLs, respectively. The strong and narrow (FWHM ~ 38 – 46 meV) QW PL spectra with no trace of deep level emission are indications of excellent material quality of the samples. ISB absorption spectra for the three samples with different well thicknesses are shown in Figure 16. The absorption peaks are relatively narrow ($\Delta E/E_{peak}$ ~12 - 15%) with linewidths typical for a bound-to-bound transition in a QW. With decreasing well width from 15 MLs to 11 MLs the ISB transition energy shifts from 4.9 μm to 3.35 μm.

The peak ISB transition energy as a function of QW width was plotted in Figure 17. The energy separation between the conduction band levels, E1 and E2, was calculated using the envelope function approximation with the conduction band profile shown in the inset. The nonparabolicity of the conduction band was taken into account for the calculation of the energy levels in the QW. Excellent agreement between the experimental results and theoretical calculation was obtained. The calculations suggest that the wavelength of the ISB transition can be extended to the near IR region with further decrease of the QW width. It should be possible to obtain an ISB transition with wavelength as low as 1.65 μm with a well width of ~ 5 MLs (the minimum well thickness to ensure two bound states in the well). With the use of strain compensation, or using coupled quantum well structures, the ISB transition wavelength could be extended further, to the 1.55 μm optical-communication wavelength.

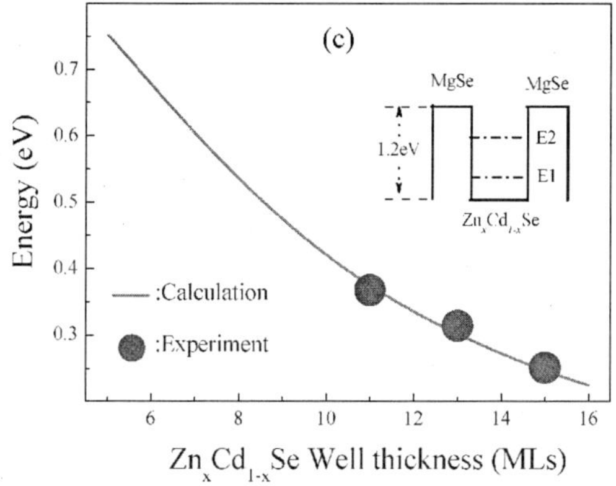

FIGURE 17: Dependency of ISB transition energy on QW thickness for ZnCdSe QWs with MgSe barriers

SUMMARY

We have presented a review of our recent results of ZnCdMgSe-based QW structures for ISB device applications. The CBO of these heterostructures were measured using a modulated spectroscopy technique known as contactless electroreflectance. QC electroluminescence at 4.8 μm was obtained from structures made from these materials. Investigations of the application of these materials to QWIP structures were also described, as well as recent studies of ISB transitions observed in CdSe self assembled QDs grown on ZnCdMgSe layers, and investigations ZnCdSe QWs with metastable MgSe barriers. Our results demonstrate that this family of wide bandgap II-VI materials are attractive for applications in mid- and near-infrared intersubband devices such as quantum cascade lasers, quantum well and quantum dot infrared photodetectors, and novel devices, such as all-optical switches, due to their large conduction band offset and excellent materials properties.

ACKNOWLEDGEMENTS

The work presented in this paper has been supported through the National Science Foundation Grant Numbers ECS-0217646 and EEC-05400832 (MIRTHE-ERC), the National Aeronautics and Space Administration Grant Number NCC-1-03009 (NASA-COSI), and Department of Defense Grant Number W911NF-04-1-0023.

REFERENCES

1. J. Faist, F. Capasso, D. L. Sivco, C. Sirtori, A. L. Hutchinson, and A. Y. Cho, Science **264**, 553 (1994).
2. *"Intersubband Transitions in Quantum Structures,"* R. Paiella, Editor, McGraw Hill, 2006.
3. H. Schnider, H. C. Liu, "Quantum Well Infrared Photodetectors; Physics and Applications (Springer, Berlin, 2007).
4. A. Lyakh, C. Pflugl, L. Diehl, Q. J. Wang, F. Capasso, X. J. Wang, J. Y. Fan, T. Tanbun-Ek, R. Maulini, A. Tsekoun, R. Go, and C. N. Patel, Applied Physics Letters **92**, 111110 (2008).
5. C. Sirtori, P. Kruck, S. Barbieri, H. Page, J. Nagle, M. beck, J. Faist and U. Oesterle, Applied Physics Letters **73**, 3486 (1998).
6. R, Teissier, D. Barate, D. Bour, A. Vicet, C. Alibert, A. N. Baranov, Applied Physics Letters **85**, 167 (2004).
7. J. Devenson, O. Cathabard, R. Teissier, and A. N. Baranov, Applied Physics Letters **91**, 251102 (2007).
8. S. P. Guo and M. C. Tamargo in *"II-VI Semiconductor Materials and Their Applications,"* M. C. Tamargo, Editor (Taylor and Francis, Ann Arbor, 2002).
9. M. Sohel, X. Zhou, H. Lu, M. N. Perez Paz, M. C. Tamargo, and M. Muñoz, Journal of Vacuum Science and Technology **B 23** (2005).
10. B. S. Li, R. Akimoto, K. Akita, and T. Hasama Applied Physics Letters **88**, 221915 (2006).
11. L. Zeng, B. X. Yang, M. C. Tamargo, E. Snoeks and L. Zhao, Applied Physics Letters **72**, 1317 (1998).
12. M. Sohel, M. Muñoz, M. C. Tamargo, Applied Physics Letters **85**, 2794 (2004).
13. F. H. Pollak and H. Shen, Materials Science and Engineering, R. **10**, 275 (1993).
14. F. H. Pollak in *"Group III Nitride Semiconductor Compounds,"* B. Gil, Editor (Clarendon, Oxford, 1998) p. 158.
15. Martín Muñoz, H. Lu, X. Zhou, M. C. Tamargo, and F. H. Pollak, Applied Physics Letters **83**, 1995 (2003).
16. O. J. Glembocki and B. V. Shanabrook, in Semiconductors and Semimetals, eds. R. K. Willardson and A. C. Beer (Academic, New York, 1992) Vol. **36**, p. 221.
17. T. Holden, P. Ram, F. H. Pollak, J. L. Freeouf, B. X. Yang, and M. C. Tamargo, Physical Reviews B **56**, 4037 (1997).
18. G. Bastard and J. A. Brum, IEEE Journal of Quantum Electronics **QE-22**, 1625 (1986).

19. G. Bastard, "*Wave Mechanics applied to Semiconductor Heterostructures,*" (Les editions de Physique, Cedex 1988).
20. H. Lu, A. Shen, M. C. Tamargo, W. Charles, I. Yokomizo, M. Muñoz, K. J. Franz, C. Gmachl, C. Y. Song and H. C. Liu, Journal of Vacuum Science and Technology B **25**, 1103 (2007).
21. F. Capasso, J. Faist, C. Sirtori, A. Y. Cho, Solid State Communications **102,** 231 (1997).
22. Q. Yang, C. Manz, W. Bronner, K. Kohler and J. Wagner, Applied Physics Letters **88**, 121127 (2006).
23. W. O. Charles, A. Shen, K. J. Franz, C. Gmachl, Q. Zhang, Y. Gong, G. F. Neumark and M. C. Tamargo, Journal of Vacuum Science and Technology B **26,** 1171 (2008).
24. M. Munoz, H. Lu, X. Zhou, M. C. Tamargo and F. H. Pollak, Applied Physics Letters **83**, 1995 (2003).
25. K. J. Franz, W. O. Charles, A. Shen, A. J. Hoffman, M. C. Tamargo and C. Cmachl, Applied Physics Letters **92**, 121105 (2008).
26. H. Lu, A. Shen, M. C. Tamargo, W. Charles, I. Yokomizo, M. Muñoz, Y. Gong, G. F. Neumark, K. J. Franz, C. Gmachl, C. Y. Song,and H. C. Liu, Journal of Vacuum Science and Technology **B 25**, 1103 (2007).
27. "*Intersubband Transitions in Quantum Wells: Physics and Device Applications I & II,*" Semiconductors and Semimetals Vols. 62 and 66, edited by H.C. Liu and F. Capasso (Academic Press, San Diego, 2000).
28. M. N. Perez Paz, H. Lu, A. Shen, F. Jean-Mary, D. Akins, M. C. Tamargo, Jouran of Crystal Growth **294,** 296–303 (2006).
29. A. Shen, H. Lu, W. Charles, I. Yokomizo, M. C. Tamargo, K. J. Franz, C. Gmachl, S. K. Zhang, X. Zhou, R. R. Alfano, and H. C. Liu, Applied Physics Letters **90**, 071910 (2007).
30. B.S. Li, A. Shen, W. O. Charles, Q. Zhang, M. C. Tamargo, Applied Physics Letters **92**, 261104 (2008).
31. M. Rabah, B. Abbar, Y. Al-Douri, B. Bouhafs, and B. Sahraoui, Materials Science & Engineering B **100**, 163 (2003).
32. H.M. Wang, J. H. Chang, T. Hanada, K. Arai, and T. Yao, Journal of Crystal Growth **208**, 253 (2000).

What determines the magnitude of Tc in HTSC?

R. Baquero

Department of Physics, Cinvestav, Av. IPN 2508, 07300 México, D.F.

Abstract. Neither Eliashberg equations nor the BCS one are predictive to the critical temperature, Tc, of a superconductor. A certain amount of phenomenological equations have been constructed with the aim of predicting the experimental values found for Tc. They preserve the form of the BCS-Tc equation but introduce different combinations of the electron-phonon interaction parameter, λ, and the electron-electron repulsion parameter, μ^*, that appear in the Eliashberg-Tc equations. The agreement with experiment is, in general, rather poor. But since these are the only instrument that exists to predict Tc, they are widely used. A criterion that "the higher the λ, the higher the Tc" has emerged from those equations and the value of this parameter became the accepted criterion to discard or accept the electron-phonon interaction as a possible mechanism in HTSC. In this paper we analyze the theoretical foundations of this criterion and the validity of the criterion itself and compare to the results of Eliashberg-Migdal theory to conclude, first, that they contradict each other whenever we are dealing with HTSC and, second, that a low electron-phonon interaction parameter, λ , is not an argument solid enough to discard the e-ph interaction as a mechanism in HTSC.

INTRODUCTION

BCS theory [1] gives rise to an equation for the critical temperature of a superconductor, Tc, that depends on the attraction parameter, V, which cannot be accurately neither calculated nor measured. So, it cannot be used as a predictive equation, as it is well known. Eliashberg gap equations [2] allow the calculation of the critical temperature of a superconductor, Tc, ounce the Eliashberg function, $\alpha^2 F(\omega)$, and some other parameters are known or set . They are not, even then, predictive as well. This arises mainly because the electron-electron repulsion parameter, μ^*, on which they depend, cannot be accurately neither measured nor calculated. When the Eliashberg function is unknown still a series of equations have been developed to predict Tc. In many cases, what is known is the electron-phonon interaction parameter, λ, which can be deduced from experiment or can be calculated as a normal state property. It is also related to the Eliashberg function as its first inverse moment. The magnitude of λ has been widely used as an indicator of the magnitude of the critical temperature of a superconductor on the bases of a criterion derived from the just-mentioned set of empirical equations that preserve the exponential dependence of the BCS one and introduce different combinations of the electron-phonon interaction and the electron-electron repulsion parameters that enter in Eliashberg-Migdal theory (EMT). Others are approximations that derive from EMT and that contain, in addition, in certain cases, parameters as the integral of the Eliashberg function, A, which is more complicated to establish without knowing the Eliashberg function itself.. There is no precise theory on the adequate magnitude of the parameter, A, for a particular superconductor, but a kind of general agreement seems to exist that it should be kept, in any case, at reasonably low values [3]. There is a whole series of papers that deal with these approximate empirical equations [4]. They all support the criterion that "the higher the λ, the higher

CP1077, *Advanced Summer School in Physics 2008, Frontiers in Contemporary Physics—EAV08*
edited by L. M. Montaño Zetina, G. Torres Vega, M. García Rocha, L. F. Rojas Ochoa, and R. López Fernández

the Tc" and actually set a limit to Tc by taking λ to infinity to obtain the highest possible value of the exponential term.. As a corollary, the impossibility emerges for a superconductor with a low value of λ to be a candidate for a high Tc. That a high-Tc material requires compulsorily a high value of λ, is the common "wisdom" nowadays. State-of-the-art calculations find systematically low values for λ in HTSC [5] and the important question is whether or not this is an argument solid enough to discard the e-ph interaction as the mechanism responsible for the superconducting phase transition in these systems.

In this paper, we analyze the theoretical foundations of this criterion to show that, first, the correspondence between λ and Tc is rather poor even for low-Tc materials; second, that it contradicts EMT, in some sense, whenever a high-temperature electron-phonon (e-ph) superconductor for which EMT holds is concerned (if it exists) and third, that to discard the e-ph mechanism on the basis of a low value of λ, is not theoretically well-founded. A low value of λ, as we show below, helps keeping the value of the parameter A at a "reasonable" magnitude. Furthermore, the results from EMT seem to indicate that for high critical temperature superconductors, the parameter λ could not contain anymore the crucial information on the factors that lead to the phase transition.

The rest of the paper is organized as follows. In the next section II, we deal with the Tc equations to recall that none of them is really predictive and to introduce a few of the empirical equations developed to predict Tc and to artificially set limits on it. In section III, we examine the criterion "the higher the λ, the higher the Tc" in some detail. Section IV is devoted to the functional derivative of Tc with the Eliashberg function, $\alpha^2F(\omega)$. In this section, we draw most of the conclusions that sustain our arguments. In section V, we use the experimental and theoretical results to analyze the actual correspondence that exists between λ and Tc. Our section VI answers to the question whether a low λ value is such a bad result for an e-ph HTSC for which EMT holds (if it exists). In the final section VII, we summarize our arguments and draw our conclusions.

THE Tc EQUATIONS AND THE LIMITS TO Tc

BCS theory [1] gives rise to Eq. (1) for the critical temperature, Tc.

$$K_B T_c = 1.13\hbar\omega_D e^{-\frac{1}{N(0)V}}. \qquad (1)$$

K_B is the Boltzmann constant, N(0) the density of electrons at the Fermi level, $\hbar$ is Planck's constant, ω_D the Debye frequency and V the attraction potential. This last parameter cannot be neither calculated nor determined experimentally. For that reason the BCS Tc-equation cannot actually be used to predict the critical temperature of a superconductor. Eliashberg gap equations [2] can be solved numerically to give an exact account of the thermodynamics of an electron-phonon (e-ph) conventional superconductors. We do not need to spell them down here. The interested reader is referred to the abundant existing literature (see the references quoted above, for example). These equations can be linearized at T = Tc and can be solved ounce the needed data are given. The so-called Eliashberg function, $\alpha^2F(\omega)$, contains all the information on the existing phonons, the electron-phonon interaction and the included information on the conduction electrons in the material. One needs further to know accurately the electron-electron coulomb repulsion parameter, μ^*. Nevertheless, this parameter cannot be obtained neither theoretically nor experimentally with enough accuracy to be useful. So, in practice, the Eliashberg linear equations valid at Tc are used to fit a proper value for μ^* to the experimental value of Tc. The exact value of μ^* depends also on the choice of the cut-off frequency that is necessary in order to end the infinite sum over the Matsubara frequencies. It is only after this fit is done that it is useful to proceed (using the same parameters) to solve the non-linear Eliashberg equations valid below Tc from which the thermodynamics of the corresponding superconductor can be calculated, resulting, in general, in very good agreement with experiment. Several codes have been produced with this purpose since long ago and they are widely known [6]. But, then, it is clear that Eliashberg equations cannot predict Tc even in the cases when the Eliashberg function, $\alpha^2F(\omega)$, is known. Actually, there is no reliable predictive equation for Tc, nowadays.

This problem has been addressed replacing the unknown product N(0)V in the BCS Tc-equation (1) by different combinations of the electron-phonon interaction parameter, λ, and the electron-electron coulomb repulsion parameter, μ^*, in an effort to reproduce the experimental results. The BCS-Tc equation can be transformed, for example, from Eq.(1) to

$$K_B T_c = 1.13\hbar\omega_D e^{-\frac{1}{\lambda-\mu^*}} \tag{2}$$

where the denominator on the exponential N(0)V has been replaced by λ-μ^*. The idea is that the net attraction can be approximately described by both expressions. A BCS-like equation can be obtained from the linear Eliashberg gap equations by performing some replacements and assuming some equivalencies. The details can be found, for example, in ref. [3]. It results in Eq. (3)

$$K_B T_c = 1.13\hbar\omega_D e^{-\frac{1+\lambda}{\lambda-\mu^*}} \tag{3}$$

which might actually be seen as the result of replacing $N(0)V \rightarrow \dfrac{1+\lambda}{\lambda-\mu^*}$.

From Eq. (1), one can extract the criterion "the higher N(0)V, the higher the Tc" which might be translated, on one hand, into "the higher the number of electrons at the Fermi energy, the higher the Tc" which is understandable as more cooper pairs can be formed from a higher number of electrons at the Fermi level and superconductivity is expected to be a more robust state in this case. On the other hand, the same condition can be interpreted as "the higher the pairing potential the higher the Tc" which is also understandable since a high pairing potential gives rise to a stronger bound in the pair state and it also translates in a more robust superconducting state. One might think that a stronger bound can be associated to a higher energy of the intermediate boson. This interpretation, as we shall see below, approaches BCS theory to the conclusions that derive directly from Eliashberg-Migdal theory.(EMT)..

The criterion that "the higher the λ, the higher the Tc" follows from Eqs. (2) and (3), and several others of the kind which are derived from or related to Eliashberg gap equations in one way or another. There is a long series of papers [4] investigating different versions of similar equations which, in essence, have all the exponential dependence of the BCS Tc-equation but with coefficients and parameters that aim to improve the agreement with experiment. It is important to stress that these equations are related to but do not follow from EMT. They are not a direct consequence of it.

A very well known formula of this sort is the one first developed by Mc Millan [7] and later refined by Allen and Dynes [8]. It reads

$$K_B T_c = \frac{\hbar \omega_{\ln}}{1.2} e^{-\left[\frac{1.04(1+\lambda)}{\lambda-\mu^*(1+0.62\lambda)}\right]} \tag{4}$$

with

$$\omega_{\ln} = \exp\{+\frac{2}{\lambda}\int_0^\infty \ln(\omega)\ \frac{\alpha^2 F(\omega)}{\omega}\ d\omega\} \tag{4a}$$

This very popular formula deviates as much as 20% from the experimental value, in the case of Pb. A point for discussion here is the adequate value for μ^* to be used in this formula that, on the other hand, requires the knowledge of the Eliashberg function, $\alpha^2F(\omega)$, itself. To use Eq.(4), μ^* is to be guessed. But, then, it is not clear whether the same guess for μ^* does not give a better approximation to Tc, when it is used to solve directly the linear Eliashberg equations, a task which, in such conditions, is almost as trivial nowadays as to use Eq. (4).

Limits to the value of Tc are found by taking λ to infinity in Eqs. (3) or (4) or in similar equations. These equations give a finite, usually, different number for the highest possible critical temperature. In other words, they set a definite limit to the critical temperature of a superconductor. EMT sets actually no limit on the possible values for a high temperature superconductor. We will come back to this point below. These limits should obviously exist but might have to come from outside EMT as from lattice instability or other arguments.

"THE HIGHER THE λ, THE HIGHER THE Tc"

The electron-phonon interaction parameter, λ, is a property of the normal state that can also be obtained from the Eliashberg function, $\alpha^2F(\omega)$, through the following formula,

$$\lambda = 2\int_0^{\omega_c} \frac{\alpha^2 F(\omega)}{\omega} d\omega \tag{5}$$

λ is therefore the first inverse momentum of the Eliashberg function , $\alpha^2F(\omega)$. It is clear that if a high value of lambda would be compulsory for a high value of Tc, then the most important phonon frequencies would be the lower ones. In other words, to have a high-Tc superconductor, its Eliashberg function should have most of its intensity at the lowest possible frequencies since $1/\omega$ becomes a high weighting factor at low frequencies (see Eq. (5)). We will show below that this conclusion is in contradiction with EMT whenever a high temperature superconductor is concerned.

THE FUNCTIONAL DERIVATIVE OF Tc WITH RESPECT TO α^2F(ω), δTc/$\delta\alpha^2$F(ω).

An important question to answer nowadays is whether it is possible for a superconductor with a low electron-phonon interaction parameter, λ, to have a high critical temperature. We address this question under the assumption that the superconductor in question is appropriately described by the e-ph EMT.

Let us first recall some of the properties of the functional derivative of Tc with respect to $\alpha^2F(\omega)$, $\delta Tc/\delta\alpha^2F(\omega)$, that will be useful to our arguments in this paper.. It is important to understand the meaning of this function in detail. The answer to the question, how does the critical temperature change (ΔTc) when we induce a small change in the Eliashberg function ($\Delta\alpha^2F(\omega)$) at some particular frequency, ω, (through doping, for example), is given precisely by this function, as follows:

$$\Delta Tc = \int_0^\infty \frac{\delta Tc}{\delta\alpha^2 F(\omega)} \Delta(\alpha^2 F(\omega))d\omega \tag{6}$$

We do not need to write explicitly the exact formula for this functional derivative here. It can be found in several places [9]. We only need here to know some of its characteristics. We merely show here how it looks like qualitatively (see Fig. 1). It rises linearly from zero and shows a maximum at a certain frequency which is characteristic to each superconductor. Then it slowly goes monotonically to zero as ω goes to infinity. But its most important property is that the just mentioned maximum is universal, this means that it is located at a frequency, ω_{opt},

$$\hbar\omega_{opt} = 7K_B T_c \tag{7}$$

called the optimal frequency. The universal law that the optimal frequency, in units of the critical temperature, is equal to 7 for any superconductor for which EMT holds, says, in other words, that the Tc is determined by the optimal frequency directly.

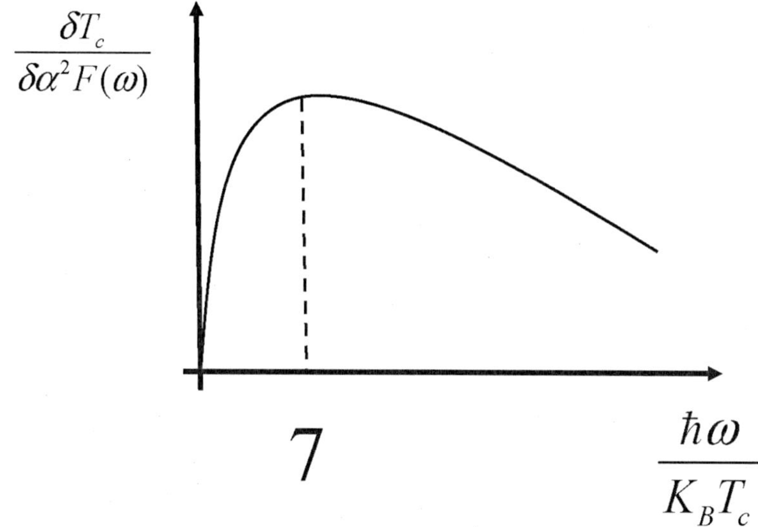

Fig. 1- The functional derivative of Tc with the Eliashberg function. Its universal maximum is at ~ω/K_BTc = 7 for all superconductors for which EMT holds (see text).

It is clear from Eq. (5) that if to get a high Tc, a high value of λ is required, the important part of the spectrum ($\alpha^2F(\omega)$), lies at low frequencies. But to get a high Tc value, according to Eq.(7) which is the result that EMT actually gives, the optimal frequency should lie at higher values and therefore the high frequencies are the important ones. When we deal with superconductors that have a high Tc, the criterion "the higher the λ, the higher the Tc" is therefore in contradiction with EMT.

An important question to answer nowadays, as we already wrote above, is whether it is possible for a superconductor with a low electron-phonon interaction parameter, λ, to have a high critical temperature. Since, according to EMT, a high critical temperature is determined by the behavior at high frequencies, what happens at low frequencies is unimportant since it does not influence the value of Tc in an essential way. So, there is nothing that can be derived from EMT against an $\alpha^2F(\omega)$ with very low weight at low frequencies (with a low λ) and a high optimal frequency which would determine a high Tc As an example, an Eliashberg function as the one sketched in Fig. 2 could represent an e-ph superconductor with a high Tc and a low λ under the condition that its optimal frequency lies at higher frequencies.

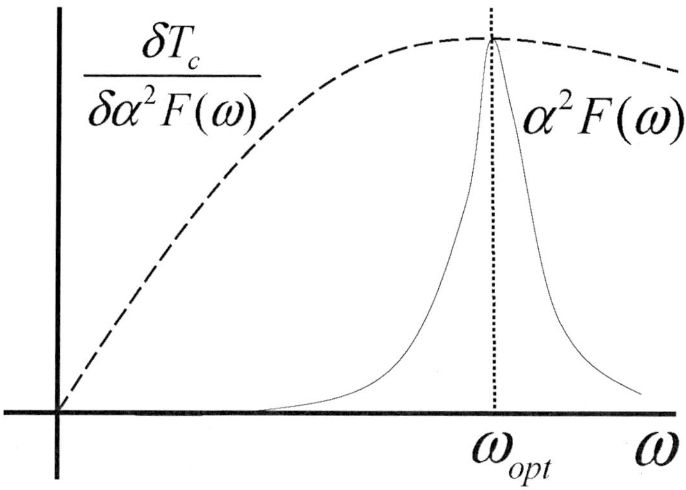

Fig.2- An hypothetical Eliashberg function that represents a superconductor for which the EMT holds with very little weight at low frequencies (low λ) and an optimal frequency lying at a very high frequency (say 70 meV) which would result in a Tc value of 10 meV, *i.e.* around 110K).

It is not possible to predict Tc nowadays but, nevertheless, the functional derivative $\delta Tc/\delta\alpha^2F(\omega)$ can be successfully used to answer more modest questions as, for example, whether a system is optimized or not. An analysis of this sort for Nb_3Ge can be found in ref [10]. This work shows explicitly that what is important is the energy at which the optimal frequency lies. For low-Tc materials, ω_{opt} lies at low frequencies and therefore the important part of the spectrum is also "in the picture captured" by λ. This will not be the case for high-Tc materials. We will come back to this point below.

THE EXPERIMENTAL VALUES FOR Tc AND λ: DO THEY REALLY CORRELATE?

If "the higher the λ, the higher the Tc" a certain correlation among the two values is expected. We saw from Eq.(5) that to get a high λ, the spectrum ($\alpha^2F(\omega)$) should have some important weight at low frequencies. On the other hand, Eq. (7) reveals that the optimal frequency, ω_{opt},, determines Tc. One expects therefore that both should then tend to agree better for low temperature superconductors where the optimal frequency, ω_{opt},, lies at low frequencies . This also gives some theoretical foundation to the success of the idea that a correspondence exists between the magnitude of λ and the one of Tc. What is really behind it is that, for low temperature superconductors, ω_{opt} lies at low frequencies, namely the ones that contribute the most to the value of λ. We examine that possibility in Table I. The data were taken from [4].

Table I contains four columns. The first from the left contains the value of the critical experimental temperature, Tc, in meV, and in ascending order and the second column shows the corresponding material. Notice that there are three insets for Nb with the same temperature. This is due to the fact that there are three different data for the corresponding value of λ. The third and fourth columns contain the material and the corresponding value of the electron-phonon interaction parameter, λ., again in ascending order from top to bottom. Would a perfect correspondence between λ and Tc existed, the names of the elements in column three and four would correspond to each other.

Tc [meV]	Material	Material	Lambda
0.1017	Al	Al	0.43
0.2034	Tl	Ta	0.69
0.2931	In	Sn	0.72
0.3233	Sn	Tl	0.8
0.3612	Hg	V	0.8
0.3862	Ta	In	0.81
0.434	La	Mo	0.9
0.4621	V	La	0.98
0.5267	Bi	Nb(Rowell)	0.98
0.6198	Pb	Nb(Arnold)	1.01
0.7379	Ga	Nb(Butler)	1.22
0.7586	Mo	Pb	1.55
0.7931	Nb	Hg	1.62
0.7931	Nb	Ga	2.25
0.7931	Nb	Bi	2.45

Table I .The value for the experimental Tc and for λ for several elements.

It is clear from this Table that there is no general correspondence between the values of the electron-phonon interaction parameter, λ, and the value of the critical temperature, Tc. The correlation works well for Al, but extremely bad for Hg, for example. So the statement "the higher the λ, the higher the Tc" is not supported by the corresponding values presented here for several conventional superconductor elements. An exact correspondence does not exists even in the most favorable case.

The correspondence between λ and Tc is not much better for the A15 compounds. We compare in the same way as in Table I, the corresponding values in the next Table II. The same conclusion can be drawn in this case. λ and Tc are not, in general, related with each other as it should be expected if λ were the proper parameter that determines the magnitude of Tc. .

Tc [meV]	Material	Material	lambda
0.4741	V3Si_1	V3Si_1	1
1.207	Nb3Al (2)	V3Si (Kihl.)	1
1.2931	V3Ga	V3Ga	1.14
1.4132	V3Si (Kihl.)	Nb3Al (2)	1.2
1.5603	Nb3Sn	Nb3Ge (2)	1.6
1.6121	Nb3Al (3)	Nb3Al (3)	1.7
1.724	Nb3Ge (2)	Nb3Sn	1.7

Table II. The A15 compounds [4]

This conclusion throws serious doubts on the use of the lambda parameter as an indicator of the magnitude of Tc. Is seems that all the approximate equations that calculate Tc based on the knowledge of λ and μ^* alone or introduce extra sophisticated parameters as the ω_{log} in Mc Millan's equation (4) could be misleading when they are applied to HTSC since they stress the behavior at low frequencies which is actually unimportant in this case according to EMT. Furthermore, these equations are not very precise even for low temperature conventional superconductors as we already mentioned (Pb). We also stress, referring to the Mc Millan's equation (4) that if we know the Eliashberg function and have to guess the electron-electron repulsion parameter, with the same data and effort, we can solve directly the Eliashberg equations at T=Tc and there is no reason *a priori* for the result not to agree better with experiment. In the worst case, there is at least a clearly set theory behind it.

The optimal frequency is calculated from the known experimental value of the critical temperature and Eq.(7) cannot be used to predict it.

What determines that an specific superconductor has a certain optimal frequency? This is an unsolved formidable problem that has not yet been addressed in the literature at least with enough detail as it is required. To solve it is actually equivalent to finding a predictive equation for Tc. Nevertheless, Eq.(7) is very useful to shade some light on whether or not the e-ph mechanism can be discarded solely on the basis of a small value of λ as we shall see, in a model, below.

IS A LOW VALUE FOR λ SO BAD FOR A HTSC?

As a final point in this paper, we explore the effect that a low λ has on the value of an important parameter, namely, the integral of the Eliashberg function, A.

$$\int_0^{\infty} \alpha^2 F(\omega) d\omega \equiv A \tag{8}$$

Let us assume a delta-function model for the Eliashberg function. The highest Tc will occurs if all its weight is located at the optimal frequency according to Eq.(7):

$$\alpha^2 F(\omega) = A\delta(\omega - \omega_0) \text{ with } \omega_0 = \omega_{opt} \tag{9}$$

From Eqs.(5) and (9), we get

$$\lambda = \frac{2A}{\omega_{opt}}. \tag{10}$$

This equation, in a certain way, is telling us, "the higher the Tc, the lower the λ" at least in the case when the parameter A does not increase very much. This parameter should not be very high in any case [3]. It is not clear whether a too high value of the parameter A is compatible with lattice stability.

In HTSC phonons of high frequency have been identified [11] For $La_{1.85}Sr_{0.15}CuO_4$, let us consider the phonon frequency at 50 meV . The amplitude of a model Eliashberg function for this material [12] appears to be maximal at this frequency. Would this phonon act as the optimal frequency, $La_{1.85}Sr_{0.15}CuO_4$ would have, according to Eq. (7) a Tc ~ 7 meV ~ 77 K. Since its experimental Tc is around 30 K, we could conclude that, according to EMT, it is not an optimized system. We could say even more, since the Sr-content dependence of Tc is fully known for this system and the highest Tc found is quite lower that 77K, we can speculate that a higher Tc could be obtained but not through the change of the Sr-content but with some different kind of impurities. This is not surprising if we take into consideration the way how the "on-earth temperature" (186K) new superconductor [13] has been obtained. Of course, there is no proof at all that the mechanism in the $La_{1.85}Sr_{0.15}CuO_4$ system is e-ph.

If we take 50 meV as the optimal frequency and believe that a high value of λ is the proper way to get a high Tc, then we are forced to consider for λ values of the order of 3 or so (for Nb: Tc=9 K, A=9 meV, and λ=1), then we get an embarrassing high A = 75 meV. But, fortunately, the result of a state-of-the-art calculation [14] is λ of the order of 0.2, a value that brings back the parameter A to a very reasonable magnitude, namely, about 5 meV, which is even lower than the one for Nb. In some conventional superconductors, as La, for example, "higher" values of Tc are found together with lower values of A [15]. So, from all these arguments, we can conclude that it might turn out that a low λ-value is, contrary to the accepted wisdom nowadays, even convenient for an e-ph HTSC for which the EMT holds, since it helps maintaining the parameter A reasonable low.

CONCLUSIONS

In the first part of this paper, we have analyzed the basis for the criterion "the higher the electron-phonon interaction parameter, λ, the higher the superconducting critical temperature, Tc". This criterion is based on approximate phenomenological equations that are not a direct consequence of Eliashberg-Migdal theory (EMT) and they not only are not very precise even for the simple superconducting elements but also use values for the parameter μ* that are guesses not well founded neither theoretically nor experimentally. We have then proceeded to show the results of EMT concerning the functional derivative of Tc with respect to the Eliashberg function, $\alpha^2F(\omega)$, $\delta Tc/\delta\alpha^2F(\omega)$, to conclude that there is a universal law that relates Tc directly to a particular phonon frequency, the optimal frequency, ω_{opt}, that is specific to each superconductor. So, to get a high critical temperature, contrary to the conclusions of the criterion based on the parameter λ, it is the high frequency part of the spectrum that is the important one. This comes directly from EMT and throws serious doubts on the validity of the conclusions that discard the electron-phonon mechanism on the basis of a low value of λ,. Furthermore, we have shown that a low value of λ .helps to prevent an e-ph HTSC (if it exists) from having an unreasonable value of the integral of the Elaiashberg function, A. It is not clear whether a too big value for the parameter A could not be associated with the instability of the lattice. Summarizing, we argue, first, that the phenomenological equations based on the knowledge of λ and μ^* alone as well as others requiring the knowledge of data that allow the solution of the Eliashberg equations themselves (as the Mc Millan one) should not be used in high-Tc superconductivity and, second, that a low value of the electron-phonon (e-ph) interaction parameter, λ, is not an argument solid enough to discard the e-ph mechanism in HTSC. We add a final argument. According to EMT, as the critical temperature of a superconductor increases phonons of each time higher frequencies become involve in the dynamics of the phase transition and the lower frequencies do not matter so much because they do not play an essential role anymore, λ ceases to capture enough details on the essential characteristics of the superconducting state. For that reason, we conclude that it is not clear whether λ contains enough information so that it can be used as a proper criterion to accept or discard the e-ph mechanism in the case of HTSC,

REFERENCES

1. J. Bardeen, L. N. Cooper, J. R. Schrieffer, Phys. Rev. **108,** 1175 (1957). There are very many books that describe BCS theory. For example, J. R. Schrieffer, *Theory of Superconductivity*, Benjamin Eds., New York, 1964; O. Navarro and R. Baquero, *Ideas Fundamentales de la Superconductividad (TheFundamental ideas of Superconductivity,* in Spanish), edited by the University of Mexico (UNAM), 2007.
2. G. M. Eliashberg, Sov. Phys. JETP **11**, 696 (1960); Zh. Eksp. Teor. Fiz. **38**, 966 (1960).
3. J. P. Carbotte, Rev. Mod. Phys. **62**, 1027 (1990). This paper contains a review on some important part of the theoretical knowledge on superconductivity prior to the HTSC, studies the knowledge of HTSC at the time and compares with experiment at several places.
4. See ref. [3] and references therein.
5. Feliciano Giustino, Marvin L. Cohen, and Steven G. Louie, Nature, **452**, 975 (2008); Rolf Heid, Klaus-Peter Bohnen, Roland Zeyher and Dirk Manske, arXiv:0707.4429v3.
6. J.M. Daams, J.P. Carbotte and R. Baquero, J. Low Temp. Phys. **35**, 547 (1979); R. Baquero, J.M. Daams and J.P. Carbotte, J. Low Temp. Phys. **42**, 585 (1981); R. Baquero and E. López-Olazagasti, Phys. Rev. **B30**, 5019 (1984).
7. W. L. Mc Millan, Phys. Rev. **167**, 331 (1968).
8. P. B. Allen and R. C. Dynes, Phys. Rev. B **12**, 905 (1975).
9. G. Bergmann and D. Rainer, Z. Phys. **263**, 59 (1973); D. Rainer and G. Bergmann, J. low Temp. Phys. **51**, 501 (1974); C. R. Leavens, Solis State Comm. **15**, 1329 (1974)

J. M. Daams and J. P. Carbotte, Can. J. Physics **56**, 1248 (1978); B. Mitrovic and J. P. Carbotte, Solid State Comm. **40**, 249 (1981)..

10. R. Baquero, J. Gutiérrez-Ibarra, L. Meza, O. Navarro and K. Kilhstrom, Rev. Mex. Fís. **35**, 461 (1985).
11. A. Lanzara *et al.*, Nature **412,** 510 (2001).
12. H. Krakauer, W.E. Pickett, and R.E. Cohen in *Proceedings of the 2nd Cinvestav Superconductivity Symposium (manifestations of the electron-phonon interaction)*, R. Baquero, editor, World Scientific, 1994., pag. 38.
13. see www.superconductors.org.
14. Rolf Heid, Klaus-Peter Bohnen, Roland Zeyher and Dirk Manske, arXiv: 0707.4429v3.
15. R. Baquero and J.P. Carbotte, J.Low Temp. Phys. **51**, 135 (1983).

XPS and AFM Study of GaAs Surface Treatment

R. Contreras-Guerrero[a,b], R. M. Wallace[a], A. Herrera-Gómez[a,c], S. Aguirre-Francisco[a], and M. López-López[b]

[a] *Department of Electrical Engineering, University Of Texas at Dallas. Richardson, TX 75083.*
[b] *Physics Department CINVESTAV-Unidad Zacatenco, D.F, Mexico 07360,*
[c] *CINVESTAV-Unidad Queretaro, Mexico 76000.*

Abstract. Obtaining smooth and atomically clean surfaces is an important step in the preparation of a surface for device manufacturing. In this work different processes are evaluated for cleaning a GaAs surface. A good surface cleaning treatment is that which provides a high level of uniformity and controllability of the surface. Different techniques are useful as cleaning treatments depending on the growth process to be used. The goal is to remove the oxygen and carbon contaminants and then form a thin oxide film to protect the surface, which is easy to remove later with thermal desorption mechanism like molecular beam epitaxy (MBE) with minimal impact to the surface. In this study, atomic force microscopy (AFM), x-ray photoelectron spectroscopy (XPS) and secondary ion mass spectrometry (SIMS) were used to characterize the structure of the surface, the composition, as well as detect oxygen and carbon contaminant on the GaAs surface. This study consists in two parts. The first part the surface was subjected to different chemical treatments. The chemical solutions were: (a) H_2SO_4:H_2O_2:H_2O (4:1:100), (b) HCl: H_2O (1:3), (c) NH_4OH 29%. The treatments (a) and (b) reduced the oxygen on the surface. Treatment (c) reduces carbon contamination. In the second part we made MOS devices on the surfaces treated. They were characterized by CV and IV electrical measurements. They show frequency dispersion.

Keywords: X-ray photoelectron spectroscopy (XPS), secondary ion mass spectrometry (SIMS).
PACS: 81.05.Ea, 68.37.Ps, 79.60.-i, 68.49.Sf.

INTRODUCTION

There has been a rapid development in research of GaAs and other III-V semiconductors particularly on the surface preparation. One of the principal problems hindering the development of GaAs (and III–V) devices is the poor surface due to native oxide layer and carbon contaminants which reduce the optical and electronic properties[1,2]. Research shows another significant problem some processes used to clean the surface can damage the surface by thermal stresses, change of surface composition, defect generation, and increase of surface roughness. Conventional surface cleaning techniques are chemical treatments (based in NH_4OH, HCl, H_2SO_4, and HF), hydrogen cracked in UHV chamber, UV irradiation, thermal cleaning[3,4]. Chemical treatment and UV irradiation are the simplest and easy to control.

In this work we investigated by x-ray photoelectron spectroscopy chemical bonds on GaAs (100), oxygen and carbon contaminates subjected to acid and base solutions. Electrical measurements on capacitors were taken to characterize the surface/oxide interface for every sample.

CP1077, *Advanced Summer School in Physics 2008, Frontiers in Contemporary Physics—EAV08*
edited by L. M. Montaño Zetina, G. Torres Vega, M. García Rocha, L. F. Rojas Ochoa, and R. López Fernández

EXPERIMENT

Part I.

Doped GaAs (100) n-type surfaces were chemically etched in:

A) H_2SO_4:H_2O_2:H_2O;4:1:100 for 2 min

B) HCl:H_2O;1:3 for 2 min

C) NH_4OH 29% for 3 min

A and B samples were rinsed with DI water. All samples were dried with nitrogen gas. All processes were done at room temperature. After that they were put into a UHV chamber and transferred to XPS chamber (OMICRON System). The samples were annealed at 300°C and 450°C. XPS was used after each step to characterize the residual impurity concentrations and carbon contamination.

Part II

We built MOS devices to evaluate the surface subjected to chemical treatments (A, B, C). The MOS device is show in figure 1. HfO_2 was grown by sputtering in high vacuum, TaN by atomic layer deposition (ALD) and the back contacts by PVD evaporation system.

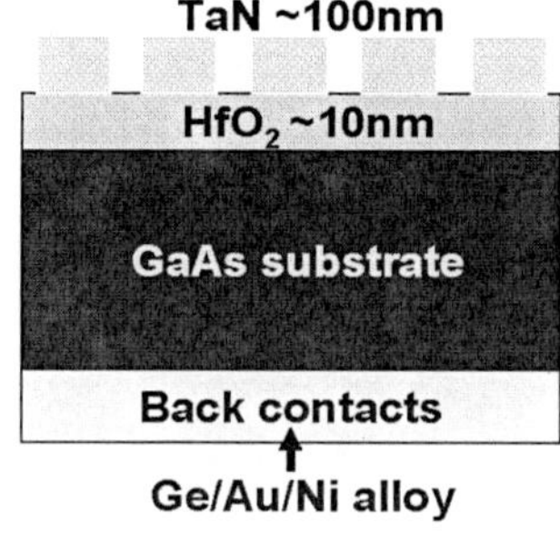

FIGURE 1. MOS device structure

Electrical measurements were taken to compare the difference between interfaces.

RESULTS

Part I. XPS data

Figure 2 shows XPS data of As-*3d* and Ga-*3d* for a sample as-received. We choice these lines because are the most surface sensitive and easier to analyze. The contributions from the oxygen bonds and GaAs peak are drawn with dashed lines.

The spectra show As-O bonds such as As^{5+}, As^{3+} commons oxides and elemental arsenic (As^{0}).

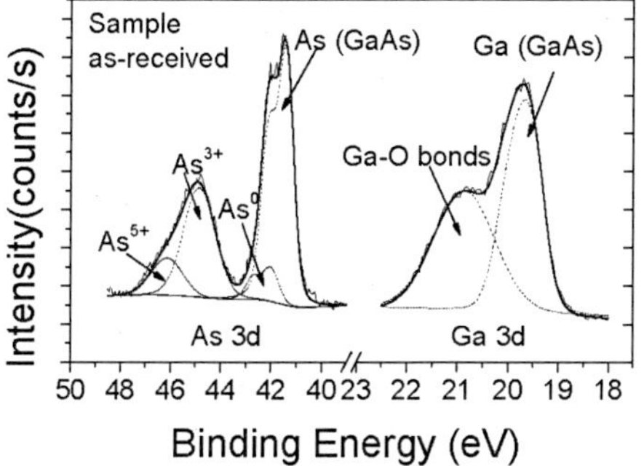

FIGURE 2. XPS data of As-*3d* and Ga-*3d* spectra for sample as-received. The spectra show native oxide

Figure 3 presents the evolution of Ga-*3d* and As-*3d* signals for increasing anneal temperatures. The HCl solution has reduced the thickness of the oxide layer. For all three samples the As-*3d* spectra at 450°C anneal show not detectable As-bonds.

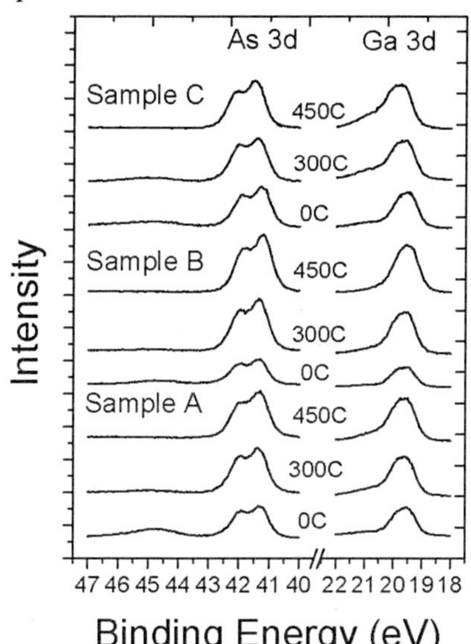

FIGURE 3. XPS data of As-*3d* and Ga-*3d* spectra for sample A, B, C. As-*3d* spectra show decrease in oxygen bonds for three treatments. Ga-*3d* spectra for sample C show oxygen bonds increase as increase the anneal temperature.

After annealing the amount of oxygen and carbon contaminates are reduced (see Figure 4). The H_2SO_4 solution has a similar behavior to HCl the carbon contamination has been increased after treatment.

The carbon contamination is strongly reduced for ammonia treatment, and after annealing carbon is not detected, but oxygen has increased.

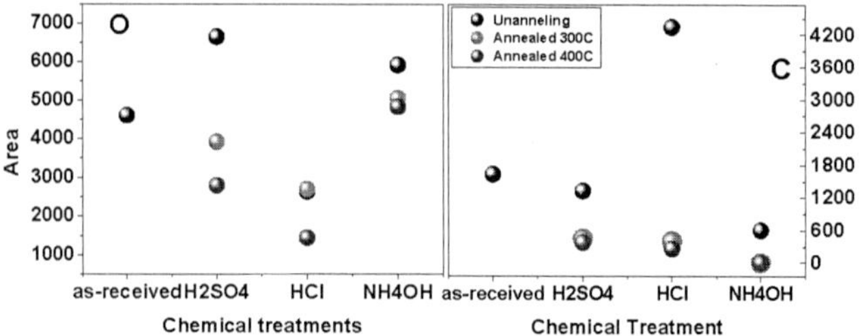

FIGURE 4. Oxygen and carbon total area from XPS data. HCl treatment reduces oxygen. Carbon is strongly reduced with NH_4OH treatment but oxygen increase.

AFM micrographs

H_2SO_4 Treatment (Sample A) strongly impacted the surface by increasing roughness (RA – 2.24nm). If we compare with sample B and C the roughness average values remain similar with the as-received sample (about ~1.4-1.6nm) (see figure 5).

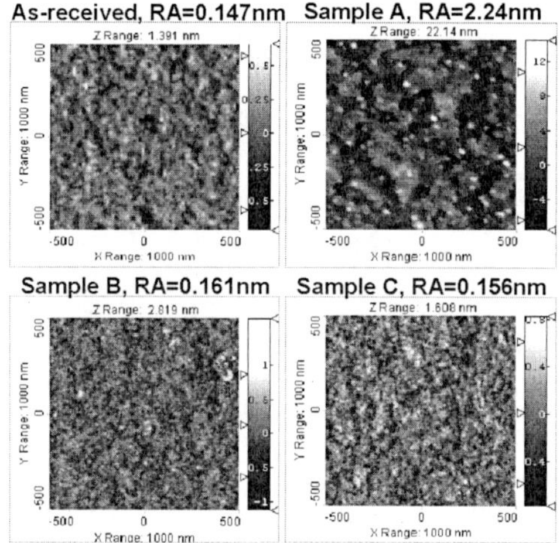

FIGURE 5. AFM Micrographs. Sample A (H_2SO_4 Treatment) strongly impact to the surface. Samples C and B (ammonia treatment) left a smooth surface.

Part II. Electrical Measurement

The capacitors show frequency dispersion. In the figure 6 we show the voltage vs. Capacitance/Area for samples B and C. We know that sample C has oxides remain on the surface and sample B has oxides and carbon contaminates. We speculate that the oxide remaining on the surface causes frequency dispersion.

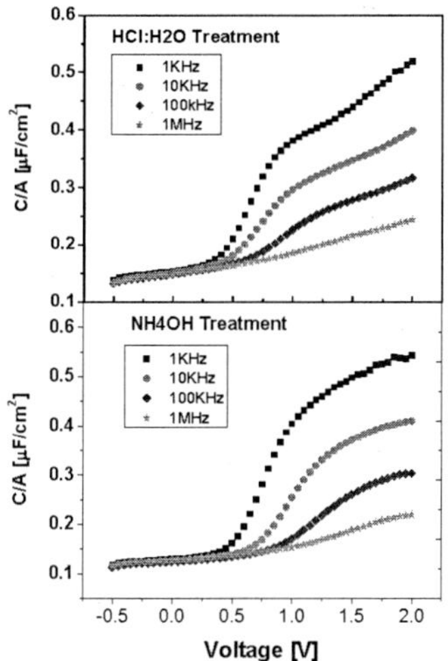

FIGURE 6. Frequency dispersion is observed in CV measurements on MOS devices for the three samples.

CONCLUSIONS

NH_4OH treatment reduces carbon concentration and left a smooth surface. HCl treatment reduces oxygen and carbon contamination.

Annealing to 450 °C almost eliminates As-O concentration and reduces the total amount of elemental arsenic for the three treatments.

The presence of oxides seems to be playing an essential role in the frequency dispersion observed in CV measurements on GaAs.

ACKNOWLEDGMENTS

This work was partially supported by CONACyT Mexico D.F. and the MARCO Focus Center on Materials.

REFERENCES

[1] Z.Liu, Y. Sun, F. Machuca, P. Pianetta, W. E. Spicer *et. al. J. Vac. Sci. Technol.* B 21(4), Jul/Aug 2003

[2] Z.Liu, Y. Sun, F. Machuca, P. Pianetta, W. E. Spicer *et. al. J. Vac. Sci. Technol.* A 21(1) Jan/Feb 2003

[3] M. V. Lebedev *et. al. Appl. Surf. Sci* 229 (2004) 226-232

[4] M.G. Kang *et al. Thin Solid Films* 308-309 (1997) 634-642

Temperature Dependence of the Photoluminescence Linewidth of $In_{0.2}Ga_{0.8}As$ Quantum Wells on GaAs(311) Substrates

J. S. Rojas-Ramírez[a], R. Goldhahn[b], P. Moser[b], J. Huerta-Ruelas[c], J. Hernández-Rosas[a], M. López-López[d]

[a]*Physics Department, Centro de Investigación y Estudios Avanzados del IPN Apartado Postal 14-740, México D.F., México 07360.*
[b]*Technical University Ilmenau, Institut of Physics, PF 100565, 98684 Ilmenau, Germany*
[c] *Centro de Investigación en Ciencia Aplicada y Tecnología Avanzada del IPN Cerro Blanco 141, Col. Colinas del Cimatario, Querétaro Qro., México 76090*
[d] *Centro de Física Aplicada y Tecnología Avanzada, Universidad Nacional Autónoma de México, Apartado Postal 1-1010, Querétaro 76000, México*

Abstract. We have studied the photoluminescence (PL) properties of strained $In_{0.2}Ga_{0.8}As$/GaAs quantum well (QW) structures grown by molecular beam epitaxy on (311)-oriented substrates. Special attention has been paid to the PL linewidth measurements in terms of the full width at half maximum (FWHM). We obtained the FWHM temperature dependence in the range of 5–250 K and compared it to measurements done in unstrained AlGaAs/GaAs(100) QW structures. The linewidth broadening of the unstrained system could be satisfactorily explained by means of a model which takes into account the exciton-acoustic phonon and exciton-LO phonon interactions. This model doesn't describe adequately the experimental data from the InGaAs/GaAs system. Remarkable differences, such as the dependence of the linewidth on the dimensions of the wells were found. We think that the additional effects to which the InGaAs QWs are subjected, namely, strain and built-in electric fields, are the origin of the completely different behavior of the linewidth broadening.

Keywords: Photoluminescence, Quantum wells, MBE, linewidth broadening
PACS: 8.67.De, 81.05.Ea, 81.15.Hi.

INTRODUCTION

Heteroepitaxy has been commonly used for the growth of semiconductor strained layers. The strain substantially modifies the semiconductor electron band structure. The use of strain has been particularly exploited in the InGaAs system[1]. High speed MODFETs and low threshold quantum well (QW) lasers based on this system are now a reality.

Many device structures have, however, been synthesized on conventional [100]-oriented GaAs substrates. Theoretical calculations have shown that large strain-induced electric fields exist in thin alternating layers of InGaAs and GaAs oriented in the [n11] crystallographic directions[2]. The presence of these fields is known to modify the optical properties of the films. It should be noted that InGaAs strained-layers structures grown on (100) surfaces exhibit no strain-induced polarization.

CP1077, *Advanced Summer School in Physics 2008, Frontiers in Contemporary Physics—EAV08*
edited by L. M. Montaño Zetina, G. Torres Vega, M. García Rocha, L. F. Rojas Ochoa, and R. López Fernández

Photoluminescence (PL) spectroscopy has been extensively used in the electronic structure characterization of the InGaAs/GaAs system. In QW structures, the spectral width at half maximum (FWHM) is customarily used as a diagnostic of the QW structure quality.

At low temperatures, PL measurements in a bulk-semiconductor crystal show the recombination of free excitons mainly in the ground state. Furthermore, by increasing the temperature, the interband recombination appears and eventually dominates the PL spectra. In high quality QW structures, the PL is governed by the recombination of free excitons from cryogenic up to room temperature[3].

In this study, the results obtained by PL as a function of temperature for $In_{0.2}Ga_{0.8}As/GaAs$ QW structures grown on (311)A-oriented GaAs substrates are reported. We emphasize the temperature dependence of the linewidth and compared it with that of AlGaAs/GaAs QW structures.

EXPERIMENTAL PROCEDURE

Three $In_xGa_{1-x}As/GaAs$ QWs of 100, 50, and 25 Å of nominal thickness were grown by molecular beam epitaxy, sequentially on a semi-insulating (311)A-oriented GaAs substrate. The substrate preparation is described in detail in Ref. 4. The nominal concentration of In was kept at 20% (x=0.2). Fig. 1(a) shows a sketch of the QWs structure, which consists of a 500 Å $Al_{0.35}Ga_{0.65}As$ barrier layer, followed by an $In_{0.2}Ga_{0.8}As$ well of 100 Å between 200 Å GaAs barriers. Similar layers were grown for the 50 Å and 25 Å QWs.

Three $Al_{0.3}Ga_{0.7}As/GaAs$ QWs were grown with nominal widths of 70, 50, and 30 Å. The experimental arrangement for the sample preparation has been described elsewhere[5]. A schematic diagram of the sample structure is shown in Fig. 2(a).

Photoluminescence spectroscopy measurements were carried out from 5 to 250 K employing a Ti:Sapphire laser as excitation source with an energy line of 1.6 eV. Detected light was in the range from 1.25 to 1.5 eV.

RESULTS AND DISCUSSION

In Fig. 1(b) we present the evolution of the PL emission from the QWs as a function of temperature between 5 and 250 K. Analyzing the PL spectrum at 5 K we observe that the three peaks localized at 1.363, 1.408, and 1.453 eV, correspond to the emission of the 100, 50, and 25 Å QWs, respectively. The energy position of a PL peak E_{PL} is determined by the energy of the transition of carriers between QWs levels in the conduction and valence bands:

$$E_{PL} = E_g + E_{1e} + E_{1hh} - E_b + \Delta_\sigma - \Delta_1 \quad (1)$$

We calculated all the terms of Eq. 1. A detailed discussion is found in Ref. 6. The 5K band gap of bulk $In_{0.2}Ga_{0.8}As$ is $E_g(5K) = 1.2210$ eV. The change in the band gap energy due to the stress in the InGaAs well layers is $\Delta_\sigma = 151.8$ meV. Once the stressed InGaAs band gap energy was obtained, the energies of the first subbands of

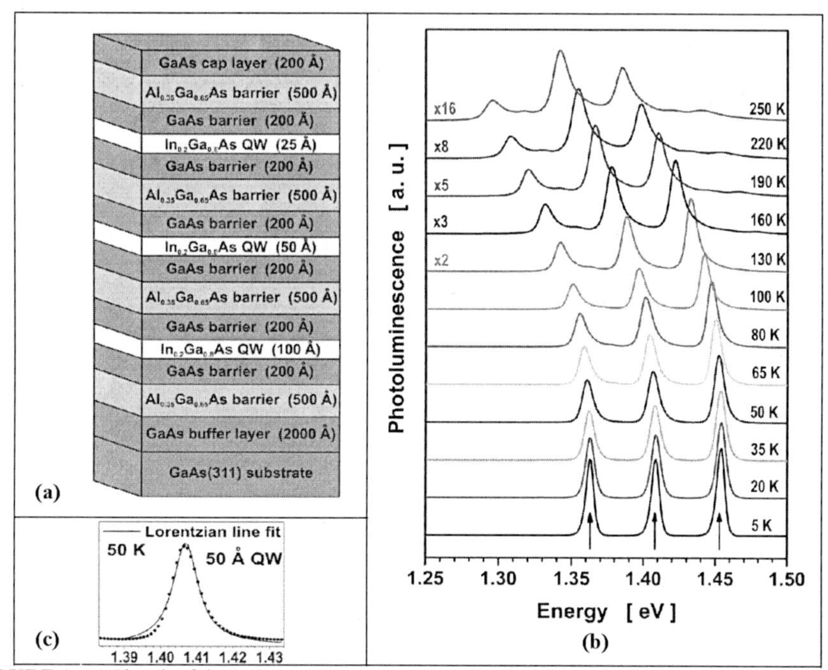

FIGURE 1. (a) Sketck of the InGaAs QWs structure studied in this work. (b) PL spectra as a function of temperature in the range of 5–250 K. (c) PL free-exciton of the 50 Å QW at 50 K.

electrons and heavy holes in the QWs are $E_{1e}(5K) = 69.8$ and $E_{1hh}(5K) = 30.4$ meV respectively. The exciton binding energy is $E_b = 8$ meV. The change of the E_{1e} and E_{1hh} levels in the presence of an electric field for the 25 Å QW is $\Delta_1 = 13.5$ meV. Our theoretical calculations for the energy position of PL peaks from the QWs at 5K are marked by arrows in the bottom of Fig. 1(b). We note that our results reproduce quite well the experimental PL peaks position.

We notice that PL peaks shift to smaller energies as the temperature is increased. We expected this behavior from the fact that the band gap energy of most bulk semiconductors decreases with increasing temperature. We fitted each PL peak to a Lorentzian function in order to accurately determine the PL features of the QWs emission: peak position, full width at half maximum (FWHM), and intensity. As an example of the fitting, Fig. 1(c) shows the PL free-exciton peak of the 50 Å QW at 50 K and a Lorentzian line fit.

Figs. 2(b) and 3(a) show the PL linewidth for both, the AlGaAs/GaAs and InGaAs/GaAs QWs systems. We will refer to the former as the unstrained system and the latter as the strained one. They exhibit a nonlinear and monotonic temperature dependence, and become broader with temperature increasing from 5 to 250 K. The FWHM behavior of the strained system is more complex. We fitted the experimental data to a five degree polynomial and determined that the temperatures at which the curvature of the fitted curves changed are 98, 110, and 62 K corresponding to the 100, 50, and 25 Å QWs, respectively.

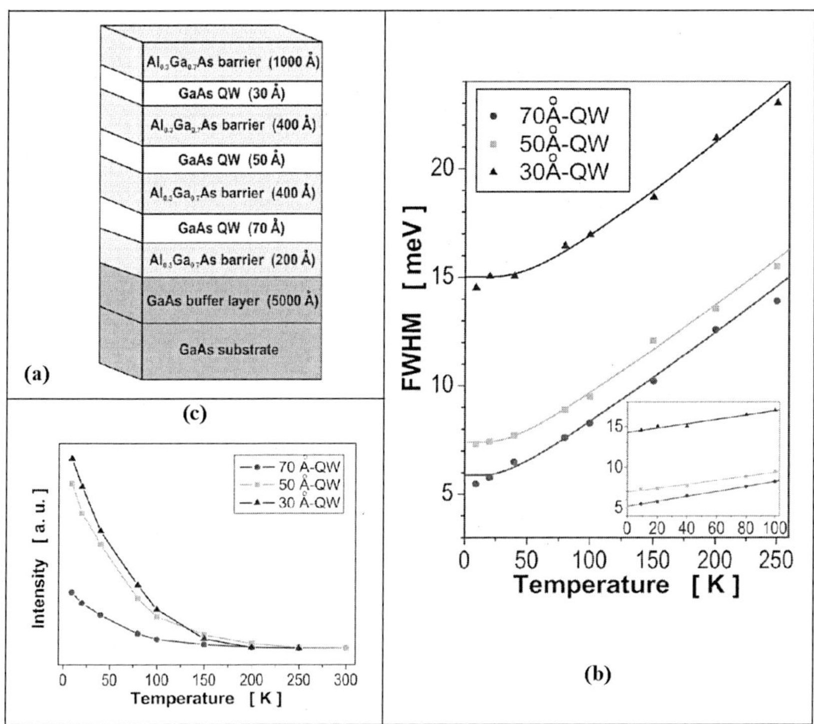

FIGURE 2. (a) Sketch of the AlGaAs/GaAs QWs structure. (b) Experimental PL linewidths in the temperature range of 5–250 K of the AlGaAs/GaAs QWs. The solid line is a least-square fit to the LO phonon contribution term in Eq. 2. The inset show the least-square fit to Eq. 2 in the low temperature region where the acoustical phonon contribution is dominant. (c) PL intensity versus temperature.

Figs. 2(c) and 3(b) show the PL intensity versus temperature for the AlGaAs/GaAs and InGaAs/GaAs QWs systems, respectively. An overall decrease of the luminescence intensity with increasing temperature can be attributed to a change in the dominant recombination mechanism, namely, from radiative to non-radiative recombinations. As well as the FWHM, the PL intensity behavior of the strained system is more complex than the unstrained one. We notice that the experimental data of the AlGaAs/GaAs system can be fitted by smooth curves with positive curvature in the whole temperature range. On the other hand, the function describing the PL intensity of the InGaAs/GaAs system present several inflection points. In particular, the inflexions located at 102, 113, and 58 K for the 100, 50, and 25 Å QWs, respectively, are very close to those found in the FWHM curves mentioned above. Some of the physical processes determining the temperature evolution of the PL intensity and FWHM must be related.

The temperature dependence of the linewidth of the direct gap of zinc-blende-type semiconductors can be expressed as follows[7,8]:

$$\Gamma_0(T) = \Gamma_0(0) + \gamma_{AC}T + \frac{\Gamma_{LO}}{\exp(\theta_{LO}/T) - 1}. \qquad (2)$$

The first term on the right-hand side of Eq. 2 is due to intrinsic effects: electron-electron interaction, impurity, dislocation, and alloy scattering at T = 0K. The second term corresponds to lifetime broadening due to the electron (exciton)-acoustic phonon interaction; γ_{AC} is the acoustical phonon coupling constant. The third term is caused by the electron (exciton)-longitudinal optical (LO) phonon interaction whose coupling strength is represented by the quantity Γ_{LO}. θ_{LO} is the LO phonon temperature.

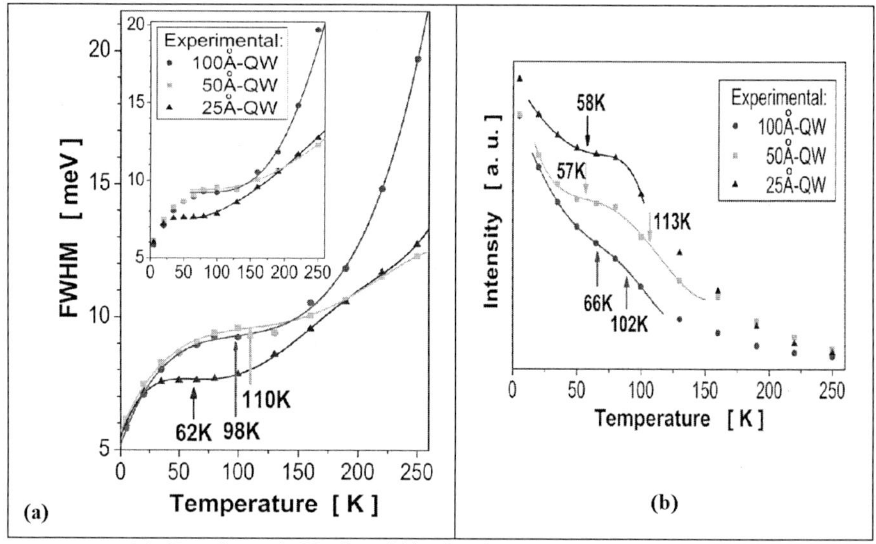

FIGURE 3. (a) Experimental PL linewidths in the temperature range of 5–250 K for the InGaAs/GaAs QWs. The solid lines are fittings to a five degree polynomial. The arrows indicate the inflection points. The inset show a least-square fit to the LO phonon contribution term in Eq. 2 in the high temperature region. (b) PL intensity as a function of temperature.

TABLE 1. Values of the parameters that describes the temperature dependence of Γ in terms of FWHM.

System	Parameter			
AlGaAs/GaAs	$\Gamma_0(0)$ [meV]	γ_{AC} [$\times 10^{-6}$ eV/K]	Γ_{LO} [meV]	θ_{LO} [K]
30 Å QW	14.29 ± 0.17	26.3 ± 2.8	7.7 ± 2.6	161 ± 39
50 Å QW	6.95 ± 0.12	24.4 ± 2.0	5.0 ± 2.2	116 ± 41
70 Å QW	5.202 ± 0.038	30.7 ± 0.6	4.4 ± 1.5	102 ± 29
InGaAs/GaAs	$\Gamma_0(0)$ [meV]	γ_{AC} [$\times 10^{-6}$ eV/K]	Γ_{LO} [meV]	θ_{LO} [K]
25 Å QW	–	–	106 ± 8.6	655 ± 130
50 Å QW	–	–	19.5 ± 1.4	499 ± 151
100 Å QW	–	–	16.8 ± 2.2	357 ± 26

As the temperature is raised, the linewidth of the AlGaAs/GaAs system first increases from its zero-temperature value because of acoustical phonon scattering, see the inset in Fig. 2(b). Above 100 K, the LO phonon contribution becomes important and eventually dominant in the linewidth. By means of a least-square fit, the values of $\Gamma_0(0)$, γ_{AC}, Γ_{LO}, and θ_{LO} are obtained and listed in Table 1. Regarding the InGaAs/GaAs system, we could not fit successfully the experimental data to Eq. 2 in the whole temperature range. Assuming that the LO phonon contribution dominates at high temperatures, in the same way that the unstrained system do, we estimated the Γ_{LO} and θ_{LO} parameters (see Table 1) by means of a least-square fit to the LO phonon contribution term in Eq. 2. See inset in Fig. 3(a).

Now we discuss the FWHM dependence on the QW thickness. In AlGaAs/GaAs QWs, the FWHM increases as the QW becomes narrower. This behavior is observed in the whole temperature range. However, the dependence of the linewidth on the dimensions of the InGaAs/GaAs QWs is not understood. The additional effects to which such a system is subjected, namely, strain and piezoelectric effects, change completely the linewidth broadening. Interesting features can be pointed out. In the low temperature region, below 30 K, the FWHM is not sensitive to the QW thickness. In the middle temperature region, (in the vicinity of the inflexions at 98, 110, and 62 K for the 100, 50, and 25 Å QWs, respectively) the FWHM remains constant under a temperature change.

ACKNOWLEDGMENTS

This work was partially supported by CONACYT-Mexico and ICTDF. The authors thank the technical assistance of R. Fragoso, A. García, and A. Guillén. MLL would like to thank CONACYT for the support during the sabbatical leave.

REFERENCES

1. P. Bhattacharya, *Properties of lattice-matched and strained Indium Gallium Arsenide*, Emis, datareviews series, no. 8. An INSPEC publication.
2. D. Sun and E. Towe, *Jpn. J. Appl. Phys.* **33**, 702-708 (1994).
3. R. Pässler and G. Oelgart, *J. Appl. Phys.* **82**, 2611-2616 (1997).
4. C. M. Yee-Rendón, A. Pérez-Centeno, M. Meléndez-Lira, G. González de la Cruz, and M. López-López, *J. Appl. Phys.* **96**, 3702-3708 (2004).
5. M. López-López, J. Luyo-Alvarado, M. Meléndez-Lira, O. Cano-Aguilar, C. Megía-García, J. Ortíz-López, G. Contreras-Puente, *J. Vac. Sci. Technol. B* **18**, 1553-1556 (2000).
6. J. S. Rojas-Ramírez, R. Goldhahn, P. Moser, J. Huerta-Ruelas, J. Hernández-Rosas, M. López-López, *J. Appl. Phys.* (2008). In revision.
7. S. Rudin, T. L. Reinecke, and B. Segall, *Phys. Rev. B* **42**, 11218-11231 (1990).
8. L. Malikova, Wojciech Krystek, Fred H. Pollak, N. Dai, A. Cavus, and M. C. Tamargo, *Phys. Rev. B* **54**, 1819-1824 (1996).

LIST OF PARTICIPANTS

Aguirre-Francisco, S. (Department of Electrical Engineering, University Of Texas at Dallas. USA)
Alva Castañeda, Yew (UPIITA)
Arciniega Castro, Marcelino (UNAM)
Ávila Romero, Jorge A. (ESIME)
Aviles Eusebio, Jair E. (ESIME)
Barro Quintero, Arturo (UABC)
Baquero, Rafael (Cinvestav)
Berdeja Sotelo, Irai A. (UAGuerrero)
Bermeo, Alberto (UAEMex)
Bojorquez Sánchez, Rubí (UABC)
Bolaños, Azucena (Cinvestav)
Cadena Rodríguez, Miguel Z. (ESIME)
Carbajal Tinoco, Mauricio D. (Cinvestav)
Carmona Loaiza, Juan M. (UNAM)
Cervantes, Aldrin (Cinvestav)
Chapa Corzo, Yussel (ESIME)
Contreras Guerrero, Rocío (Department of Electrical Engineering, University Of Texas at Dallas, USA/Cinvestav)
Cortes Cuautli, Luz del Carmen (ESIME)
Cruz Becerra, M. A. (Cinvestav)
Cruz y Cruz, Sara G. (UPIITA/Cinvestav)
Cruz Hernández, Juan C. (UAEH)
de Arcia Solís, Roberto C. (UMSNH)
de la Cruz Burelo, Eduard (Cinvestav)
de la Cruz Trujillo, Leonardo R. (UNAM)
Díaz Huerta, Claudia Celia (Cinvestav)
D'oleire Díaz, Marco A. (UAEMex)
Escobar Ortega, Julio C. (ESIME)
Espíndola Rodríguez, Moisés (ESFM)
Fernández-García, Nicolás (Cinvestav)
Ferrer Jiménez, Sandybell G. (ESFM)
Filio López, Ernesto (UPIITA)
Flores Carrillo, Diego A. (ESIME)
Flores, David C. (ESIME)
Galván Palacios, Alfredo (ESIME)
García Hernández, Luis A. (UNAM)
Goldhahn, R. (Technical University Ilmenau, Institut of Physics, Germany)
Gómez Rodríguez, Guadalupe L. (ESIME)
González Gutiérrez, Jorge (BUAP)
González Zacarías, Clio (BUAP)
Gracia Linares, Miguel (BUAP)
Gutiérrez Fosado, Yair (UAEH)
Gutiérrez Medina, Fabiola A. (ESFM)
Hernández, Eduardo R.(Institut de Ciència de Materials de Barcelona, Spain)
Hernández-Contreras, M. (Cinvestav)
Hernández González, Miguel A. (ESIME)
Hernández, Roger (Cinvestav)

Herrera Corral, Gerardo (Cinvestav)
Herrera García, Rigoberto (UABC)
Herrera-Gómez, A. (Department of Electrical Engineering, University Of Texas at Dallas, USA/Cinvestav-Querétaro)
Herrera Ramírez, Alejandra S. (UPIITA)
Hernández-Rosas, J. (Cinvestav)
Honorato Colín, Miguel A. (UAEMex)
Huerta-Ruelas, J. (CICATA-IPN)
Hurtado Torres, Raúl E. (ESIME)
Ibañez Sandoval, Araceli (ESIME)
Iriarte Hernández, Adán I. (ESIME)
Ledesma Motolinia, Mónica (BUAP)
López-López, M. (Cinvestav)
López Mendoza, Luis A. (UV)
López Muñoz, Gerardo (UPIITA)
Luna Rensendis, Aldo (ESIME)
Magaña Zacarías, Abraham M. (ESFM)
Martínez Ara, Luis A. (ESFM)
Martínez Huerta, Humberto (BUAP)
Medina García, Julieta (UPIITA)
Mejía Gil, María G. (UAEMex)
Mejía López, Carlos (ESIME)
Melchor Rodríguez, Ángel (UABC)
Méndez Chávez, Francisco J. (ESFM)
Méndez Zavaleta, Julio A. (UV)
Mendoza Castrejón, Arturo (ESFM)
Menk, Ralf Hendrik (Sincrotrone Trieste & INFN Trieste, Italy)
Meneces, Amilcar (Cinvestav-Computación)
Mercado Osorio, Brenda (UAEMex)
Meza Benítez, Mauricio (ESIME)
Moctezuma Salgado, Claudia (ESFM)
Mondragón Leal, Ricardo D. (ESIME)
Montiel Piña, Enrique (Ing. BUAP)
Morales Cortés, Humberto (ESFM)
Moser, P. (Technical University Ilmenau, Institut of Physics, Germany)
Navarro Rodríguez, Luis G. (UV)
Ochoa y Zamora, Francisco P. (Inv. (Hospital))
Olmedo Flores, Arturo (ESIME)
Ovando Vázquez, Césare (Cinvestav)
Patiño Martínez, Didier A. (UMSNH)
Ponce Loeza, Juan C. (ESIME)
Procopio, Lorenzo M. (Cinvestav)
Raffelt, Georg G. (Max-Planck-Institut für Physik (Werner-Heisenberg-Institut), Germany)
Ríos Soto, Julia (IFUG)
Rivera Sandoval, Liliana E. (BUAP)
Rocha, Luisa L. (Departamento de Farmacobiología, Cinvestav)
Rodríguez Tzompantzi, Omar (BUAP)
Rodríguez, Carlos (CICESE-UNAM)
Rojas-Ramírez, Juan Salvador (Cinvestav)

Rosas Ortiz, Oscar (Cinvestav)
Ruiz Barrera, Yuritzi (UMSNH)
Sánchez Hernández, Lidia (UAEMex)
Sánchez Tufiño, Pedro R. (ESIME)
Sánchez, Carolina (Cinvestav)
Silva Quiroz, Rafael (BUAP)
Tamargo, Maria C. (
Tecpa Tovar, Mónica (ESIME)
Téllez Limón, Ricardo (BUAP)
Torres Vargas, Gamaliel (UAEH)
Trujillo López, Luisa N. (UV)
Uribe García, Jonathan (UPIITA)
Vázquez Báez, Víctor M. (BUAP)
Vázquez Rodríguez, O. (Cinvestav)
Villalobos Mora, Sandra (ESFM)
Wallace, R. M. (Department of Electrical Engineering, University Of Texas at Dallas, USA)

AUTHOR INDEX